Viral Replication: A Comprehensive Study

Viral Replication: A Comprehensive Study

Editor: Audrey Houle

www.fosteracademics.com

www.fosteracademics.com

Cataloging-in-Publication Data

Viral replication : a comprehensive study / edited by Audrey Houle.
p. cm.
Includes bibliographical references and index.
ISBN 978-1-64646-574-3
1. Viruses--Reproduction. 2. Reproduction. 3. Virology. I. Houle, Audrey.
QR470 .V57 2023
579.2--dc23

Foster Academics,
118-35 Queens Blvd., Suite 400,
Forest Hills, NY 11375, USA

ISBN 978-1-64646-574-3 (Hardback)

Contents

Preface

Viral replication refers to the creation of biological viruses in the target host cells during an infection. Viruses should first enter the cell before viral replication could take place. The virus continuously infects new hosts by producing numerous copies of its genome and packaging these copies. The replication of viruses varies significantly and is determined by the type of genes involved. Majority of DNA viruses accumulate in the nucleus whereas majority of RNA viruses grow only in the cytoplasm. Viruses can only replicate in living cells. The host cell must supply the synthetic machinery, energy and low-molecular-weight precursors, which are required for the synthesis of nucleic acids and viral proteins. The nucleic acid polymerases are behind the replication of genome of viruses and they are also involved in producing genetic variety, which is necessary for evading the host's defenses. Polymerases and other replication machinery components could be used as anti-viral targets. This book explores all the important aspects of viral replication in the present day scenario. It is appropriate for students seeking detailed information in this area of study as well as for experts.

The information shared in this book is based on empirical researches made by veterans in this field of study. The elaborative information provided in this book will help the readers further their scope of knowledge leading to advancements in this field.

Finally, I would like to thank my fellow researchers who gave constructive feedback and my family members who supported me at every step of my research.

Editor

The Effect of Oral Administration of dsRNA on Viral Replication and Mortality in *Bombus terrestris*

Niels Piot *, Simon Snoeck, Maarten Vanlede, Guy Smagghe and Ivan Meeus

Laboratory of Agrozoology, Department of Crop Protection, Faculty of Bioscience Engineering, Ghent University, 9000 Ghent, Belgium; E-Mails: simonp.snoeck@ugent.be (S.S.); maarten.vanlede@gmail.com (M.V.); guy.smagghe@ugent.be (G.S.); ivan.meeus@ugent.be (I.M.)

* Author to whom correspondence should be addressed; E-Mail: niels.piot@ugent.be;

Academic Editors: Elke Genersch and Sebastian Gisder

Abstract: Israeli acute paralysis virus (IAPV), a single-stranded RNA virus, has a worldwide distribution and affects honeybees as well as other important pollinators. IAPV infection in honeybees has been successfully repressed by exploiting the RNA interference (RNAi) pathway of the insect's innate immune response with virus-specific double stranded RNA (dsRNA). Here we investigated the effect of IAPV infection in the bumblebee *Bombus terrestris* and its tissue tropism. *B. terrestris* is a common pollinator of wild flowers in Europe and is used for biological pollination in agriculture. Infection experiments demonstrated a similar pathology and tissue tropism in bumblebees as reported for honeybees. The effect of oral administration of virus-specific dsRNA was examined and resulted in an effective silencing of the virus, irrespective of the length. Interestingly, we observed that non-specific dsRNA was also efficient against IAPV. However further study is needed to clarify the precise mechanism behind this effect. Finally we believe that our data are indicative of the possibility to use dsRNA for a broad range viral protection in bumblebees.

Keywords: IAPV; dsRNA; bumblebee; virus; immune response

1. Introduction

Managed bees are used worldwide as pollinators to aid food production in open field as well as in greenhouses [1]. Honeybees and bumblebees are the two most commonly used species for commercial pollination purposes. Domesticated honeybees are hosts to multiple viruses [2–4], but also different bumblebee species have been reported to carry viruses [5,6]. Managed bees could act as reservoirs for pathogens, and spillover of pathogens from managed bees toward wild pollinators could disrupt natural host-pathogen interactions [7]. Mainly for protozoan parasites from reared bumblebees this phenomena has been studied [8–12], but also disturbance of host-pathogen interaction in wild bumblebees by domesticated honeybees could be happening [6,13]. The risks associated with pathogen spillover could in certain geographic regions be one of the drivers responsible for the decline of pollinators [7,14]. Pathogen eradication in managed bees can already prevent initial spillover events. For honeybees the administration of dsRNA has been successfully applied to lower infection levels of several pathogens, for example Israeli acute paralysis virus (IAPV), Deformed wing virus (DWV), Chinese sacbrood virus (CSBV) and *Nosema ceranae* [15–19]. For bumblebees oral administration of dsRNA has remained unexplored, although genome analysis revealed the presence of the RNAi pathway [20] responsible for processing dsRNA into short interfering RNA (siRNA). These siRNAs target RNA with the corresponding sequence which leads to cleavage of the targeted RNA with the help of Dicer-2. Indeed injection of gene-specific dsRNA can induce gene silencing in bumblebees [21] and also IAPV-specific silencing was obtained after injection of virus-specific dsRNA [22]. This RNAi mechanism has long been reasoned to be the main antiviral immune response in invertebrates as they would lack a non-specific antiviral response. However more recent reports also suggest the presence of a non-specific antiviral response in invertebrates triggered by nucleic acids, as dsRNA, that are recognized as pathogen-associated molecular patterns (PAMPs) [23–25]. Still, more research is needed to clarify the precise pathways of this response.

In contrast to honeybees, bumblebees are reared in a closed environment, therefore it should be feasible to produce pathogen-free bumblebees via a combination of screening and preventative measures [26]. dsRNA could help to achieve this goal. Although caution is needed, as in honeybees treatment with dsRNA cannot eliminate the infections [15,17,18], it can significantly lower the viral titer, and thereby perhaps reduce it to titers below the threshold for horizontal transmission. Another option is the application of dsRNA as a preventive treatment. Here we hypothesize that treatment with dsRNA, prior to infection, could prime the bumblebee, making virus infection less likely. This priming could not only eliminate viral spread within breeding facilities, but could also immunize bumblebees when they are used in the field. Indeed aside from the above discussed pathogen spillover mechanism, spillback principles are often overlooked [27]. In spillback the introduced species, in our case a managed bumblebee, could be picking up local parasites or viruses and act as a new host disturbing natural host-pathogen interactions. Here we reported on how different dsRNA treatments can lower infection with IAPV, a virus present in honeybees and bumblebees, causing damage in both species and possibly other pollinators.

2. Materials and Methods

2.1. Viral Stock

For the production of the viral stock we obtained 240 adult bumblebees from virus-free colonies (Biobest, Westerlo, Belgium). One hundred twenty bumblebees were injected with 2 μL of IAPV solution containing 200 virus particles. The starting virus solution originated from honeybees and was kindly provided by Joachim de Miranda (Swedish University of Agricultural Sciences, Uppsala, Sweden), and quantified by transmission electron microscopy (CODA-CERVA, Brussels, Belgium) as described by Meeus *et al.* [28]. The other 120 bumblebees were injected with nuclease free water and served as a control. After injection bumblebees were kept in micro-colonies of 10 workers at 30 °C and 60% RH (MLR-352 incubator, Sanyo/Panasonic, Osaka, Japan) and *ad libitum* access to 50% sugar water (Biogluc®, Biobest) [29]. Four days after injection the bumblebees were stored at −80 °C until the virus purification. Bumblebees were crushed in 0.01 M phosphate buffer (pH 7.0) 0.02% diethyl dithiocarbamate. The exoskeletons were discarded and the remaining liquid was centrifuged (20 min 800× *g*). Supernatant was collected and centrifuged at 40,000× *g* for 4 h at 4 °C. The pellet was suspended in nuclease free water and stored at −80 °C. We performed a quality control on both the viral stock and the control inoculum. Only the virus stock contained IAPV, checked with primers described by Cox-Foster *et al.* [30]. The virus titer was checked with transmission electron microscopy (CODA-CERVA) resulting in 1×10^8 virus particles/μL. We screened for possible contaminating viruses in both viral stock and control inoculum, we tested for slow bee paralysis virus, Kashmir bee virus, acute bee paralysis virus and DWV with PCR and all were negative. Finally, the infectivity of the prepared viral stock and control inoculum solution was tested by injection of 2 μL control inoculum and viral stock. Four days after injection the infection status was determined with PCR [30]. The viral stock solution could successfully infect, and no infection was detected when 2 μL of the control inoculum was injected.

2.2. IAPV Infectivity and Mortality

Bumblebees of the same age were kept in micro-colonies per treatment, in an incubator (MLR-352, Sanyo/Panasonic) at 30 °C, 60% RH and constant darkness. Bees had *ad libitum* access to 50% sugar water (Biogluc®, Biobest, Westerlo, Belgium) [29]. Prior to feeding bumblebees were starved for 4 h. Bumblebees were inoculated individually in separate boxes (∅ 9 cm) with water ($n = 17$), or with 1×10^7 ($n = 16$), 2×10^7 ($n = 16$) and 1×10^8 ($n = 18$) virus particles in a 20 μL droplet of 50% sugar water (Biogluc®, Biobest). Feeding was done under red light, to minimize disturbance of the bees. After inoculation behavior and mortality were followed for 20 days. After 20 days, the infection status of the bees was checked by PCR as described by Cox-Foster *et al.* [30]. Survival analysis was done in SPSS Statistics (version 22.0, IBM Corp, Armonk, NY, USA) with the Mantel-Cox (log rank) test. At day 4 and day 7 post infection 2 bumblebees fed with either 1×10^7 or 1×10^8 were sacrificed for tissue tropism analysis, these bumblebees were not included in the survival analysis.

2.3. dsRNA Administration and Effects on Infectivity and Mortality of IAPV Infected Bees

To test the effect of dsRNA on IAPV infectivity and mortality, we fed dsRNA on a daily basis for 6 consecutive days to bumblebees of the same age. Each day a 4h-starved bee received 2 μg of dsRNA in 20 μL of 50% sugar water (Biogluc®, Biobest). On the third day the bees also received 1×10^7 virus particles suspended in 20 μL of 50% sugar water (Biogluc®, Biobest) [29]. Bees were fed individually as described earlier. In the first experiment bees were put in micro-colonies per treatment after each feeding session and behavior and mortality were scored for 22 days. In the second experiment bees were kept individually during the whole course of the experiment. Five days after the last dsRNA feeding day bumblebees were sacrificed and stored at −80 °C until extraction to determine the viral titer.

2.4. Extraction and Virus Quantification

The gut, fat body and head of the bumblebees was dissected and crushed in 350 μL RLT buffer (supplied with the Qiagen RNeasy Mini kit). Further RNA extraction proceeded according manufacturer's protocol (RNeasy Mini Kit, Qiagen, Venlo, The Netherlands) and were stored at −80 °C until further use. RNA concentrations and purity were determined using a NanoDrop ND-1000 spectrophotometer. A two-step real time RT-PCR was used for quantification of IAPV. First cDNA was made (SuperScript® II Reverse Transcriptase, Life Technologies, Carlsbad, CA, USA) using 500 ng of RNA and oligo-dT primers, according to manufacturer's protocol. The cDNA was 10 times diluted and IAPV was quantified using a CFX96™ Real-Time PCR Detection system (Bio-Rad, Hercules, CA, USA), performing each reaction in duplicate. The total reaction volume of 20 μL contained 10 μL GoTaq® qPCR Master Mix, (Promega, Madison, WI, USA), 1 μL (10 μM) Forward and 1 μL (10 μM) Reverse primer (see Table 1) and 8 μL diluted cDNA. Nuclease free water was used as a negative control (NTC). Peptidylprolyl isomerase A (PPIA), reported as a stable reference gene in *B. terrestris* with IAPV infection was used for normalization as described by Niu *et al.* [31]. The qPCR protocol for both IAPV (E = 92.6%) and PPIA (E = 79.3%) reference gene was as follows: 95 °C for 5 min, 40 cycles of 95 °C for 30 s and 60 °C for 1 min. The relative normalized IAPV titer was determined as the IAPV titer after normalization with the reference gene PPIA. All qPCR data was processed with the Bio-Rad CFX Manager 3.0 Software (Bio-Rad).

Table 1. Primers used for Israeli acute paralysis virus (IAPV) detection, for virus quantification and for dsRNA synthesis, T7 promotor-sequence not included.

Target	Use	Forward Primer (5′→3′)	Reverse Primer (5′→3′)	Reference
IAPV	PCR	CGAACTTGGTGACTTGAAGG	GCATCAGTCGTCTTCCAGGT	[30]
PPIA	RT-qPCR	TCGTAATGGAGTTGAGGAGTGA	CTTGGCACATGAAGTTTGGAAT	[31]
IAPV	RT-qPCR	CCATGCCTGGCGATTCAC	CTGAATAATACTGTGCGTATC	[31]
dsVP586	dsRNA	ACCTGGAAGACGATTGATGC	CTGCCCACTTCCAAACAACT	This study
dsVP443	dsRNA	TATAGATGCCGCTCCATGTG	CTGCCCACTTCCAAACAACT	This study
dsVP293	dsRNA	ACCTGGAAGACGATTGATGC	GTGGGTTTGACGGGTATCAC	This study

2.5. dsRNA Synthesis

The RNA extract from the IAPV stock solution was subjected to a DNase treatment and subsequently to RT-PCR with oligo-dT primers, as described earlier. The fragments corresponding to the structural protein-coding region, dsVP586 (bases 7986-8571; accession: NC_009025.1) were PCR amplified with primers (Table 1) carrying a T7 promotor-sequence at the $5'$ end, dsRNA for this target was synthesized for three different lengths, 293 bp (dsVP293); 443 bp (dsVP443); and 586 bp (dsVP586). Primers were designed using Primer3 software [32], dsRNA was further produced according manufacturer's protocol (MEGAscript® RNA kit, Life Technologies), except in the elution step where nuclease free water was used to elute the dsRNA instead of elution buffer. Eluted dsRNA was subjected to a DNAse treatment. dsRNA for GPF (dsGFP, 455 bp) was produced in the same way, starting from a plasmid containing the nucleotide sequence for GFP. The sequences of all dsRNA fragments were compared to the genome of *B. terrestris* and did not contain any similarity longer than 20 bp [20].

3. Results

3.1. Infectivity of IAPV in Bumblebees and Its Effect on Survival

Previous studies showed that 0.5×10^7 IAPV particles is near the minimum amount required for oral infection in bumblebees, and reported no significant effect on mortality with this dose [28]. Here we inoculated bumblebee workers with three different doses of IAPV (*i.e.*, 1×10^7, 2×10^7 and 1×10^8) and tested the infection status ($n = 2$, per dose per time point) in different body parts (*i.e.*, head, gut and fat body) after four and seven days. For all treatments we also followed mortality and bee behavior.

After a single oral administration of 1×10^8 virus particles, we were able to detect IAPV in the head, gut and fat body both four and seven days post infection (p.i.). Feeding of 1×10^7 virus particles had a slower infection progression as IAPV was only detectable in the gut at 4 days p.i., while it took until day seven for virus to appear in the other tested body parts. This reduced infection progression was also apparent in the survival scoring of the bumblebees. Figure 1 shows that feeding of 1×10^8 IAPV virus particles resulted in significant increased mortality after 20 days ($p = 0.001$, Mantel-Cox) compared to the no virus control. Feeding of 2×10^7 virus particles resulted in reduced survival, although not significant ($p = 0.09$, Mantel-Cox). When 1×10^7 virus particles were fed, there was nearly no increase in mortality, which is consistent with previous reports showing only sub-lethal effects [28].

IAPV therefore seems to have a similar pathology in bumblebees as in honeybees. Indeed high mortality within 10 days after feeding of IAPV was reported for honeybees [33]. Experiments comparing doses in honeybees and bumblebees and the naturally occurring titers of IAPV are needed to draw further conclusions.

Looking at the behavior of the bumblebees we noticed disorientated movements: the bumblebees moved around lifting their paws high in the air and tipping over frequently. These observations were not recorded before, and were only seen when high amounts (1×10^8 or 2×10^7) of virus particles were fed and shortly before dying. In honeybees infected with IAPV similar symptoms of spasms have been reported shortly before dying [33]. Darkening of the abdomen is described for IAPV infected honeybees as an early symptom [33], while in bumblebees no such symptoms were recorded in our experiments.

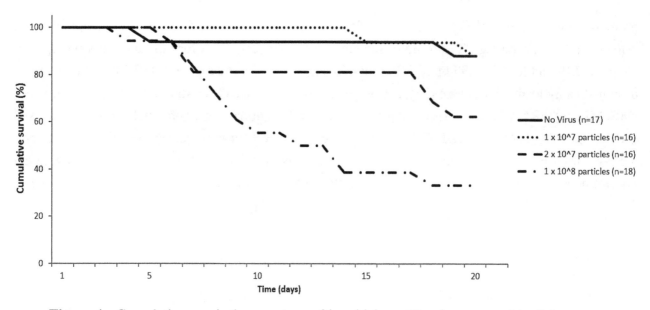

Figure 1. Cumulative survival percentage of bumblebees (*Bombus terrestris*) of the same age fed with different Israeli acute paralysis virus (IAPV) concentrations.

Although there is no knowledge on IAPV titers in naturally infected wild bumblebees, titers of Acute bee paralysis virus, a close related virus have been found to range from 10^4 to 10^{11} genome copies per bumblebee [6].

3.2. Effect of Oral Application of dsRNA on IAPV Infectivity, Relative Virus Titer and Bumblebee Survival

The aim of this experiment was to evaluate if oral administration of dsRNA specific for IAPV could prevent infection in bumblebees. We opted to administer 1×10^7 virus particles to measure the effect of dsRNA on the infection success. Administration of dsVP586 ($n = 10$) and dsGFP ($n = 10$) three days before and three days after inoculation could not prevent infection. After eight days all tested body parts (*i.e.*, head, gut and fat body) ($n = 5$, each treatment) were infected.

In Figure 2b, the relative IAPV titers of both the dsVP586 and dsGFP treatment are compared at eight days p.i. in different body parts. We detected a markedly lower relative normalized virus titer in the head ($p < 0.001$), the gut ($p = 0.015$) and the fat body ($p = 0.014$) in the dsGFP-treated samples compared to the dsVP586 treatment. This result was somewhat unexpected, but actually the impact of non-specific dsRNA (*i.e.*, dsGFP) and the IAPV-specific dsRNA (*i.e.*, dsVP586) on mortality followed the same pattern. The dsGFP control had a significantly higher survival compared to the virus-specific treatment (Mantel-Cox, $p = 0.009$) (Figure 2a). The dsGFP control showed a survival of 60% after 22 days compared to only 10% survival for the dsVP586 treatment. Due to the design of the experiment, where we wanted to test if virus-specific dsRNA could prevent infection, dsGFP served as a control and no virus-only control was used, as infection success with no treatment was proven in the first experiment. Therefore we cannot draw conclusions on the silencing efficiency, however the observed effect of dsGFP compared to the virus-specific dsRNA treatment remains noteworthy. This effect contradicts with previous studies on the use of dsRNA on IAPV infected honeybees, where the non-specific dsRNA control, also dsGFP, had no effect on the survival compared to the virus-specific dsRNA [15]. For DWV infected adult honeybees a slight positive effect of dsGFP on the survival was reported, although the specific dsRNA was significantly better than the dsGFP treatment [18]. Feeding of dsRNA specific for CSBV could prevent infection in larvae as opposed to the unrelated dsGFP, for which there was also no effect on survival [17].

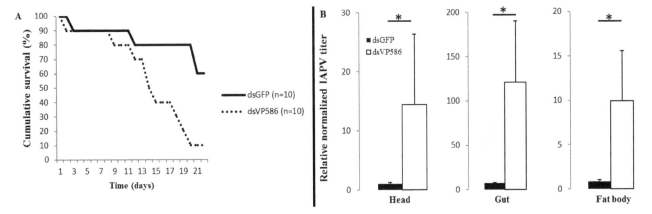

Figure 2. (**A**) Cumulative survival percentage of bumblebees (*Bombus terrestris*) infected with 1×10^7 IAPV particles and fed with virus-specific dsVP586 ($n = 10$), and non-specific dsGFP ($n = 10$); (**B**) Relative normalized IAPV titer in the head, gut and fat body of bumblebees fed with the virus-specific dsVP586 ($n = 5$) and the non-specific dsGFP ($n = 5$) relative to the dsGFP treatment in the head; asterisks indicate significant differences ($p < 0.05$).

Additionally, the effect of dsGFP in bumblebees on the relative viral titer seems much higher than reported in honeybees. Maori *et al.* [15] reported no effect of dsGFP on the relative viral titer of IAPV infected honeybees in contrast to a significant reduction after virus-specific dsRNA was administered. Nevertheless the positive effect of administering non-specific dsRNAs on virus titers has been reported in invertebrates. A study, although working with an artificial virus (GFP recombinant Sindbis virus), reported a positive effect of non-specific dsRNA on the viral titer in honeybees [24]. In shrimp similar results have been reported where non-specific dsRNAs had a positive effect on survival after infection with Taura syndrome virus (TSV) and White spot syndrome virus (WSSV) [25].

3.3. Confirmation of Lower Relative Virus Titers after Non-Specific and Specific dsRNA Treatment

The lower effect observed for the virus-specific dsRNA (dsVP586) could be due to the differences in length of the dsRNA. The used dsVP586 is much longer than the dsGFP fragment. Indeed size can be a limiting factor in uptake of the dsRNA molecules [34]. To elucidate if the observed effect was due to the length of the dsRNA molecules, the same experiment was repeated. In this second experiment dsRNA molecules specific for IAPV were synthesized in three different lengths, all targeting the same region of the IAPV genome: dsVP293 shorter than dsGFP, dsVP443 with approximately the same length, and dsVP586 the same molecule used in the first experiment. A virus-only control was also used to assess the silencing efficiency of the different treatments. Here only the head was used to compare the impact of the dsRNA on the virus titers, as extraction of the head resulted in less variation in quantity and quality of the RNA extracts; in less variation between biological replicates; and in higher significant effects in the first experiment. Figure 3 gives the relative virus titer in the head of a virus-only control (oral inoculation of IAPV and later on sugar water instead of dsRNA), dsGFP administration as well as all three different IAPV-specific dsRNA treatments. Here all virus-specific treatments showed a significant drop in virus titers compared to the virus-only control ($p < 0.05$). The length of the dsRNA molecule did not have a significant effect on the virus silencing with the IAPV-specific fragment, as there was no significant difference in virus titer between all three virus-specific fragments ($p > 0.10$). The viral titer of the dsGFP control was significantly different from the only-virus control ($p < 0.05$) but not from the virus-specific treatments ($p > 0.10$). These findings confirm the positive effect of dsGFP found in the first experiment.

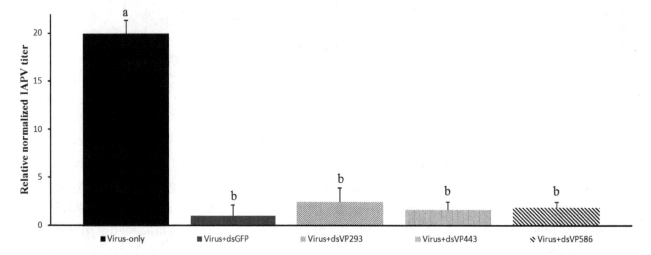

Figure 3. Relative normalized IAPV titer in the head of bumblebees (*Bombus terrestris*) of the same age fed with: only virus; unrelated dsRNA (dsGFP); virus-specific dsRNA of three different lengths, dsVP293, dsVP443 and dsVP586; $n = 5$ for each treatment; relative to the dsGFP treatment in the head. Different letters indicate significant difference ($p < 0.05$).

4. Discussion

4.1. Pathology and Tissue Tropism of IAPV in Bumblebees

IAPV and other dicistroviruses have been reported in several key pollinators as well as in other insects belonging to different orders [35]. Active replication of IAPV has also been demonstrated in the bumblebee *Bombus impatiens* by detection of the negative strand [36]. However there is still a lack of information concerning the pathology of these dicistroviruses in wild non-*Apis* pollinators [35]. As these viruses have been shown to be able to transmit between honeybees and bumblebees [5], the impact of infections needs to be addressed for these wild non-*Apis* pollinators. We reported here on the pathology and tissue distribution of IAPV infection of *B. terrestris*. Infection with IAPV seems to have similar effects on mortality as reported for honeybees when high doses are fed [15,33]. The tissue tropism also seems similar, where in honeybees IAPV can also be found in the head, the gut and the fat body [37]. At colony level there is however no information on the impact of IAPV in bumblebees, except for artificial micro-colonies consisting of only workers [28]. This is the next step that needs to be taken, as we provide the first evidence of the impact of IAPV on single workers, the impact on queens and colony development is still missing. Indeed wild nest are very hard to find, but studying the impact on colony level is also of importance as wild bees often go through more harsh conditions compared to honeybees [38]. The biological relevance of feeding such high doses of virus particles to look at pathology remains to be tested for IAPV. On the other hand, viral loads as high as 1×10^9 RNA copies have been detected in bee feces for chronic bee paralysis virus (CBPV) [39], giving some idea about biological relevant doses that can be a source for infection. IAPV has also been detected in honeybee feces although there are no quantitative data so far [37]. Indeed fecal-oral transmission is suggested as one of the main transmission routes between taxa [9,40]. This can happen through shared flower use where feces could contaminate the exterior of the bee and thereafter the nectar as well as the pollen of flowers [5]. The impact of IAPV on wild bees needs further investigation as we provide initial evidence that this virus may also be detrimental to bumblebees as is reported for honeybees [41,42].

4.2. dsRNA As Possible Treatment to Prevent Spillback Principles

The risk of pathogen spillover associated with the use of commercial bumblebees is well described [8,40,43,44]. However, the use of pathogen free commercial bees can also disturb the natural host-pathogen equilibrium, as these bees can be competent hosts for the native pathogens and so increase the pathogen prevalence, the so called spillback mechanism [27,45]. Here we investigated if oral administration of dsRNA could be of help to prevent IAPV infection in bumblebees, this in an effort to prevent spillback of viruses in pathogen free managed bees. In honeybees this technique successfully lowered the viral titer and prolonged the lifespan of infected bees [15–18]. Administration of dsRNA was not able to prevent infection with IAPV in bumblebees. Feeding of dsRNA could significantly prolong the lifespan of the infected bumblebees and significantly lower the virus titer. It therefore seems that initial infection of naturally occurring viruses will not be prevented, but the reduced viral replication holds promising results. Indeed dsRNA treatment of managed bees could in this way prevent massive infection loads of these bees, avoiding spillback of viruses into the environment. If these lower virus titers will indeed diminish the possibility of spreading infectious doses of virus and prevent spillback mechanisms needs to be further investigated.

4.3. Does Non-Specific dsRNA Works as A PAMP in Bumblebees?

The effect of IAPV-specific dsRNA molecules on the virus titer suggests the presence of a functional specific antiviral RNAi immune response in bumblebees, as all the genes required for the RNAi machinery are present in the genome of *B. terrestris* [20].

Our results, done in bumblebees, demonstrate that a non-specific dsRNA molecule has the same or even higher potency to silence IAPV. In honeybees, virus-specific dsRNA proofed to be more effective to silence viruses compared to non-specific dsRNA [15,17,18]. This suggests that the role of a virus-unspecific immune response might be of greater importance in bumblebees. It remains speculative if the massive domestication and long term selection towards high honey productive honeybees had its trade-offs toward diminished efficiency of certain innate viral immune pathways. On the other hand no effect of inbreeding was observed for encapsulation and phenoloxidase activity, in honeybees [46]. However, the impact of inbreeding and selection on viral immune responses in honeybees still remains unexplored.

The effect of non-specific dsRNA molecules on virus titer and survival has been reported in several invertebrates including honeybees [23,24], thus a pronounced presence in bumblebee is possible. Actually the amount of evidence supporting the hypothesis that there is also a non-specific antiviral response in invertebrates triggered by nucleic acid molecules, as dsRNA, is rising [23]. Flenniken and Andino [24] persued an attempt to find the immune pathways involved in the virus silencing by non-specific dsRNA. Microarray analysis and qPCR in honeybees after virus infection or administration of non-specific dsRNA, diminishing viral loads, revealed that most up-regulated genes could not be linked with previously reported immune pathways. They suggested the involvement of uncharacterized signal transduction cascades in virus control by non-specific dsRNA [24]. Although it could be that known pathways are still involved but regulation is missed by microarray and qPCR, as these techniques are limited to see regulation on mRNA level. While these techniques miss post-transcriptional regulation like post-translational modification. Furthermore, it could also be that the precise time point of up-regulation of the involved cascade genes was missed, further study is needed to clarify this.

Although the exact immune response pathway is unknown, it seems that non-specific dsRNA acts as a PAMP triggering the anti-viral defense. It could be that PAMP recognition strengthens the conventional RNAi pathway, making it more efficient when the actual virus infects and triggers the virus-specific RNAi machinery. But it could also be that in bumblebees other viral defense pathways like the JAK-STAT, Imd, Toll or Jun-*N*-terminal kinase immune pathways are more or equally important and triggered by non-specific dsRNA molecular patterns.

Feeding experiments in honeybees, all show a clear effect of virus-specific dsRNAs [15,16,18]; injection experiments with Sindbis virus in honeybees on the other hand demonstrated an effect of non-specific dsRNA [24] similar to our results in bumblebees after feeding. We can speculate that the siRNA response in different host and against different viruses is not clear-cut. Aside from the influence of the host and virus on antiviral responses it seems the method of delivery can also have an influence. We hypothesize that the biological importance of a certain immune pathway, be it the siRNA pathway (triggered by virus-specific dsRNA) or other pathways, even for the same virus within the same host can be different, depending on the mode of infection and severity of the infection. Furthermore also the actual fitness of the host, its immunologic competence, and genetic background are at play. This would explain the differences reported between non-specific dsRNA and virus-specific dsRNA in viral silencing efficiency [15,17,18,24,25], as well as the differences observed in our experiments. In both our experiments there was a positive effect of the non-specific dsRNA on the virus titer. In the first experiment this effect was significantly higher compared to the virus-specific dsRNA treatment, however in the second experiment there was no significant difference in the effect between the virus-specific and non-specific dsRNA. As both experiments were conducted at different times, the bumblebees used for the experiments originated from different colonies. During the course of both experiments the environmental conditions were kept constant. However, the genetic and physiological background of the bees could be different. It is known that both food stress [47] and genotype [48,49] can influence the innate immune response of bumblebee workers.

Aside from which pathways is triggered, the fact that non-specific dsRNA can silence IAPV opens avenues to prevent replication of multiple viruses in bumblebees. However, the effect of random dsRNA on other viruses in bumblebees first need to be tested. Although our preventive treatment did not lower the infection success of 1×10^7 virus particles, it could significantly lower the virus titer. This treatment could possibly protect bumblebees from high infections when placed in a natural environment and prevent possible spillback.

Acknowledgments

This research was supported by the Fund for Scientific Research Flanders (FWO-Vlaanderen) and the Special Research Fund of the Ghent University (BOF) to Guy Smagghe. The authors would like to thank Biobest (Westerlo, Belgium) for providing bumblebees and Joachim de Miranda (Swedish University of Agricultural Sciences, Uppsala, Sweden) for providing the virus.

Author Contributions

Niels Piot, Ivan Meeus and Guy Smagghe conceived and designed the experiments. Niels Piot, Maarten Vanlede and Simon Snoeck conducted the experiments. Analysis and interpretation of the results was done by Niels Piot, Simon Snoeck, Maarten Vanlede and Ivan Meeus. The manuscript was written by Niels Piot and critically read by Ivan Meeus and Guy Smagghe

References

1. Velthuis, H.H.W.; van Doorn, A. A century of advances in bumblebee domestication and the economic and environmental aspects of its commercialization for pollination. *Apidologie* **2006**, *37*, 421–451. [CrossRef]

2. Chen, Y.P.; Pettis, J.S.; Collins, A.; Feldlaufer, M.F. Prevalence and transmission of honeybee viruses. *Appl. Environ. Microbiol.* **2006**, *72*, 606–611. [CrossRef] [PubMed]

3. Ribière, M.; Ball, B.V.; Aubert, M. Natural History And Geographical Distribution Of Honey Bee Viruses. In *Virology and the Honey Bee*; Aubert, M., Ball, B., Fries, I., Moritz, R., Milani, N., Bernardinelli, I., Eds.; EC Publications: Luxembourg, Luxembourg, 2008; pp. 15–84.

4. De Miranda, J.R.; Cordoni, G.; Budge, G. The Acute bee paralysis virus-Kashmir bee virus-Israeli acute paralysis virus complex. *J. Invertebr. Pathol.* **2010**, *103*, S30–S47. [CrossRef] [PubMed]

5. Singh, R.; Levitt, A.L.; Rajotte, E.G.; Holmes, E.C.; Ostiguy, N.; Vanengelsdorp, D.; Lipkin, W.I.; Depamphilis, C.W.; Toth, A.L.; Cox-Foster, D.L. RNA viruses in hymenopteran pollinators: Evidence of inter-taxa virus transmission via pollen and potential impact on non-Apis hymenopteran species. *PLoS ONE* **2010**, *5*, e14357. [CrossRef] [PubMed]

6. McMahon, D.P.; Fürst, M.A.; Caspar, J.; Theodorou, P.; Brown, M.J.F.; Paxton, R.J. A sting in the spit: Widespread cross-infection of multiple RNA viruses across wild and managed bees. *J. Anim. Ecol.* **2015**, in press. [CrossRef] [PubMed]

7. Meeus, I.; Brown, M.J.F.; de Graaf, D.C.; Smagghe, G. Effects of invasive parasites on bumble bee declines. *Conserv. Biol.* **2011**, *25*, 662–671. [CrossRef] [PubMed]

8. Murray, T.E.; Coffey, M.F.; Kehoe, E.; Horgan, F.G. Pathogen prevalence in commercially reared bumble bees and evidence of spillover in conspecific populations. *Biol. Conserv.* **2013**, *159*, 269–276. [CrossRef]

9. Graystock, P.; Goulson, D.; Hughes, W.O.H. The relationship between managed bees and the

prevalence of parasites in bumblebees. *Peer J. Prepr.* **2014**, *2*, 1–24. [CrossRef] [PubMed]

10. Schmid-Hempel, R.; Eckhardt, M.; Goulson, D.; Heinzmann, D.; Lange, C.; Plischuk, S.; Escudero, L.R.; Salathé, R.; Scriven, J.J.; Schmid-Hempel, P. The invasion of southern South America by imported bumblebees and associated parasites. *J. Anim. Ecol.* **2014**, *83*, 823–837. [CrossRef] [PubMed]

11. Arbetman, M.P.; Meeus, I.; Morales, C.L.; Aizen, M.A.; Smagghe, G. Alien parasite hitchhikes to Patagonia on invasive bumblebee. *Biol. Invasions* **2012**, *15*, 489–494. [CrossRef]

12. Maharramov, J.; Meeus, I.; Maebe, K.; Arbetman, M.; Morales, C.; Graystock, P.; Hughes, W.O.H.; Plischuk, S.; Lange, C.E.; de Graaf, D.C.; *et al.* Genetic variability of the neogregarine Apicystis bombi, an etiological agent of an emergent bumblebee disease. *PLoS ONE* **2013**, *8*, e81475. [CrossRef] [PubMed]

13. Fürst, M.A; McMahon, D.P.; Osborne, J.L.; Paxton, R.J.; Brown, M.J.F. Disease associations between honeybees and bumblebees as a threat to wild pollinators. *Nature* **2014**, *506*, 364–366. [CrossRef] [PubMed]

14. Goulson, D.; Nicholls, E.; Botias, C.; Rotheray, E.L. Bee declines driven by combined stress from parasites, pesticides, and lack of flowers. *Science* **2015**, *347*. [CrossRef] [PubMed]

15. Maori, E.; Paldi, N.; Shafir, S.; Kalev, H.; Tsur, E.; Glick, E.; Sela, I. IAPV, a bee-affecting virus associated with colony collapse disorder can be silenced by dsRNA ingestion. *Insect Mol. Biol.* **2009**, *18*, 55–60. [CrossRef] [PubMed]

16. Hunter, W.; Ellis, J.; Vanengelsdorp, D.; Hayes, J.; Westervelt, D.; Glick, E.; Williams, M.; Sela, I.; Maori, E.; Pettis, J.; *et al.* Large-scale field application of RNAi technology reducing Israeli acute paralysis virus disease in honey bees (*Apis mellifera*, hymenoptera: Apidae). *PLoS Pathog.* **2010**, *6*, e1001160. [CrossRef] [PubMed]

17. Liu, X.; Zhang, Y.; Yan, X.; Han, R. Prevention of chinese sacbrood virus infection in apis cerana using rna interference. *Curr. Microbiol.* **2010**, *61*, 422–428. [CrossRef] [PubMed]

18. Desai, S.D.; Eu, Y.J.; Whyard, S.; Currie, R.W. Reduction in deformed wing virus infection in larval and adult honey bees (*Apis mellifera* L.) by double-stranded RNA ingestion. *Insect Mol. Biol.* **2012**, *21*, 446–455. [CrossRef] [PubMed]

19. Paldi, N.; Glick, E.; Oliva, M.; Zilberberg, Y.; Aubin, L.; Pettis, J.; Chen, Y.; Evans, J.D. Effective gene silencing in a microsporidian parasite associated with honeybee (*Apis mellifera*) colony declines. *Appl. Environ. Microbiol.* **2010**, *76*, 5960–5964. [CrossRef] [PubMed]

20. Sadd, B.M.; Barribeau, S.M.; Bloch, G.; de Graaf, D.C.; Dearden, P.; Elsik, C.G.; Gadau, J.; Grimmelikhuijzen, C.J.; Hasselmann, M.; Lozier, J.D.; *et al.* The genomes of two key bumblebee species with primitive eusocial organization. *Genome Biol.* **2015**, *16*. [CrossRef] [PubMed]

21. Deshwal, S.; Mallon, E.B. Antimicrobial peptides play a functional role in bumblebee anti-trypanosome defense. *Dev. Comp. Immunol.* **2013**, *42*, 240–243. [CrossRef] [PubMed]

22. Piot, N.; Meeus, I.; Smagghe, G. Injection of Israeli acute paralysis virus-specific dsRNA reduces virus replication and mortality in *Bombus terrestris*. Submitted.

23. Wang, P.H.; Weng, S.P.; He, J.G. Nucleic acid-induced antiviral immunity in invertebrates: An evolutionary perspective. *Dev. Comp. Immunol.* **2014**, *48*, 291–296. [CrossRef] [PubMed]

24. Flenniken, M.L.; Andino, R. Non-specific dsRNA-mediated antiviral response in the honey bee.

PLoS ONE **2013**, *8*, e77263. [CrossRef] [PubMed]

25. Robalino, J.; Bartlett, T.C.; Chapman, R.W.; Gross, P.S.; Browdy, C.L.; Warr, G.W. Double-stranded RNA and antiviral immunity in marine shrimp: Inducible host mechanisms and evidence for the evolution of viral counter-responses. *Dev. Comp. Immunol.* **2007**, *31*, 539–547. [CrossRef] [PubMed]

26. Meeus, I.; Mosallanejad, H.; Niu, J.; de Graaf, D.C.; Wäckers, F.; Smagghe, G. Gamma irradiation of pollen and eradication of Israeli acute paralysis virus. *J. Invertebr. Pathol.* **2014**, *121*, 10–13. [CrossRef] [PubMed]

27. Strauss, A.; White, A.; Boots, M. Invading with biological weapons: The importance of disease-mediated invasions. *Funct. Ecol.* **2012**, *26*, 1249–1261. [CrossRef]

28. Meeus, I.; de Miranda, J.R.; de Graaf, D.C.; Wäckers, F.; Smagghe, G. Effect of oral infection with Kashmir bee virus and Israeli acute paralysis virus on bumblebee (*Bombus terrestris*) reproductive success. *J. Invertebr. Pathol.* **2014**, *121*, 64–69. [CrossRef] [PubMed]

29. Mommaerts, V.; Sterk, G.; Smagghe, G. Hazards and uptake of chitin synthesis inhibitors in bumblebees Bombus terrestris. *Pest Manag. Sci.* **2006**, *62*, 752–758. [CrossRef] [PubMed]

30. Cox-Foster, D.L.; Conlan, S.; Holmes, E.C.; Palacios, G.; Evans, J.D.; Moran, N.A.; Quan, P.-L.; Briese, T.; Hornig, M.; Geiser, D.M.; *et al.* A metagenomic survey of microbes in honey bee colony collapse disorder. *Science* **2007**, *318*, 283–287. [CrossRef] [PubMed]

31. Niu, J.; Cappelle, K.; de Miranda, J.R.; Smagghe, G.; Meeus, I. Analysis of reference gene stability after Israeli acute paralysis virus infection in bumblebees Bombus terrestris. *J. Invertebr. Pathol.* **2014**, *115*, 76–79. [CrossRef] [PubMed]

32. Koressaar, T.; Remm, M. Enhancements and modifications of primer design program Primer3. *Bioinformatics* **2007**, *23*, 1289–1291. [CrossRef] [PubMed]

33. Maori, E.; Lavi, S.; Mozes-Koch, R.; Gantman, Y.; Peretz, Y.; Edelbaum, O.; Tanne, E.; Sela, I. Isolation and characterization of Israeli acute paralysis virus, a dicistrovirus affecting honeybees in Israel: Evidence for diversity due to intra- and inter-species recombination. *J. Gen. Virol.* **2007**, *88*, 3428–3438. [CrossRef] [PubMed]

34. Huvenne, H.; Smagghe, G. Mechanisms of dsRNA uptake in insects and potential of RNAi for pest control: A review. *J. Insect Physiol.* **2010**, *56*, 227–235. [CrossRef] [PubMed]

35. Manley, R.; Boots, M.; Wilfert, L. Emerging viral disease risk to pollinating insects: Ecological, evolutionary and anthropogenic factors. *J. Appl. Ecol.* **2015**, *52*, 331–340. [CrossRef] [PubMed]

36. Levitt, A.L.; Singh, R.; Cox-Foster, D.L.; Rajotte, E.; Hoover, K.; Ostiguy, N.; Holmes, E.C. Cross-species transmission of honey bee viruses in associated arthropods. *Virus Res.* **2013**, *176*, 232–240. [CrossRef] [PubMed]

37. Chen, Y.P.; Pettis, J.S.; Corona, M.; Chen, W.P.; Li, C.J.; Spivak, M.; Visscher, P.K.; DeGrandi-Hoffman, G.; Boncristiani, H.; Zhao, Y.; *et al.* Israeli acute paralysis virus: Epidemiology, pathogenesis and implications for honey bee health. *PLoS Pathog.* **2014**, *10*, e1004261. [CrossRef] [PubMed]

38. Goulson, D.; Lye, G.C.; Darvill, B. Decline and conservation of bumble bees. *Annu. Rev. Entomol.* **2008**, *53*, 191–208. [CrossRef] [PubMed]

39. Ribière, M.; Lallemand, P.; Iscache, A.L.; Schurr, F.; Celle, O.; Blanchard, P.; Olivier, V.;

Faucon, J.P. Spread of infectious chronic bee paralysis virus by honeybee (*Apis mellifera* L.) feces. *Appl. Environ. Microbiol.* **2007**, *73*, 7711–7716. [CrossRef] [PubMed]

40. Otterstatter, M.C.; Thomson, J.D. Does pathogen spillover from commercially reared bumble bees threaten wild pollinators? *PLoS ONE* **2008**, *3*, e2771. [CrossRef] [PubMed]

41. Hou, C.; Rivkin, H.; Slabezki, Y.; Chejanovsky, N. Dynamics of the presence of israeli acute paralysis virus in honey bee colonies with colony collapse disorder. *Viruses* **2014**, *6*, 2012–2027. [CrossRef] [PubMed]

42. Ribière, M.; Drajnudel, P.; Faucon, J.-P. The collapse of bee colonies: The CCD case ("Colony collapse disorder") and the IAPV virus (Israeli acute paralysis virus). *Virology* **2008**, *12*, 319–322.

43. Colla, S.R.; Otterstatter, M.C.; Gegear, R.J.; Thomson, J.D. Plight of the bumble bee: Pathogen spillover from commercial to wild populations. *Biol. Conserv.* **2006**, *129*, 461–467. [CrossRef]

44. Graystock, P.; Yates, K.; Evison, S.E.F.; Darvill, B.; Goulson, D.; Hughes, W.O.H. The Trojan hives: Pollinator pathogens, imported and distributed in bumblebee colonies. *J. Appl. Ecol.* **2013**, *50*, 1207–1215. [CrossRef]

45. Jones, C.M.; Brown, M.J.F. Parasites and genetic diversity in an invasive bumblebee. *J. Anim. Ecol.* **2014**, *83*, 1428–1440. [CrossRef] [PubMed]

46. Lee, G.M.; Brown, M.J.F.; Oldroyd, B.P. Inbred and outbred honey bees (*Apis mellifera*) have similar innate immune responses. *Insectes Soc.* **2012**, *60*, 97–102. [CrossRef]

47. Moret, Y.; Schmid-Hempel, P. Survival for immunity: The price of immune system activation for bumblebee workers. *Science* **2000**, *290*, 1166–1168. [CrossRef] [PubMed]

48. Schlüns, H.; Sadd, B.M.; Schmid-Hempel, P.; Crozier, R.H. Infection with the trypanosome Crithidia bombi and expression of immune-related genes in the bumblebee Bombus terrestris. *Dev. Comp. Immunol.* **2010**, *34*, 705–709. [CrossRef] [PubMed]

49. Wilfert, L.; Gadau, J.; Baer, B.; Schmid-Hempel, P. Natural variation in the genetic architecture of a host-parasite interaction in the bumblebee Bombus terrestris. *Mol. Ecol.* **2007**, *16*, 1327–1339. [CrossRef] [PubMed]

2

Relevance of Viroporin Ion Channel Activity on Viral Replication and Pathogenesis

Jose L. Nieto-Torres [1,†], Carmina Verdiá-Báguena [2,†], Carlos Castaño-Rodriguez [1], Vicente M. Aguilella [2,*] and Luis Enjuanes [1,*]

[1] Department of Molecular and Cell Biology, National Center of Biotechnology (CNB-CSIC), Campus Universidad Autónoma de Madrid, 28049 Madrid, Spain; E-Mails: jlnieto@cnb.csic.es (J.L.N.-T.); ccastano@cnb.csic.es (C.C.-R.)

[2] Laboratory of Molecular Biophysics, Department of Physics, Universitat Jaume I, 12071 Castellón, Spain; E-Mail: verdia@uji.es

[†] These authors contributed equally to this work.

[*] Authors to whom correspondence should be addressed; E-Mails: aguilell@uji.es (V.M.A.); L.Enjuanes@cnb.csic.es (L.E.).

Academic Editors: Luis Carrasco and José Luis Nieva

Abstract: Modification of host-cell ionic content is a significant issue for viruses, as several viral proteins displaying ion channel activity, named viroporins, have been identified. Viroporins interact with different cellular membranes and self-assemble forming ion conductive pores. In general, these channels display mild ion selectivity, and, eventually, membrane lipids play key structural and functional roles in the pore. Viroporins stimulate virus production through different mechanisms, and ion channel conductivity has been proved particularly relevant in several cases. Key stages of the viral cycle such as virus uncoating, transport and maturation are ion-influenced processes in many viral species. Besides boosting virus propagation, viroporins have also been associated with pathogenesis. Linking pathogenesis either to the ion conductivity or to other functions of viroporins has been elusive for a long time. This article summarizes novel pathways leading to disease stimulated by viroporin ion conduction, such as inflammasome driven immunopathology.

Keywords: viroporins;   virus;   ion   channel;   protein-lipid   pore;   replication; pathogenesis; inflammasome

1. Introduction

Cells maintain optimum subcellular compartment ionic conditions, different to those of the extracellular media by controlling ion transport through lipid membranes. Different cell organelles present particular ion compositions. These asymmetric distributions of ions among biological membranes generate electrochemical gradients, essential for the proper cell functioning [1]. Crucial aspects of the cell are governed by the membrane potential, Ca^{2+} stores in the endoplasmic reticulum (ER) and the Golgi apparatus, and different pH conditions found in the organelles of the secretory pathway, which benefit from those ion gradients. The coordinate action of a multitude of ion channels and transporters generates and tightly controls these ionic milieus found within cells.

It is well known that viruses exploit and modify host-cell ion homeostasis in favor of viral infection. To that purpose, a wide range of viruses encode viroporins [2]. Viroporins constitute a large family of multifunctional proteins broadly distributed in different viral families, and are mainly concentrated in RNA viruses [2]. Highly pathogenic human viruses, such as influenza A virus (IAV), human immunodeficiency virus 1 (HIV-1), hepatitis C virus (HCV), several picornaviruses, respiratory syncytial virus (RSV), and coronaviruses (CoVs), such as the one responsible for the severe acute respiratory syndrome (SARS-CoV), and the etiologic agent of Middle East respiratory syndrome (MERS-CoV), encode at least one viroporin [3–9]. These are transmembrane proteins that stimulate crucial aspects of the viral life cycle through a variety of mechanisms. Noticeably, these proteins oligomerize in cell membranes to form ion conductive pores, which generally display mild ion selectivity, indicating that viroporins do not show preference for particular ionic species. The measurements of channel conductance are in accordance with the formation of relatively wide pores, supporting the non-specificity of viroporins. The influence of the lipid charge in channel function is a distinctive feature of some viroporins, as reported in the case of SARS-CoV E protein [10,11]. Ion channel (IC) activity is relevant for virus propagation and may have a great impact on host-cell ionic milieus and physiology [2,12,13]. Once inserted on cell membranes, viroporins tune ion permeability at different organelles to stimulate a variety of viral cycle stages that will be described below. IC activity ranges from almost essential, to highly or moderately necessary for viruses to yield properly.

Besides modifying cellular processes to favor virus propagation, the loss of ion homeostasis triggered by viral IC activity may have deleterious consequences for the cell, from stress responses to apoptosis [2,14,15]. That is why cells have evolved mechanisms to sense the ion imbalances caused by infections and elaborate immune responses to counteract viruses. Interestingly, the IC activity of several viroporins triggers the activation of a macromolecular complex called the inflammasome, key in the stimulation of innate immunity [16–21]. Inflammasomes control pathways essential in the resolution of viral infections. However, its disproportionate stimulation can lead to disease. In fact, disease worsening in several respiratory viruses infections is associated with inflammasome-driven immunopathology [22,23].

Taking into consideration the relevance of IC activity in viral production, and its direct effect in pathology and disease, ion conductivity and its pathological stimulated pathways can represent targets for combined therapeutic interventions.

2. Ion Channels Formed by Viroporins

In general, viroporins are small proteins (less than 100 amino acids) with at least one amphipathic helix that constitutes its transmembrane domain, spanning lipid membranes [2]. Larger viroporins have also been described in CoVs. This is the case of SARS-CoV 3a protein, porcine epidemic diarrhea virus (PEDV) 3 protein, or human coronavirus 229E (HCoV-229E) 4a protein [24–26]. To form the ion conductive pore, viroporins self-assemble and oligomerize, which is a key feature of this family of proteins. Structural studies for either the transmembrane domain or for full-length viroporins have revealed the molecular architecture of these viral ion channels, which can present different oligomerization statuses. IAV M2, picornavirus 2B and Chlorella virus Kcv form tetrameric structures [27–30], whereas pentamers have been described for HIV-1 Vpu, SARS-CoV and MERS-CoV E proteins, and RSV SH protein [9,31–34]. HCV p7 and human papillomavirus E5 proteins form hexameric channels [35,36]. In addition, most measurements of channel conductance are in accordance with the formation of relatively wide pores, again in agreement with the non-specificity of viroporins [6,10,37,38].

Ion selectivity of ion channels indicates the preference of the pore for a specific ion and defines the functional roles that the channel may display. It is known that IAV M2 protein channels are highly selective for protons and ion conductance is activated at low pH [39–41]. Likewise the IAV M2 channel, the Kcv protein of Chlorella virus is another highly selective channel, which contains a conserved K^+ selectivity filter [29]. Nevertheless, most viroporins usually show mild ion selectivity, which means that, in general terms, these channels do not display preference for a particular ion. HIV-1 Vpu protein displays a mild cationic selectivity in NaCl and KCl electrolyte solutions [4]. Similarly, HCV p7 channels are selective for monovalent cations (Na^+ and K^+) over monovalent anions (Cl^-) [42]. Moreover, functionally relevant H^+ transport has also been identified for this protein in cell culture [12]. ORF4a protein of HCoV-229E forms a channel that prefers cations over anions but does not show a clear specificity for a particular type of cation [25]. Still, a lot of electrophysiology experiments remain to be done for a proper characterization of viroporin selectivity. Setting aside a few highly proton selective viroporins, the vast majority display a weak selectivity, either cationic or anionic, which is strongly dependent on the lipid charge of their host membrane, as discussed below in detail.

Interestingly, under some circumstances the selectivity of these channels can be modulated. SH protein exhibits a poor cationic selectivity at neutral pH, which turns into anionic at acidic pH [43]. This is consistent with the titration of histidines, the only titratable residues of the SH protein. The presence of Ca^{2+} in selectivity experiments performed in KCl solutions reduced the cationic preference of p7 channels, which may indicate that Ca^{2+} affect the selectivity filter [42]. Probably, the most striking mechanism influencing IC selectivity and conductance has been recently reported for SARS-CoV E protein [10,11]. It was observed that the lipids are an integral component of the pore structure because electrophysiological measurements proved that the lipid charge modulates the ion transport properties of the SARS-CoV E protein. These findings suggested that viroporins can assemble into alternative complex structures, forming protein-lipid pores (Figure 1).

PROTEIN PORE   **PROTEIN-LIPID PORE**

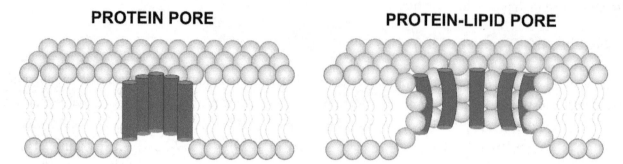

Figure 1. Ion channels formed by viroporins. Left depiction represents a channel exclusively formed by protein monomers (blue cylinders) inserted on a lipid membrane. Schematic on the right shows a protein-lipid pore. In this latter case, the lipid head groups (cyan circles) are oriented towards the channel pore, modulating ion conductance and selectivity.

The IC activity of a number of transmembrane proteins, as well as of small peptides and antimicrobial peptides, is strongly dependent on the lipid environment. The case of SARS-CoV E protein is a clear example of how lipid membrane charge influences the main ion transport properties of an IC. E protein behaves quite differently when reconstituted in neutral or negatively-charged membranes (Figure 2).

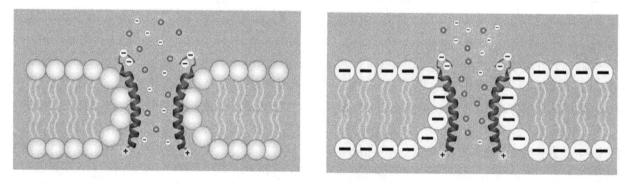

Figure 2. Functional involvement of lipid head groups in the protein-lipid pore formed by SARS-CoV E protein. Depictions represent E protein channels inserted in non-charged membranes (left) or negatively-charged membranes (right), under low solute concentrations and neutral pH. In these circumstances, each E protein monomer presents two negative charges provided by glutamic acid residues, and a positive charge conferred by an arginine. When reconstituted in fully or partially negatively-charged membranes, lipid head groups provide additional negative charges to the pore, which makes E protein channel more selective for cations and more conductive.

Conductance experiments showed that in non-charged phosphatidylcholine membranes, the E protein channel acts as a non-selective neutral pore, since the channel conductance changes linearly with bulk solution conductivity [11]. In contrast, in negatively charged phosphatidylserine membranes the variation of channel conductance with solution concentration follows different regimes depending on the salt concentration; a characteristic trait of charged pores [10,11,44]. These experiments, together with ion selectivity measurements [11], suggest that E protein behaves as a charged pore in negatively-charged membranes. Parallel observations are made when reconstituting SARS-CoV E protein in membranes containing a small percentage of charged lipids and, in particular, in membranes containing the same amount of charged lipid ($\sim$20%) as the ERGIC-Golgi membranes, where E protein channel is presumably localized [10]. These observations support the view that some polar lipid heads line the pore lumen together with the protein transmembrane helices and that both, protein and lipids, contribute to channel selectivity. An additional proof reinforcing this hypothesis is provided by electrophysiology experiments using lipids with negative intrinsic curvature. It is known that the protein-lipid pore formation is favored in membranes with positive curvature [45]. In fact, when E protein is reconstituted in dioleoyl phosphatidylethanolamine membranes, a lipid with intrinsic negative curvature, it is much more difficult to achieve channel insertion because the membrane negative curvature becomes energetically unfavorable for the assembly of a proteolipidic structure [10]. The E protein-lipid channel may not constitute a unique case, as poliovirus 2B permeabilization displayed a strict requirement for anionic phospholipids in the membrane composition [28], which may indicate that lipid molecules are involved in the pore formation.

Regulatory mechanisms, such as gating, operate in some viroporins. Histidine residues present in the IAV M2 transmembrane domain become protonated when encountering low pH conditions, and experience a conformational change that opens the channel pore allowing H^+ flux [46]. The influence of the histidine residues in the IC is also observed in the SH protein. In this protein the conductance is pH-dependent and consistent with the titration of these residues [43].

Viroporins could be regulating the transport capacity of cellular ion transporters, besides showing their intrinsic IC activity. A well-defined case is that of human papillomavirus oncoprotein E5, which interacts with vacuolar H^+ ATPase (V-ATPase) modulating its function and leading to cell transformation [47,48]. Interestingly, other viroporins interact with cellular ion pumps. SARS-CoV E protein binds the alpha subunit of Na^+/K^+ ATPase in infected cells [49]. IAV M2 protein binds Na^+/K^+ ATPase beta1 subunit [50]. It is well known that the ion transport capacity of cellular pumps such as Na^+/K^+ ATPase and sarcoendoplasmic reticulum Ca^{2+} ATPase (SERCA) may be influenced by the interaction with other proteins. FXYD proteins and phospholamban are well-characterized examples [51,52]. These regulatory proteins are small transmembrane proteins that oligomerize and eventually form ICs when reconstituted in artificial lipid membranes [53], as happens for viroporins. Some examples of viroporins modifying the activity of cellular ion channels have also been reported. HIV-1 Vpu interacts with the K^+ channel TASK-1 promoting viral release [54]. In addition, indirect inhibition of epithelial sodium channels has been reported for IAV M2 and SARS-CoV E proteins [55,56]. The possibility of a dual role of ion modulation by viroporins, depending both on their intrinsic transport properties and on their possible regulatory activity on key cellular ion transporters, increases the potential relevance of this family of proteins.

3. Ion Channel Activity and Virus-Host Interaction

3.1. Virus Production

Viroporins are involved in processes relevant for virus production. In general, these proteins do not affect viral genome replication, but stimulate other key aspects of the viral cycle such as entry, assembly, trafficking and release of viral particles [2,13]. As a consequence, partial or total deletion of viroporins usually leads to significant decreases in viral yields. HCV p7 viroporin is indispensable for virus propagation, as no infectious viruses are recovered when p7 protein is eliminated [57]. Other viruses are more tolerant to viroporin deletion. Thus, IAV or HIV-1 lacking M2 and Vpu, respectively, can be efficiently rescued. Nevertheless, in general, viroporin defective viruses show significantly lower virus yields, ranging from 10- to 100-fold [58,59]. The relevance of E viroporin in coronavirus production seems to be species-specific. Deletion of E gene in transmissible gastroenteritis virus (TGEV) and MERS-CoV generates replication-competent propagation-deficient viruses [60,61]. E gene is not essential for virus production in SARS-CoV and mouse hepatitis virus (MHV), although in its absence viral titters are reduced from 100- to 1000-fold, respectively [62,63]. Silencing of SARS-CoV 3a protein, PEDV 3 protein, and HCoV-229E 4a protein expression in virus infection cause a reduction in virus titers [24,25,64], supporting that these viroporins are involved in virus production. In the case of SARS-CoV 3a protein, this has also been shown in mice infected with a SARS-CoV variant missing 3a protein (C. Castaño-Rodriguez, J.L. Nieto-Torres and L. Enjuanes, unpublished results). In some cases, viroporin deletion induces a growth restriction that can be cell or tissue specific. Thus, RSV lacking the SH gene, shows an efficient growth in cell culture, but a limited production in the nasal turbinates of infected mice or chimpanzees, as compared with the wild type virus [65,66].

Collectively, previous data support that viroporins stimulate virus propagation. A key issue is to understand the relevance of their IC conductance on this activity. Both viroporin IC activity and other functions not related to ion conductivity apparently affect virus propagation. Some non-conductive viroporin domains are important players in viral morphogenesis. The cytoplasmic tail of several viroporins actively participates in virus production. A few examples are now briefly described. CoV E protein C-terminal domain interacts with the viral membrane (M) protein during the morphogenesis process [67]. In fact, the last nine amino acids of SARS-CoV E protein C-terminus stimulate virus growth [68]. This protein sequence includes a PDZ-binding motif involved in interactions with cellular proteins and virulence. IAV M2 cytoplasmic domain interacts with the M1 protein to favor virus assembly at the site of budding [69]. An amphipathic helix located in the cytoplasmic tail of the IAV M2 protein induces bending of cellular membranes, necessary for budding, and excision of the nascent particles [70]. This functional domain works independently to the M2 ion conductive properties. HIV-1 Vpu also facilitates budding termination of newly formed viruses in an IC independent manner. Vpu deleted viruses are less efficiently released and remain attached to the plasma membrane of infected cells [59]. In this case, the transmembrane domain of Vpu accounts for this phenotype by establishing critical interactions with cellular proteins such as one with the interferon-stimulated protein tetherin. This binding prevents tetherin from inhibiting the release of viral particles at the plasma membrane of infected cells [71,72]. Both IC activity and tetherin inhibitory property are concentrated in the Vpu transmembrane domain, nevertheless these functions work independently [73].

Accumulating reports support that viroporin ion conductivity favors virus yields. Inhibition of viroporin IC properties, through subtle mutations that may not interfere with other functions of the protein, affects viral growth to different extents. Point mutations that blocked HCV p7 IC activity either completely abrogated virus production or resulted in a 100-fold restriction in virus yields, depending on the virus strain [12,57]. Influenza viruses lacking M2 ion conductivity, presented either a 15-fold reduction of viral titer in tissue culture [74], or showed a standard production in cell culture but a restricted growth in the nasal turbinates of infected mice [75]. SARS-CoV E protein ion conductivity also supports viral production and fitness. E protein IC activity was knocked down by introducing point mutations in key residues of its transmembrane domain [10]. Although not essential for virus production, E protein IC activity stimulated viral propagation. Viruses lacking E protein ion conductivity were outcompeted by others displaying this function [22]. In fact, IC defective viruses tended to restore ion conductance by introducing compensatory mutations in the transmembrane domain of the protein both in cell cultures and in mice [22]. In agreement with these data, IAV lacking M2 IC activity showed fitness defects, as it was outgrowth by the parental virus, in an even faster manner than that observed in SARS-CoV [74]. The relevance of ion conductivity in boosting virus production remains unknown for the moment in other viral systems. Initial findings argued that scrambling of the HIV-1 Vpu transmembrane domain inactivated ion conductivity and partially inhibited viral production [76]. However, this alteration of the transmembrane domain affected the Vpu-tetherin interaction and IC activity. Recent reports showed that mutations that specifically inhibited ion conductivity, but not Vpu-tetherin interaction, did not affect VLP production; therefore, this interaction with tetherin seems to be responsible for the observed phenotype [73].

Pharmacological inhibition of viroporin ion conductance further supports the role of IC in virus propagation. Several compounds inhibit viroporin ion conductivity in artificial lipid membranes, and some of them efficiently reduce viral growth when administered to infected cells. Inhibition of M2 protein IC activity mediated by amantadine prevents or decreases viral growth in a strain specific manner [39,77]. Amantadine also inhibits HCV growth [13]. HCoV-229-E and MHV growth is restricted by hexamethylene amiloride, which blocks E protein ion conductivity [78]. The viral growth inhibitory properties of these compounds encourage its use as potential antivirals. This aspect will be further addressed below.

Widely conserved pathways influenced by viroporin ion conductivity, and others that seem to be species-specific are now described and graphically summarized to facilitate their overview (Figure 3).

3.1.1. Viral Entry

Early key aspects of virus cycle, such as viral entry, can be boosted through ion conductivity. Non-enveloped viral particles lack viroporins as structural components, and these proteins are poorly represented in the membrane of enveloped viruses [63,79–81]. Nevertheless, the viroporin molecules embedded in the viral envelope can actively participate in viral entry; IAV M2 protein is a well characterized example. IAV is internalized within the cell through an endocytic process. Endosomes containing viral particles fuse with acidic organelles to form endolysosomes. The H^+ ATPase of these organelles pumps H^+ inside their lumen lowering the pH. The few M2 copies present in the IAV envelope are activated by the low pH conditions and allow the flux of H^+ inside the viral particle. The acidification of the endocytosed virion drives a series of conformational changes in the viral hemagglutinin (HA) protein, leading to the exposure of a fusion peptide [82,83]. In parallel, the M1 protein layer, which underlies the viral envelope, and protects the viral ribonucleoproteins, is disassembled under these pH conditions [84]. The fusion peptide finally triggers the fusion of the viral and endosomal membranes, resulting in the release of viral ribonucleoproteins in the cell cytoplasm. Inhibition of M2 protein

IC activity by amantadine results in incomplete uncoating of the virion in some IAV strains [39,77]. This mechanism does not seem relevant for HCV virus, as its entry is independent of p7 IC activity [12]. Whether or not this pathway, or related ones, participate in the entrance of other viral species remains to be established.

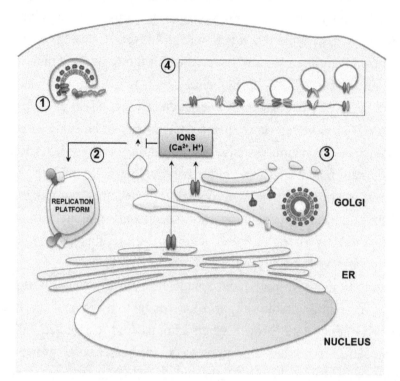

Figure 3. Pathways stimulated by viroporin ion channel activity leading to virus production. Viral-membrane embedded viroporins (red ellipses) transport H^+ inside the endocytosed virion. This causes structural changes in fusion and matrix proteins facilitating the uncoating of viral ribonucleoproteins (**1**); Viroporin-mediated ions leak from intracellular organelles such as the endoplasmic reticulum (ER) or the Golgi apparatus towards the cytoplasm causes a blockade of vesicle transport and/or hijacking of autophagic membranes. These processes finally result in the accumulation of membranous structures that will serve as platforms for viral replication and morphogenesis. Blue, red and green structures show the viral replicase (**2**); In addition, equilibration of Golgi and secretory pathway organelles' pH protect both viral proteins involved in entry (blue structures and green ellipses) and newly formed virions that can be sensitive to acidic environments (**3**); Viroporins (red and blue ellipses), locate in the budding neck of some enveloped viruses. These proteins may interact and oligomerize, rearranging the formation of channels, which additionally could facilitate virion scission (**4**).

3.1.2. Remodeling of Cell Organelles

Modification of the ionic content of intracellular compartments can ultimately result in functional and morphological transformations. The ion conductivity of several viroporins affects protein transport. Alteration of the ionic content of the Golgi apparatus and the endosomes interfere with cellular protein processing and sorting [85]. IAV M2 causes a lag in protein transport through the Golgi apparatus [86]. This delay on transport can be inhibited by the addition of amantadine, an IC activity inhibitor of

M2 protein. Furthermore, monensin, an H^+/Na^+ antiporter, causes similar defects. Both M2 and monensin induce dilation of Golgi cisternae, although to different extents. Infectious bronchitis virus (IBV) E protein also affects protein transport through the secretory pathway, whereas the mutant predicted to inhibit IC activity does not [87]. Coxsackievirus 2B protein disturbs pH and calcium homeostasis in the Golgi and the ER, thereby inhibiting protein transport. Mutations inhibiting 2B ion conductivity restored proper protein trafficking [88]. Ca^{2+} is involved in membrane fusion events, and its leakage to the cell cytoplasm causes anterograde vesicle trafficking blockage. SARS-CoV 3a protein is both necessary and sufficient for SARS-CoV Golgi fragmentation and promotes the accumulation of intracellular vesicles that may be used for virus formation or for non-lytic release of virus particles [89]. The exact relevance of protein transport delay on virus production remains to be established. It has been speculated that alteration of ionic homeostasis in transport vesicles affects anterograde trafficking leading to conglomeration of cell membranes that will serve as the platforms for viral replication and budding [88]. In addition, ionic efflux from intracellular organelles can lead to pathways triggering the accumulation of autophagic vacuoles as platforms for virus replication. In fact, rotavirus nsp4 viroporin allows Ca^{2+} efflux form ER thereby triggering autophagy in favor of viral infection [90].

3.1.3. Protection of the Viral Progeny

In many cases, viral proteins, and even virions, have to progress through the secretory pathway, encountering the low pH found within the Golgi apparatus lumen [91]. These acidic conditions may inactivate forming virions, which eventually are acid-sensitive [12]. Viroporin ion conductivity is critical in viral progeny protection and, occasionally, it can be trans-complemented by heterologous viroporins. IAV M2 IC activity is thought to keep the pH of the Golgi apparatus and its associated vesicles above a threshold, in order to avoid conformational changes in HA which may lead to its premature activation rendering non-infectious viruses. In fact, blocking of M2 ion conductivity by amantadine induces irreversible activation of HA [92]. This effect can be overcome by the treatment of infected cells with monensin that alkalizes the Golgi in an analogous manner to M2 protein. Similarly to what has been shown for M2 protein, HCV p7 protein mediates a H^+ leak from intracellular organelles, and this alkalinization is required for the production of infectious viruses [12]. Mutant viruses lacking p7 ion conductivity were not rescued, and treatment with amantadine greatly diminished the production of wild type infectious virions. Furthermore, in the absence of p7 IC activity, alternative approaches to balance the pH of intracellular organelles result in partial recovery of infectious viruses. Either Bafilomycin A1 treatment, an inhibitor of a H^+ vacuolar ATPase (V-ATPase), or expression of IAV M2, but not the M2 IC mutant, rescued virus production [12].

3.1.4. Release of Newly Formed Virions

The increase of cellular permeability by viroporins, IC activity may ultimately result in cell lysis, a requisite for the egress of non-enveloped viruses [2]. It has also been speculated that disruption of ion gradients may trigger the membrane fusion events necessary for budding termination and release of enveloped viruses [2]. Nevertheless, in the latter case other non-ion conductive viroporin functions, as those previously described for M2 and Vpu, may also be relevant in the release process. In addition, we propose that viroporins, located in the budding neck of forming viruses [70], could oligomerize to form an IC facilitating membrane fusion and budding termination.

3.2. Pathogenesis

Besides their key role in virus propagation, viroporins are also virulence factors in different viral systems. Total or partial deletion of non-essential viroporins usually leads to attenuated phenotypes. Interestingly, these viroporin-deleted viruses have been successfully used as potential vaccine candidates. Mice infected with an IAV lacking M2 protein showed no weight-loss, nor pathology, and were protected against the wild type virus [93]. An RSV defective for SH protein showed growth attenuation in the upper respiratory tracts of mice and chimpanzees and caused mild disease [65,66]. A classical swine fever virus (CSFV) missing full-length or partial-length p7 protein, similar to HCV p7 viroporin, lacked virulence [94]. A SARS-CoV lacking the full-length E protein (SARS-CoV-ΔE) was attenuated in hamsters, transgenic mice expressing the human SARS-CoV receptor (hACE2), and in both young and elderly mice. SARS-CoV-ΔE induced increased stress and limited NF-κB driven inflammation [62,95,96]. In addition, elimination of defined E protein regions of 6–12 amino acids in its carboxy-terminus leads to viral attenuation [68,97]. Several of these mutants are promising potential vaccine candidates for SARS-CoV. SARS-CoV 3a protein regulates host-cellular responses involved in the activation of pro-inflammatory genes such as C-Jun and NF-κB [98–100], and in the production of pro-inflammatory cytokines and chemokines such as IL-8 or CCL5 [99]. Furthermore, 3a protein IC activity has been linked to its pro-apoptotic function [101]. PEDV 3 protein has also been involved in pathogenesis as a PEDV with a 49-nucleotide deletion in 3 gene lacks IC activity and is attenuated [24].

Viroporins are frequently associated with virus propagation, as their deletion from the viral genome usually causes viral titer drops that, by themselves, could contribute to the attenuated viral phenotype. However, viroporin removal can modulate cellular signaling pathways leading to virulence. It has been proposed that the protein transport blockage exerted by the IC activity of viroporins, such as that from coxsackievirus 2B and coronavirus E proteins causes a delay in the transport of major histocompatibility complex class I (MHC-I) molecules towards the plasma membrane [87,88,102]. This MHC-I trafficking defect allows the evasion of adaptive immune responses, leading to more productive infections (Figure 4).

Over stimulation of immune responses can also lead to pathogenesis. Indeed, IC activity is an important trigger of immunopathology, as demonstrated for SARS-CoV E protein. Mutant viruses lacking ion conductivity, which showed proficient replication in mice lungs, presented an attenuated phenotype. Animals infected with the SARS-CoVs lacking E protein IC activity showed a reduced mortality in comparison with those inoculated with the parental virus [22]. Interestingly, mice infected with the IC defective viruses presented much less edema accumulation in the lung airways, the ultimate cause of acute respiratory distress syndrome (ARDS) [103]. Flooded bronchioles and alveoli fail to interchange gases, leading to hypoxemia and eventually to death. Noticeably, there was a good correlation between edema accumulation and the disassembly of the airway epithelia which, when intact, drive edema resolution through an ion mediated water reabsorption. Edema accumulation and airway epithelia damage is accompanied by an exacerbated proinflammatory response in the lung parenchyma [103]. Animals infected with the IC deficient viruses presented decreased levels of IL-1β, TNF and IL-6 in the lung airways, key mediators of the ARDS progression. IL-1β is a key orchestrator of the inflammatory response and plays a critical role in counteracting invading viruses, however overstimulation of this pathway can lead to unwanted deleterious effects for the organism. In fact, over production of IL-1β is related with important pathologies such as gout, atherosclerosis, diabetes, ARDS and asthma [104–107]. As a consequence, IL-1β production is tightly regulated in the organism, through the inflammasome multiprotein complex.

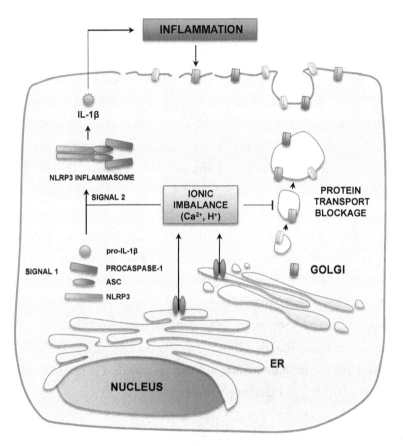

Figure 4. Pathways stimulated by viroporin ion channel activity leading to pathology. Molecular patterns associated with viral infections are recognized by cellular sensors (signal 1), which activate the transcription and translation of the NLRP3 inflammasome components (NLRP3, ASC and procaspase-1) and the inactive pro-IL-1β. Viroporins inserted in the intracellular organelles, such as the endoplasmic reticulum (ER) or the Golgi apparatus, favor the leak of Ca^{2+} and H^+ ions that move following their electrochemical gradient into cell cytoplasm. This ionic imbalance (signal 2) induces the assembly of the inflammasome complex, which triggers the maturation of pro-IL-1β into IL-1β through the action of caspase-1. Secreted IL-1β mediates a potent pro-inflammatory response that can be deleterious for the cell and the organism, when overstimulated. In addition, alteration of ionic milieus in intracellular compartments comes along with a protein transport delay or blockage. This results in a decrease of the levels of MHC-I molecules (blue rectangles) at the plasma membrane, preventing the infected cell to be recognized by the immune system. Protein transport blockage also diminishes the levels and activity in the cell surface of ion channels and transporters, crucial in the resolution of edema accumulation. Epithelial sodium channels (green structure) and Na^+/K^+ ATPase (purple rectangles) impairment have been related to the worsening of viral respiratory diseases such as those caused by SARS-CoV, IAV or RSV.

How can viral IC activity stimulate this exacerbated inflammatory response leading to pathology and disease? Inflammasomes participate in pathogen recognition by sensing disturbances in cellular milieus, including intracellular ionic concentrations [108,109]. The NLRP3 inflammasome is one of the most extensively studied [110]. Two signals are generally required to activate this complex: the first one is usually triggered by molecular patterns associated to the viral infection, such as double-stranded RNA. This leads to the transcription and translation of the different components of the inflammasome and the immature pro-IL-1β [16]. Only when a second signal is simultaneously present, the components of the inflammasome are assembled, leading to the cleavage and activation of caspase-1. This protein processes pro-IL-1β into its active form IL-1β that is released to the extracellular media to promote proinflammation [16]. Interestingly, viroporin IC activity represents the second signal required for inflammasome activation, inducing the release of active IL-1β (Figure 4). In general, viroporins induce an efflux of ions, such as H^+ and Ca^{2+} that move from their intracellular stores into the cytoplasm following strong electrochemical gradients, triggering the NLRP3 inflammasome. IAV M2 was the first noticed ion conducting viral protein activating this pathway [16]. Nowadays, several other viroporins stimulating the inflammasome have been identified. RSV SH, human rhinovirus 2B, encephalomyocarditis virus 2B, HCV p7, and IAV PB1-F2, among others, are additional examples [17–20,23,111]. This mechanism of overstimulation of IL-1β production seems to have key consequences in pathogenesis. The severity of human rhinovirus infection is linked to overinflammation and release of cytokines, including IL-1β. Rhinovirus 2B protein causes a Ca^{2+} leakage from the ER and Golgi apparatus essential for inflammasome activation and IL-1β production [19]. Similarly, RSV SH activates this pathway resulting in immunopathology [18].

Given the wide variety of effects that viroporin IC activity has on viral propagation and its influence on pathogenesis, inhibition of IC conductivity has been a promising target for therapeutic interventions. A growing list of compounds interfering with viroporin IC activity of several viruses such as IAV, HCV, HIV-1, coronaviruses, and RSV, has been described [43,78,112–114]. Despite being active in artificial lipid bilayers, and sometimes in cell culture, the pharmacological use of antagonists of IC activity in humans is still limited to a reduced number of cases. Amantadine was the first inhibitor of viroporin IC activity approved for use in humans, and it has been utilized in clinics for around 20 years in the treatment of IAV [113]. Amantadine binds M2 protein at the N-terminal lumen of the channel and at the C-terminal surface of the protein with high and low affinities, respectively, blocking ion conductance and interfering with viral replication [115]. However, IAV variants containing mutations in the M2 transmembrane domain that conferred resistance to amantadine emerged [115,116]. Amantadine and hexamethylene amiloride are effective inhibitors of HCV p7 IC activity as shown using *in vitro* systems [42,112]. HCV shows a genotype-dependent sensitivity to amantadine, when the drug is applied at high doses, which may explain its inefficacy in infected patients, limiting the application of amantadine in HCV disease treatment [117]. Similarly, high concentrations of hexamethylene amiloride are required to inhibit p7 conductivity in cell culture, which increased its toxicity, making this drug unsuitable for clinical administration [117]. However, drug screenings have identified other promising compounds interfering p7 IC activity, such as long alkyl iminosugars derivatives and BIT225, with the latter showing modest but successful restriction of HCV load in infected patients [6,118]. These compounds may represent the basis for a therapeutic treatment of hepatitis C.

Taking into consideration the fast selection of drug-resistant viruses under drug pressure, the simultaneous inhibition of IC activity, and other signaling pathways stimulated by ion conductivity leading to pathology, may represent a better treatment option. Interfering with these pathological pathways constitutes a valuable approach, with the advantage that it is independent of the appearance of drug resistant viruses. Targeting exacerbated inflammatory responses, such as those triggered by IC activity leading to inflammasome activation in various respiratory infections, may reduce disease worsening and progression [23]. Novel specific inhibitors of NLRP3 inflammasome showing promising results for inflammatory diseases could be applied for these purposes [119].

4. Summary and Future Prospects

Viroporins have been known for a long time as key contributors to virus propagation and stimulators of pathogenesis. Several reports have dissected the crucial impact that IC activity has in these processes. It is remarkable how these small channels have high implications at the cell level favoring virus production and, at the same time, causing disturbances at a tissue or organism level eventually leading to pathology. Understanding the molecular and physicochemical structure of these ion pores may constitute the basis for rational design of specific IC activity inhibitors and strategies to counteract the pathology mediated by IC activity.

Acknowledgments

The work done by the authors was supported by grants from the government of Spain (BIO2013-42869-R, FIS2013-40473-P), Generalitat Valenciana (Prometeo 2012/069), Fundació Caixa Castelló-Bancaixa (P1-1B2012-03) and a U.S. National Institutes of Health (NIH) project (5P01AI060699). JLN-T received a contract from NIH. CC-R received a contract from Fundacion La Caixa.

Author Contributions

Jose L. Nieto-Torres, Carmina Verdiá-Báguena, Carlos Castaño-Rodriguez, Vicente M. Aguilella and Luis Enjuanes, wrote and revised the manuscript. Jose L. Nieto-Torres and Carmina Verdiá-Báguena illustrated the figures.

References

1. Dubyak, G.R. Ion homeostasis, channels, and transporters: An update on cellular mechanisms. *Adv. Physiol. Educ.* **2004**, *28*, 143–154. [CrossRef] [PubMed]
2. Nieva, J.L.; Madan, V.; Carrasco, L. Viroporins: Structure and biological functions. *Nat. Rev. Microbiol.* **2012**, *10*, 563–574. [CrossRef] [PubMed]
3. De Jong, A.S.; Visch, H.J.; de Mattia, F.; van Dommelen, M.M.; Swarts, H.G.; Luyten, T.; Callewaert, G.; Melchers, W.J.; Willems, P.H.; van Kuppeveld, F.J. The coxsackievirus 2B protein increases efflux of ions from the endoplasmic reticulum and Golgi, thereby inhibiting protein trafficking through the Golgi. *J. Biol. Chem.* **2006**, *281*, 14144–14150. [CrossRef] [PubMed]

4. Ewart, G.D.; Sutherland, T.; Gage, P.W.; Cox, G.B. The Vpu protein of human immunodeficiency virus type 1 forms cation-selective ion channels. *J. Virol.* **1996**, *70*, 7108–7115. [PubMed]

5. Henkel, M.; Mitzner, D.; Henklein, P.; Meyer-Almes, F.J.; Moroni, A.; Difrancesco, M.L.; Henkes, L.M.; Kreim, M.; Kast, S.M.; Schubert, U.; *et al.* The proapoptotic influenza A virus protein PB1-F2 forms a nonselective ion channel. *PLoS ONE* **2010**, *5*, e11112. [CrossRef] [PubMed]

6. Pavlovic, D.; Neville, D.C.; Argaud, O.; Blumberg, B.; Dwek, R.A.; Fischer, W.B.; Zitzmann, N. The hepatitis C virus p7 protein forms an ion channel that is inhibited by long-alkyl-chain iminosugar derivatives. *Proc. Natl. Acad. Sci. USA* **2003**, *100*, 6104–6108. [CrossRef] [PubMed]

7. Pinto, L.H.; Holsinger, L.J.; Lamb, R.A. Influenza virus M2 protein has ion channel activity. *Cell* **1992**, *69*, 517–528. [CrossRef]

8. Wilson, L.; McKinlay, C.; Gage, P. SARS coronavirus E protein forms cation-selective ion channels. *Virology* **2004**, *330*, 322–331. [CrossRef] [PubMed]

9. Surya, W.; Li, Y.; Verdia-Baguena, C.; Aguilella, V.M.; Torres, J. MERS coronavirus envelope protein has a single transmembrane domain that forms pentameric ion channels. *Virus Res.* **2015**, *201*, 61–66. [CrossRef] [PubMed]

10. Verdia-Baguena, C.; Nieto-Torres, J.L.; Alcaraz, A.; Dediego, M.L.; Torres, J.; Aguilella, V.M.; Enjuanes, L. Coronavirus E protein forms ion channels with functionally and structurally-involved membrane lipids. *Virology* **2012**, *432*, 485–494. [CrossRef] [PubMed]

11. Verdia-Baguena, C.; Nieto-Torres, J.L.; Alcaraz, A.; Dediego, M.L.; Enjuanes, L.; Aguilella, V.M. Analysis of SARS-CoV E protein ion channel activity by tuning the protein and lipid charge. *Biochim. Biophys. Acta* **2013**, *1828*, 2026–2031. [CrossRef] [PubMed]

12. Wozniak, A.L.; Griffin, S.; Rowlands, D.; Harris, M.; Yi, M.; Lemon, S.M.; Weinman, S.A. Intracellular proton conductance of the hepatitis C virus p7 protein and its contribution to infectious virus production. *PLoS Pathog.* **2010**, *6*, e1001087. [CrossRef] [PubMed]

13. Steinmann, E.; Pietschmann, T. Hepatitis C virus p7-a viroporin crucial for virus assembly and an emerging target for antiviral therapy. *Viruses* **2010**, *2*, 2078–2095. [CrossRef] [PubMed]

14. Madan, V.; Castello, A.; Carrasco, L. Viroporins from RNA viruses induce caspase-dependent apoptosis. *Cell. Microbiol.* **2008**, *10*, 437–451. [CrossRef] [PubMed]

15. Bhowmick, R.; Halder, U.C.; Chattopadhyay, S.; Chanda, S.; Nandi, S.; Bagchi, P.; Nayak, M.K.; Chakrabarti, O.; Kobayashi, N.; Chawla-Sarkar, M. Rotaviral enterotoxin nonstructural protein 4 targets mitochondria for activation of apoptosis during infection. *J. Biol. Chem.* **2012**, *287*, 35004–35020. [CrossRef] [PubMed]

16. Ichinohe, T.; Pang, I.K.; Iwasaki, A. Influenza virus activates inflammasomes via its intracellular M2 ion channel. *Nat. Immunol.* **2010**, *11*, 404–410. [CrossRef] [PubMed]

17. Ito, M.; Yanagi, Y.; Ichinohe, T. Encephalomyocarditis virus viroporin 2B activates NLRP3 inflammasome. *PLoS Pathog.* **2012**, *8*, e1002857. [CrossRef] [PubMed]

18. Triantafilou, K.; Kar, S.; Vakakis, E.; Kotecha, S.; Triantafilou, M. Human respiratory syncytial virus viroporin SH: A viral recognition pathway used by the host to signal inflammasome activation. *Thorax* **2013**, *68*, 66–75. [CrossRef] [PubMed]

19. Triantafilou, K.; Kar, S.; van Kuppeveld, F.J.; Triantafilou, M. Rhinovirus-induced calcium flux

triggers NLRP3 and NLRC5 activation in bronchial cells. *Am. J. Respir. Cell Mol. Biol.* **2013**. [CrossRef] [PubMed]

20. McAuley, J.L.; Tate, M.D.; MacKenzie-Kludas, C.J.; Pinar, A.; Zeng, W.; Stutz, A.; Latz, E.; Brown, L.E.; Mansell, A. Activation of the NLRP3 inflammasome by IAV virulence protein PB1-F2 contributes to severe pathophysiology and disease. *PLoS Pathog.* **2013**, *9*, e1003392. [CrossRef] [PubMed]

21. Zhang, K.; Hou, Q.; Zhong, Z.; Li, X.; Chen, H.; Li, W.; Wen, J.; Wang, L.; Liu, W.; Zhong, F. Porcine reproductive and respiratory syndrome virus activates inflammasomes of porcine alveolar macrophages via its small envelope protein E. *Virology* **2013**, *442*, 156–162. [CrossRef] [PubMed]

22. Nieto-Torres, J.L.; Dediego, M.L.; Verdia-Baguena, C.; Jimenez-Guardeño, J.M.; Regla-Nava, J.A.; Fernandez-Delgado, R.; Castaño-Rodriguez, C.; Alcaraz, A.; Torres, J.; Aguilella, V.M.; *et al.* Severe acute respiratory syndrome coronavirus envelope protein ion channel activity promotes virus fitness and pathogenesis. *PLoS Pathog.* **2014**, *10*, e1004077. [CrossRef] [PubMed]

23. Triantafilou, K.; Triantafilou, M. Ion flux in the lung: Virus-induced inflammasome activation. *Trends Microbiol.* **2014**, *22*, 580–588. [CrossRef] [PubMed]

24. Wang, K.; Lu, W.; Chen, J.; Xie, S.; Shi, H.; Hsu, H.; Yu, W.; Xu, K.; Bian, C.; Fischer, W.B.; *et al.* PEDV ORF3 encodes an ion channel protein and regulates virus production. *FEBS Lett.* **2012**, *586*, 384–391. [CrossRef] [PubMed]

25. Zhang, R.; Wang, K.; Lv, W.; Yu, W.; Xie, S.; Xu, K.; Schwarz, W.; Xiong, S.; Sun, B. The ORF4a protein of human coronavirus 229E functions as a viroporin that regulates viral production. *Biochim. Biophys. Acta* **2014**. [CrossRef] [PubMed]

26. Lu, W.; Zheng, B.J.; Xu, K.; Schwarz, W.; Du, L.; Wong, C.K.; Chen, J.; Duan, S.; Deubel, V.; Sun, B. Severe acute respiratory syndrome-associated coronavirus 3a protein forms an ion channel and modulates virus release. *Proc. Natl. Acad. Sci. USA* **2006**, *103*, 12540–12545. [CrossRef] [PubMed]

27. Acharya, R.; Carnevale, V.; Fiorin, G.; Levine, B.G.; Polishchuk, A.L.; Balannik, V.; Samish, I.; Lamb, R.A.; Pinto, L.H.; DeGrado, W.F.; *et al.* Structure and mechanism of proton transport through the transmembrane tetrameric M2 protein bundle of the influenza A virus. *Proc. Natl. Acad. Sci. USA* **2010**, *107*, 15075–15080. [CrossRef] [PubMed]

28. Agirre, A.; Barco, A.; Carrasco, L.; Nieva, J.L. Viroporin-mediated membrane permeabilization. Pore formation by nonstructural poliovirus 2B protein. *J. Biol. Chem.* **2002**, *277*, 40434–40441. [CrossRef] [PubMed]

29. Plugge, B.; Gazzarrini, S.; Nelson, M.; Cerana, R.; Van Etten, J.L.; Derst, C.; DiFrancesco, D.; Moroni, A.; Thiel, G. A potassium channel protein encoded by Chlorella virus PBCV-1. *Science* **2000**, *287*, 1641–1644. [CrossRef] [PubMed]

30. Shim, J.W.; Yang, M.; Gu, L.Q. *In vitro* synthesis, tetramerization and single channel characterization of virus-encoded potassium channel Kcv. *FEBS Lett.* **2007**, *581*, 1027–1034. [CrossRef] [PubMed]

31. Parthasarathy, K.; Ng, L.; Lin, X.; Liu, D.X.; Pervushin, K.; Gong, X.; Torres, J. Structural

flexibility of the pentameric SARS coronavirus envelope protein ion channel. *Biophys. J.* **2008**, *95*, 39–41. [CrossRef] [PubMed]

32. Pervushin, K.; Tan, E.; Parthasarathy, K.; Lin, X.; Jiang, F.L.; Yu, D.; Vararattanavech, A.; Soong, T.W.; Liu, D.X.; Torres, J. Structure and inhibition of the SARS coronavirus envelope protein ion channel. *PLoS Pathog.* **2009**, *5*, e1000511. [CrossRef] [PubMed]

33. Park, S.H.; Mrse, A.A.; Nevzorov, A.A.; Mesleh, M.F.; Oblatt-Montal, M.; Montal, M.; Opella, S.J. Three-dimensional structure of the channel-forming trans-membrane domain of virus protein "u" (Vpu) from HIV-1. *J. Mol. Biol.* **2003**, *333*, 409–424. [CrossRef] [PubMed]

34. Gan, S.W.; Ng, L.; Lin, X.; Gong, X.; Torres, J. Structure and ion channel activity of the human respiratory syncytial virus (hRSV) small hydrophobic protein transmembrane domain. *Protein Sci.* **2008**, *17*, 813–820. [CrossRef] [PubMed]

35. Luik, P.; Chew, C.; Aittoniemi, J.; Chang, J.; Wentworth, P., Jr.; Dwek, R.A.; Biggin, P.C.; Venien-Bryan, C.; Zitzmann, N. The 3-dimensional structure of a hepatitis C virus p7 ion channel by electron microscopy. *Proc. Natl. Acad. Sci. USA* **2009**, *106*, 12712–12716. [CrossRef] [PubMed]

36. Wetherill, L.F.; Holmes, K.K.; Verow, M.; Muller, M.; Howell, G.; Harris, M.; Fishwick, C.; Stonehouse, N.; Foster, R.; Blair, G.E.; *et al.* High-risk human papillomavirus E5 oncoprotein displays channel-forming activity sensitive to small-molecule inhibitors. *J. Virol.* **2012**, *86*, 5341–5351. [CrossRef] [PubMed]

37. Tosteson, M.T.; Pinto, L.H.; Holsinger, L.J.; Lamb, R.A. Reconstitution of the influenza virus M2 ion channel in lipid bilayers. *J. Membr. Biol.* **1994**, *142*, 117–126. [CrossRef] [PubMed]

38. Marassi, F.M.; Ma, C.; Gratkowski, H.; Straus, S.K.; Strebel, K.; Oblatt-Montal, M.; Montal, M.; Opella, S.J. Correlation of the structural and functional domains in the membrane protein vpu from HIV-1. *Proc. Natl. Acad. Sci USA* **1999**, *96*, 14336–14341. [CrossRef] [PubMed]

39. Wang, C.; Lamb, R.A.; Pinto, L.H. Direct measurement of the influenza a virus M2 protein ion channel activity in mammalian cells. *Virology* **1994**, *205*, 133–140. [CrossRef] [PubMed]

40. Vijayvergiya, V.; Wilson, R.; Chorak, A.; Gao, P.F.; Cross, T.A.; Busath, D.D. Proton conductance of influenza virus M2 protein in planar lipid bilayers. *Biophys. J.* **2004**, *87*, 1697–1704. [CrossRef] [PubMed]

41. Lin, T.I.; Schroeder, C. Definitive assignment of proton selectivity and attoampere unitary current to the M2 ion channel protein of influenza A virus. *J. Virol.* **2001**, *75*, 3647–3656. [CrossRef] [PubMed]

42. Premkumar, A.; Wilson, L.; Ewart, G.D.; Gage, P.W. Cation-selective ion channels formed by p7 of hepatitis C virus are blocked by hexamethylene amiloride. *FEBS Lett.* **2004**, *557*, 99–103. [CrossRef]

43. Li, Y.; To, J.; Verdia-Baguena, C.; Dossena, S.; Surya, W.; Huang, M.; Paulmichl, M.; Liu, D.X.; Aguilella, V.M.; Torres, J. Inhibition of the human respiratory syncytial virus small hydrophobic protein and structural variations in a bicelle environment. *J. Virol.* **2014**, *88*, 11899–11914. [CrossRef] [PubMed]

44. Malev, V.V.; Schagina, L.V.; Gurnev, P.A.; Takemoto, J.Y.; Nestorovich, E.M.; Bezrukov, S.M. Syringomycin e channel: A lipidic pore stabilized by lipopeptide? *Biophys. J.* **2002**, *82*,

1985–1994. [CrossRef]

45. Sobko, A.A.; Kotova, E.A.; Antonenko, Y.N.; Zakharov, S.D.; Cramer, W.A. Effect of lipids with different spontaneous curvature on the channel activity of colicin E1: Evidence in favor of a toroidal pore. *FEBS Lett.* **2004**, *576*, 205–210. [CrossRef] [PubMed]

46. Wang, C.; Lamb, R.A.; Pinto, L.H. Activation of the M2 ion channel of influenza virus: A role for the transmembrane domain histidine residue. *Biophys. J.* **1995**, *69*, 1363–1371. [CrossRef]

47. Goldstein, D.J.; Schlegel, R. The E5 oncoprotein of bovine papillomavirus binds to a 16 kDa cellular protein. *EMBO J.* **1990**, *9*, 137–145. [PubMed]

48. Andresson, T.; Sparkowski, J.; Goldstein, D.J.; Schlegel, R. Vacuolar H(+)-ATPase mutants transform cells and define a binding site for the papillomavirus E5 oncoprotein. *J. Biol. Chem.* **1995**, *270*, 6830–6837. [PubMed]

49. Nieto-Torres, J.L.; DeDiego, M.L.; Alvarez, E.; Jimenez-Guardeño, J.M.; Regla-Nava, J.A.; Llorente, M.; Kremer, L.; Shuo, S.; Enjuanes, L. Subcellular location and topology of severe acute respiratory syndrome coronavirus envelope protein. *Virology* **2011**, *415*, 69–82. [CrossRef] [PubMed]

50. Mi, S.; Li, Y.; Yan, J.; Gao, G.F. Na(+)/K (+)-ATPase beta1 subunit interacts with M2 proteins of influenza A and B viruses and affects the virus replication. *Sci. China Life Sci.* **2010**, *53*, 1098–1105. [CrossRef] [PubMed]

51. Geering, K. Function of FXYD proteins, regulators of Na, K-ATPase. *J. Bioenerg. Biomembr.* **2005**, *37*, 387–392. [CrossRef] [PubMed]

52. Glaves, J.P.; Trieber, C.A.; Ceholski, D.K.; Stokes, D.L.; Young, H.S. Phosphorylation and mutation of phospholamban alter physical interactions with the sarcoplasmic reticulum calcium pump. *J. Mol. Biol.* **2011**, *405*, 707–723. [CrossRef] [PubMed]

53. Minor, N.T.; Sha, Q.; Nichols, C.G.; Mercer, R.W. The gamma subunit of the Na, K-ATPase induces cation channel activity. *Proc. Natl. Acad. Sci. USA* **1998**, *95*, 6521–6525. [CrossRef] [PubMed]

54. Hsu, K.; Seharaseyon, J.; Dong, P.; Bour, S.; Marban, E. Mutual functional destruction of HIV-1 Vpu and host Task-1 channel. *Mol. Cell* **2004**, *14*, 259–267. [CrossRef]

55. Lazrak, A.; Iles, K.E.; Liu, G.; Noah, D.L.; Noah, J.W.; Matalon, S. Influenza virus M2 protein inhibits epithelial sodium channels by increasing reactive oxygen species. *FASEB J.* **2009**, *23*, 3829–3842. [CrossRef] [PubMed]

56. Ji, H.L.; Song, W.; Gao, Z.; Su, X.F.; Nie, H.G.; Jiang, Y.; Peng, J.B.; He, Y.X.; Liao, Y.; Zhou, Y.J.; *et al.* SARS-CoV proteins decrease levels and activity of human ENaC via activation of distinct PKC isoforms. *Am. J. Physiol. Lung Cell. Mol. Physiol.* **2009**, *296*, L372–L383. [CrossRef] [PubMed]

57. Steinmann, E.; Penin, F.; Kallis, S.; Patel, A.H.; Bartenschlager, R.; Pietschmann, T. Hepatitis C virus p7 protein is crucial for assembly and release of infectious virions. *PLoS Pathog.* **2007**, *3*, e103. [CrossRef] [PubMed]

58. Cheung, T.K.; Guan, Y.; Ng, S.S.; Chen, H.; Wong, C.H.; Peiris, J.S.; Poon, L.L. Generation of recombinant influenza A virus without M2 ion-channel protein by introduction of a point mutation at the 5′ end of the viral intron. *J. Gen. Virol.* **2005**, *86*, 1447–1454. [CrossRef] [PubMed]

59. Strebel, K.; Klimkait, T.; Martin, M.A. A novel gene of HIV-1, Vpu, and its 16-kilodalton product. *Science* **1988**, *241*, 1221–1223. [CrossRef] [PubMed]

60. Ortego, J.; Escors, D.; Laude, H.; Enjuanes, L. Generation of a replication-competent, propagation-deficient virus vector based on the transmissible gastroenteritis coronavirus genome. *J. Virol.* **2002**, *76*, 11518–11529. [CrossRef] [PubMed]

61. Almazan, F.; DeDiego, M.L.; Sola, I.; Zuñiga, S.; Nieto-Torres, J.L.; Marquez-Jurado, S.; Andres, G.; Enjuanes, L. Engineering a replication-competent, propagation-defective Middle East respiratory syndrome coronavirus as a vaccine candidate. *mBio* **2013**, *4*, e00650-13. [CrossRef] [PubMed]

62. DeDiego, M.L.; Alvarez, E.; Almazan, F.; Rejas, M.T.; Lamirande, E.; Roberts, A.; Shieh, W.J.; Zaki, S.R.; Subbarao, K.; Enjuanes, L. A severe acute respiratory syndrome coronavirus that lacks the E gene is attenuated *in vitro* and *in vivo*. *J. Virol.* **2007**, *81*, 1701–1713. [CrossRef] [PubMed]

63. Kuo, L.; Masters, P.S. The small envelope protein E is not essential for murine coronavirus replication. *J. Virol.* **2003**, *77*, 4597–4608. [CrossRef] [PubMed]

64. Akerstrom, S.; Mirazimi, A.; Tan, Y.J. Inhibition of SARS-CoV replication cycle by small interference rnas silencing specific SARS proteins, 7a/7b, 3a/3b and S. *Antivir. Res.* **2007**, *73*, 219–227. [CrossRef] [PubMed]

65. Bukreyev, A.; Whitehead, S.S.; Murphy, B.R.; Collins, P.L. Recombinant respiratory syncytial virus from which the entire SH gene has been deleted grows efficiently in cell culture and exhibits site-specific attenuation in the respiratory tract of the mouse. *J. Virol.* **1997**, *71*, 8973–8982. [PubMed]

66. Whitehead, S.S.; Bukreyev, A.; Teng, M.N.; Firestone, C.Y.; St Claire, M.; Elkins, W.R.; Collins, P.L.; Murphy, B.R. Recombinant respiratory syncytial virus bearing a deletion of either the NS2 or SH gene is attenuated in chimpanzees. *J. Virol.* **1999**, *73*, 3438–3442. [PubMed]

67. Corse, E.; Machamer, C.E. The cytoplasmic tails of infectious bronchitis virus E and M proteins mediate their interaction. *Virology* **2003**, *312*, 25–34. [CrossRef]

68. Jimenez-Guardeño, J.M.; Nieto-Torres, J.L.; DeDiego, M.L.; Regla-Nava, J.A.; Fernandez-Delgado, R.; Castaño-Rodriguez, C.; Enjuanes, L. The PDZ-binding motif of severe acute respiratory syndrome coronavirus envelope protein is a determinant of viral pathogenesis. *PLoS Pathog.* **2014**, *10*, e1004320. [CrossRef] [PubMed]

69. Chen, B.J.; Leser, G.P.; Jackson, D.; Lamb, R.A. The influenza virus M2 protein cytoplasmic tail interacts with the M1 protein and influences virus assembly at the site of virus budding. *J. Virol.* **2008**, *82*, 10059–10070. [CrossRef] [PubMed]

70. Rossman, J.S.; Jing, X.; Leser, G.P.; Lamb, R.A. Influenza virus M2 protein mediates ESCRT-independent membrane scission. *Cell* **2010**, *142*, 902–913. [CrossRef] [PubMed]

71. Neil, S.J.; Zang, T.; Bieniasz, P.D. Tetherin inhibits retrovirus release and is antagonized by HIV-1 Vpu. *Nature* **2008**, *451*, 425–430. [CrossRef] [PubMed]

72. Van Damme, N.; Goff, D.; Katsura, C.; Jorgenson, R.L.; Mitchell, R.; Johnson, M.C.; Stephens, E.B.; Guatelli, J. The interferon-induced protein BST-2 restricts HIV-1 release and is downregulated from the cell surface by the viral Vpu protein. *Cell Host Microbe* **2008**, *3*, 245–252. [CrossRef] [PubMed]

73. Bolduan, S.; Votteler, J.; Lodermeyer, V.; Greiner, T.; Koppensteiner, H.; Schindler, M.; Thiel, G.; Schubert, U. Ion channel activity of HIV-1 Vpu is dispensable for counteraction of CD317. *Virology* **2011**, *416*, 75–85. [CrossRef] [PubMed]

74. Takeda, M.; Pekosz, A.; Shuck, K.; Pinto, L.H.; Lamb, R.A. Influenza A virus M2 ion channel activity is essential for efficient replication in tissue culture. *J. Virol.* **2002**, *76*, 1391–1399. [CrossRef] [PubMed]

75. Watanabe, T.; Watanabe, S.; Ito, H.; Kida, H.; Kawaoka, Y. Influenza A virus can undergo multiple cycles of replication without M2 ion channel activity. *J. Virol.* **2001**, *75*, 5656–5662. [CrossRef] [PubMed]

76. Schubert, U.; Ferrer-Montiel, A.V.; Oblatt-Montal, M.; Henklein, P.; Strebel, K.; Montal, M. Identification of an ion channel activity of the Vpu transmembrane domain and its involvement in the regulation of virus release from HIV-1-infected cells. *FEBS Lett.* **1996**, *398*, 12–18. [CrossRef]

77. Grambas, S.; Bennett, M.S.; Hay, A.J. Influence of amantadine resistance mutations on the pH regulatory function of the M2 protein of influenza A viruses. *Virology* **1992**, *191*, 541–549. [CrossRef]

78. Wilson, L.; Gage, P.; Ewart, G. Hexamethylene amiloride blocks E protein ion channels and inhibits coronavirus replication. *Virology* **2006**, *353*, 294–306. [CrossRef] [PubMed]

79. Zebedee, S.L.; Lamb, R.A. Influenza A virus M2 protein: Monoclonal antibody restriction of virus growth and detection of M2 in virions. *J. Virol.* **1988**, *62*, 2762–2772. [PubMed]

80. Maeda, J.; Repass, J.F.; Maeda, A.; Makino, S. Membrane topology of coronavirus E protein. *Virology* **2001**, *281*, 163–169. [CrossRef] [PubMed]

81. Raamsman, M.J.B.; Locker, J.K.; de Hooge, A.; de Vries, A.A.F.; Griffiths, G.; Vennema, H.; Rottier, P.J.M. Characterization of the coronavirus mouse hepatitis virus strain A59 small membrane protein E. *J. Virol.* **2000**, *74*, 2333–2342. [CrossRef] [PubMed]

82. Wharton, S.A.; Belshe, R.B.; Skehel, J.J.; Hay, A.J. Role of virion M2 protein in influenza virus uncoating: Specific reduction in the rate of membrane fusion between virus and liposomes by amantadine. *J. Gen. Virol.* **1994**, *75*, 945–948. [CrossRef] [PubMed]

83. Bullough, P.A.; Hughson, F.M.; Skehel, J.J.; Wiley, D.C. Structure of influenza haemagglutinin at the pH of membrane fusion. *Nature* **1994**, *371*, 37–43. [CrossRef] [PubMed]

84. Stauffer, S.; Feng, Y.; Nebioglu, F.; Heilig, R.; Picotti, P.; Helenius, A. Stepwise priming by acidic pH and a high K+ concentration is required for efficient uncoating of influenza A virus cores after penetration. *J. Virol.* **2014**, *88*, 13029–13046. [CrossRef] [PubMed]

85. Nakamura, N.; Tanaka, S.; Teko, Y.; Mitsui, K.; Kanazawa, H. Four Na+/H+ exchanger isoforms are distributed to Golgi and post-Golgi compartments and are involved in organelle pH regulation. *J. Biol. Chem.* **2005**, *280*, 1561–1572. [CrossRef] [PubMed]

86. Sakaguchi, T.; Leser, G.P.; Lamb, R.A. The ion channel activity of the influenza virus M2 protein affects transport through the Golgi apparatus. *J. Cell Biol.* **1996**, *133*, 733–747. [CrossRef] [PubMed]

87. Ruch, T.R.; Machamer, C.E. A single polar residue and distinct membrane topologies impact the function of the infectious bronchitis coronavirus E protein. *PLoS Pathog.* **2012**, *8*, e1002674.

[CrossRef] [PubMed]

88. De Jong, A.S.; de Mattia, F.; van Dommelen, M.M.; Lanke, K.; Melchers, W.J.; Willems, P.H.; van Kuppeveld, F.J. Functional analysis of picornavirus 2B proteins: Effects on calcium homeostasis and intracellular protein trafficking. *J. Virol.* **2008**, *82*, 3782–3790. [CrossRef] [PubMed]

89. Freundt, E.C.; Yu, L.; Goldsmith, C.S.; Welsh, S.; Cheng, A.; Yount, B.; Liu, W.; Frieman, M.B.; Buchholz, U.J.; Screaton, G.R.; *et al.* The open reading frame 3a protein of severe acute respiratory syndrome-associated coronavirus promotes membrane rearrangement and cell death. *J. Virol.* **2010**, *84*, 1097–1109. [CrossRef] [PubMed]

90. Crawford, S.E.; Hyser, J.M.; Utama, B.; Estes, M.K. Autophagy hijacked through viroporin-activated calcium/calmodulin-dependent kinase kinase-beta signaling is required for rotavirus replication. *Proc. Natl. Acad. Sci. USA* **2012**, *109*, E3405–E3413. [CrossRef] [PubMed]

91. Anderson, R.G.; Orci, L. A view of acidic intracellular compartments. *J. Cell Biol.* **1988**, *106*, 539–543. [CrossRef] [PubMed]

92. Takeuchi, K.; Lamb, R.A. Influenza virus M2 protein ion channel activity stabilizes the native form of fowl plague virus hemagglutinin during intracellular transport. *J. Virol.* **1994**, *68*, 911–919. [PubMed]

93. Watanabe, S.; Watanabe, T.; Kawaoka, Y. Influenza A virus lacking M2 protein as a live attenuated vaccine. *J. Virol.* **2009**, *83*, 5947–5950. [CrossRef] [PubMed]

94. Gladue, D.P.; Holinka, L.G.; Largo, E.; Fernandez Sainz, I.; Carrillo, C.; O'Donnell, V.; Baker-Branstetter, R.; Lu, Z.; Ambroggio, X.; Risatti, G.R.; *et al.* Classical swine fever virus p7 protein is a viroporin involved in virulence in swine. *J. Virol.* **2012**, *86*, 6778–6791. [CrossRef] [PubMed]

95. DeDiego, M.L.; Nieto-Torres, J.L.; Jimenez-Guardeño, J.M.; Regla-Nava, J.A.; Alvarez, E.; Oliveros, J.C.; Zhao, J.; Fett, C.; Perlman, S.; Enjuanes, L. Severe acute respiratory syndrome coronavirus envelope protein regulates cell stress response and apoptosis. *PLoS Pathog.* **2011**, *7*, e1002315. [CrossRef] [PubMed]

96. DeDiego, M.L.; Nieto-Torres, J.L.; Regla-Nava, J.A.; Jimenez-Guardeño, J.M.; Fernandez-Delgado, R.; Fett, C.; Castaño-Rodriguez, C.; Perlman, S.; Enjuanes, L. Inhibition of NF-kappaB mediated inflammation in severe acute respiratory syndome coronavirus-infected mice increases survival. *J. Virol.* **2014**, *88*, 913–924. [CrossRef] [PubMed]

97. Regla-Nava, J.A.; Nieto-Torres, J.L.; Jimenez-Guardeño, J.M.; Fernandez-Delgado, R.; Fett, C.; Castaño-Rodriguez, C.; Perlman, S.; Enjuanes, L.; DeDiego, M.L. SARS coronaviruses with mutations in E protein are attenuated and promising vaccine candidates. *J. Virol.* **2015**, *89*, 3870–3887. [CrossRef] [PubMed]

98. Obitsu, S.; Ahmed, N.; Nishitsuji, H.; Hasegawa, A.; Nakahama, K.; Morita, I.; Nishigaki, K.; Hayashi, T.; Masuda, T.; Kannagi, M. Potential enhancement of osteoclastogenesis by severe acute respiratory syndrome coronavirus 3a/X1 protein. *Arch. Virol.* **2009**, *154*, 1457–1464. [CrossRef] [PubMed]

99. Kanzawa, N.; Nishigaki, K.; Hayashi, T.; Ishii, Y.; Furukawa, S.; Niiro, A.; Yasui, F.; Kohara, M.; Morita, K.; Matsushima, K.; *et al.* Augmentation of chemokine production by severe acute

respiratory syndrome coronavirus 3a/X1 and 7a/X4 proteins through NF-kappaB activation. *FEBS Lett.* **2006**, *580*, 6807–6812. [CrossRef] [PubMed]

100. Narayanan, K.; Huang, C.; Makino, S. SARS coronavirus accessory proteins. *Virus Res.* **2008**, *133*, 113–121. [CrossRef] [PubMed]

101. Wong, S.L.; Chen, Y.; Chan, C.M.; Chan, C.S.; Chan, P.K.; Chui, Y.L.; Fung, K.P.; Waye, M.M.; Tsui, S.K.; Chan, H.Y. *In vivo* functional characterization of the SARS-coronavirus 3a protein in drosophila. *Biochem. Biophys. Res. Commun.* **2005**, *337*, 720–729. [CrossRef] [PubMed]

102. Cornell, C.T.; Kiosses, W.B.; Harkins, S.; Whitton, J.L. Coxsackievirus B3 proteins directionally complement each other to downregulate surface major histocompatibility complex class I. *J. Virol.* **2007**, *81*, 6785–6797. [CrossRef] [PubMed]

103. Matthay, M.A.; Zemans, R.L. The acute respiratory distress syndrome: Pathogenesis and treatment. *Annu. Rev. Pathol.* **2011**, *6*, 147–163. [CrossRef] [PubMed]

104. Pugin, J.; Ricou, B.; Steinberg, K.P.; Suter, P.M.; Martin, T.R. Proinflammatory activity in bronchoalveolar lavage fluids from patients with ards, a prominent role for interleukin-1. *Am. J. Respir. Crit. Care Med.* **1996**, *153*, 1850–1856. [CrossRef] [PubMed]

105. Grommes, J.; Soehnlein, O. Contribution of neutrophils to acute lung injury. *Mol. Med.* **2011**, *17*, 293–307. [CrossRef] [PubMed]

106. dos Santos, G.; Kutuzov, M.A.; Ridge, K.M. The inflammasome in lung diseases. *Am. J. Physiol. Lung Cell. Mol. Physiol.* **2012**, *303*, L627–L633. [CrossRef] [PubMed]

107. Strowig, T.; Henao-Mejia, J.; Elinav, E.; Flavell, R. Inflammasomes in health and disease. *Nature* **2012**, *481*, 278–286. [CrossRef] [PubMed]

108. Murakami, T.; Ockinger, J.; Yu, J.; Byles, V.; McColl, A.; Hofer, A.M.; Horng, T. Critical role for calcium mobilization in activation of the NLRP3 inflammasome. *Proc. Natl. Acad. Sci. USA* **2012**, *109*, 11282–11287. [CrossRef] [PubMed]

109. Munoz-Planillo, R.; Kuffa, P.; Martinez-Colon, G.; Smith, B.L.; Rajendiran, T.M.; Nunez, G. K(+) efflux is the common trigger of NLRP3 inflammasome activation by bacterial toxins and particulate matter. *Immunity* **2013**, *38*, 1142–1153. [CrossRef] [PubMed]

110. Elliott, E.I.; Sutterwala, F.S. Initiation and perpetuation of NLRP3 inflammasome activation and assembly. *Immunol. Rev.* **2015**, *265*, 35–52. [CrossRef] [PubMed]

111. Shrivastava, S.; Mukherjee, A.; Ray, R.; Ray, R.B. Hepatitis C virus induces interleukin-1beta (IL-1beta)/IL-18 in circulatory and resident liver macrophages. *J. Virol.* **2013**, *87*, 12284–12290. [CrossRef] [PubMed]

112. Griffin, S.D.; Beales, L.P.; Clarke, D.S.; Worsfold, O.; Evans, S.D.; Jaeger, J.; Harris, M.P.; Rowlands, D.J. The p7 protein of hepatitis C virus forms an ion channel that is blocked by the antiviral drug, amantadine. *FEBS Lett.* **2003**, *535*, 34–38. [CrossRef]

113. Oxford, J.S. Antivirals for the treatment and prevention of epidemic and pandemic influenza. *Influenza Other Respir. Viruses* **2007**, *1*, 27–34. [CrossRef] [PubMed]

114. Ewart, G.D.; Mills, K.; Cox, G.B.; Gage, P.W. Amiloride derivatives block ion channel activity and enhancement of virus-like particle budding caused by HIV-1 protein Vpu. *Eur. Biophys. J.* **2002**, *31*, 26–35. [CrossRef] [PubMed]

115. Cady, S.D.; Schmidt-Rohr, K.; Wang, J.; Soto, C.S.; Degrado, W.F.; Hong, M. Structure of the amantadine binding site of influenza M2 proton channels in lipid bilayers. *Nature* **2010**, *463*, 689–692. [CrossRef] [PubMed]

116. Govorkova, E.A.; Webster, R.G. Combination chemotherapy for influenza. *Viruses* **2010**, *2*, 1510–1529. [CrossRef] [PubMed]

117. Griffin, S.; Stgelais, C.; Owsianka, A.M.; Patel, A.H.; Rowlands, D.; Harris, M. Genotype-dependent sensitivity of hepatitis C virus to inhibitors of the p7 ion channel. *Hepatology* **2008**, *48*, 1779–1790. [CrossRef] [PubMed]

118. Luscombe, C.A.; Huang, Z.; Murray, M.G.; Miller, M.; Wilkinson, J.; Ewart, G.D. A novel hepatitis C virus p7 ion channel inhibitor, BIT225, inhibits bovine viral diarrhea virus *in vitro* and shows synergism with recombinant interferon-alpha-2b and nucleoside analogues. *Antivir. Res.* **2010**, *86*, 144–153. [CrossRef] [PubMed]

119. Coll, R.C.; Robertson, A.A.; Chae, J.J.; Higgins, S.C.; Munoz-Planillo, R.; Inserra, M.C.; Vetter, I.; Dungan, L.S.; Monks, B.G.; Stutz, A.; *et al.* A small-molecule inhibitor of the NLRP3 inflammasome for the treatment of inflammatory diseases. *Nat. Med.* **2015**, *21*, 248–255. [PubMed]

3

Autophagy Activated by Bluetongue Virus Infection Plays a Positive Role in its Replication

Shuang Lv, Qingyuan Xu, Encheng Sun, Tao Yang, Junping Li, Yufei Feng, Qin Zhang, Haixiu Wang, Jikai Zhang and Donglai Wu *

State Key Laboratory of Veterinary Biotechnology, Harbin Veterinary Research Institute, Chinese Academy of Agricultural Sciences, Harbin 150001, China;
E-Mails: shuanglv8728@163.com (S.L.); x_qingyuan@163.com (Q.X.); ecsunhvri@163.com (E.S.); yangtao04@126.com (T.Y.); lijunping916@163.com (J.L.); fyf_anna@163.com (Y.F.); zq_dolphin11@163.com (Q.Z.); wanghaixiu217@163.com (H.W.); 15776628009@163.com (J.Z.)

* Author to whom correspondence should be addressed; E-Mail: dlwu@hvri.ac.cn.

Academic Editor: Andrew Mehle

Abstract: Bluetongue virus (BTV) is an important pathogen of wild and domestic ruminants. Despite extensive study in recent decades, the interplay between BTV and host cells is not clearly understood. Autophagy as a cellular adaptive response plays a part in many viral infections. In our study, we found that BTV1 infection triggers the complete autophagic process in host cells, as demonstrated by the appearance of obvious double-membrane autophagosome-like vesicles, GFP-LC3 dots accumulation, the conversion of LC3-I to LC3-II and increased levels of autophagic flux in BSR cells (baby hamster kidney cell clones) and primary lamb lingual epithelial cells upon BTV1 infection. Moreover, the results of a UV-inactivated BTV1 infection assay suggested that the induction of autophagy was dependent on BTV1 replication. Therefore, we investigated the role of autophagy in BTV1 replication. The inhibition of autophagy by pharmacological inhibitors (3-MA, CQ) and RNA interference (siBeclin1) significantly decreased viral protein synthesis and virus yields. In contrast, treating BSR cells with rapamycin, an inducer of autophagy, promoted viral protein expression and the production of infectious BTV1. These findings lead us to conclude that autophagy is activated by BTV1 and contributes to its replication, and provide novel insights into BTV-host interactions.

Keywords: bluetongue virus (BTV); autophagy; interplay; replication

1. Introduction

Bluetongue is an insect-transmitted disease that infects wild and domestic ruminants, especially cattle and sheep [1]. The causative agent, bluetongue virus (BTV), a member of the *Orbivirus* genus in the *Reoviridae* family, is an architecturally complex arbovirus. The virion is a non-enveloped icosahedral particle consisting of the outermost double-capsids and ten genomic double-stranded RNA (dsRNA) segments that encode seven structural proteins (VP1 to VP7) and four nonstructural proteins (NS1 to NS4) [2,3]. Due to the impact of bluetongue on international trade and economic development, BTV has been the subject of extensive molecular virology and structural biology studies [4–7]. Even so, the underlying mechanisms guiding the interactions between the virus and host remain poorly understood.

Viruses exploit multiple mechanisms to modulate host cells, presumably for replicative advantages. Among them, autophagy is a highly conserved catabolic process that mediates the clearance of long-lived proteins and damaged organelles via a lysosomal degradative pathway, encompassing macroautophagy, microautophagy and chaperone-mediated autophagy [8,9]. Many autophagy-related genes (ATG) are involved in this intracellular degradation process to maintain the homeostasis of cells. These genes work in coordination to regulate autophagy, including the formation of autophagosomes and their fusion with lysosomes [10]. Under normal conditions, autophagy proceeds at a basal level, but it is markedly activated in response to a variety of extracellular and intracellular stimuli, including nutrient starvation, energy depletion, reactive oxygen stress and microbial infection [11–13].

Many recent studies have reported that viral infections are involved in the complex autophagic process. Although autophagy commonly serves as a defense mechanism against viral infection [14,15], some viruses appear to be unaffected by autophagy, and many viruses have evolved to escape or to exploit this mechanism to promote their survival and replication in different ways [8,16–18]. Thus, the role of autophagy in host-virus interactions is diverse for different viruses.

Regarding BTV, we speculate that autophagy is likely to be involved in BTV infection based on several experimental clues: (i) In one study, activated host autophagy appeared to inhibit post-BTV infection when cells were treated with two compound agents, C003 and C052 [19], but a more specialized and systematic verification has not been reported. In addition, another member of the *Orbivirus* genus, epizootic hemorrhagic disease virus (EHDV), can induce autophagy in cultured mammalian cells [20]; (ii) Recent research findings have described the modulation of autophagy by mammalian reovirus (MRV) and by avian reovirus (ARV), which both belong to the *Reoviridae* family and share a similar virus structure [13,21,22]; (iii) Finally, crosstalk has been observed between apoptosis and autophagy pathways [23,24]. Exploitation of the host apoptosis pathway is a critical strategy utilized by BTV for its pathological effects, as evidenced by a recent experimental system that demonstrated that BTV infection triggers apoptosis in mammalian cells and that the uncoating of BTV was required for this process [25,26]. However, to date, the role of autophagy during BTV infection and replication is unknown.

In this report, we provide evidence that autophagy is triggered in permissive BSR cells upon infection by BTV1. Notably, a similar phenomenon was observed in primary natural host cells. By regulating autophagy via pharmacological treatment and RNA interference, we demonstrate that autophagy plays a positive and critical role in BTV replication. A better comprehension of the interactions between BTV and host autophagic responses will provide new insights into viral pathogenesis and antiviral drug development.

2. Materials and Methods

2.1. Cells, Viruses and Plasmids

BSR cells were maintained in Dulbecco's modified Eagle's medium (DMEM) supplemented with 5% fetal bovine serum (FBS) and antibiotics (0.1 mg· mL^{-1} of streptomycin and 100 IU· mL^{-1} of penicillin). Primary lamb lingual epithelial cells were maintained in DMEM supplemented with 10% FBS and 1% antibiotics. BTV1 strain SZ97/1 (GenBank No. JN848767) was propagated in BHK-21 cells.

To construct pEGFP-LC3B, the LC3B gene was amplified from BSR cells with primers (LC3F: GTGAATTCTCCGTCCGAGAAGACCTTCAAG; LC3R: AAGTCGACTTACACAGCCATTGCTGTCCCGA) that were designed based on the sequence of LC3B (GenBank No. NM_026160.4) and then cloned into pEGFP-C1 to express LC3B fused with the EGFP protein at its *N*-terminus.

2.2. Virus Infection and Titration Assays

BSR cells in 6-well plates were infected with BTV1 at a multiplicity of infection (MOI) of 1 (100 μL per well) for autophagy induction, and the supernatant was removed after absorption for 1 h. The cell monolayers were rinsed three times with sterile phosphate buffered saline (PBS, pH 7.2) and incubated in fresh complete medium at 37 °C for the indicated times. Then, the extracellular viruses from cell supernatants and intracellular viruses from cells treated with multiple freeze-thaws were harvested and serially diluted 10-fold to infect confluent BSR cells in 96-well plates. Finally, cytopathic effects (CPEs) were observed and recorded at 72 h post-infection (hpi). Virus titers were calculated as TCID$_{50}$/mL using the Reed-Muench method.

2.3. SDS-PAGE and Western Blot

Following the indicated treatment or transfection, all samples were lysed using Western and IP lysis buffer (P0013, Beyotime, Beijing, China) containing PMSF and Complete Protease Inhibitor Cocktail (04693116001, Roche, Penzberg, Germany). The clarified lysates pelleted by centrifugation were boiled in 5× SDS-PAGE loading buffer for 10 min, and then were subjected to 12% SDS-PAGE gel and then electrotransferred onto NC membranes. The membranes were blocked for 2 h, and incubated with the following primary antibodies for 2 h: Rabbit anti-LC3B antibody (L7543, Sigma, St Louis, MO, USA), rabbit anti-p62/SQSTM1 antibody (P0067, Sigma), mouse anti-actin antibody (TA-09, Zhongshan Golden Bridge, Beijing, China), rabbit anti-Beclin1 antibody (3738S, CST, Danvers, MA, USA) and mouse monoclonal antibodies against the VP2 NS3 and NS1 proteins of BTV1. The membranes were then incubated with IRDye® 800 CW goat anti-mouse IgG or goat anti-rabbit IgG (LiCor BioSciences, Lincoln, NA, USA) as secondary antibodies. The blots were visualized using an Odyssey Infrared Imaging System (LiCor BioSciences).

2.4. Transmission Electron Microscopy

Transmission electron microscopy (TEM) was used to observe autophagosomes as described previously [27]. Specifically, BSR cells were collected at 12 and 24 hpi with BTV1 at an MOI of 1. The mock group and rapamycin-treated group were from 24 h samples. Then, the samples were prepared for TEM observation. Autophagosomes were judged as double-membrane vesicles with diameter of 0.5 to 2.0 μm.

2.5. Confocal Fluorescence Microscopy

For the detection of autophagosomes, BSR cells were grown to 70% confluence in 35-mm glass-bottomed culture dishes (NEST) and then transfected with 2 μg of pEGFP-LC3 using Lipofectamine 2000 (11668-27, Invitrogen, Carlsbad, CA, USA). The cells were disposed with rapamycin or virus as described above at 12 h post-transfection, and then were fixed with cold absolute ethyl alcohol at 24 h post-treatment. The formation of EGFP-LC3 fluorescence dots was observed using a Leica SP2 confocal system (Leica Microsystems, Wetzlar, Germany). The BTV infection was marked using antibody against the BTV NS1 protein, followed by the corresponding secondary antibody conjugated to TRITC. The cell nucleus was counterstained with DAPI.

To observe the effect of chloroquine (CQ) on LC3 puncta, BSR cells transfected by pEGFP-LC3 were infected with BTV in presence or absence of CQ for 24 h at 12 h post-transfection. The cell nucleus was counterstained with DAPI.

To detect autophagic flux, BSR cells were transfected with pEGFP-LC3, infected with BTV or left untreated as described above. After fixment and permeabilization, the cells were stained with 50 nM LysoTracker Red DND-99 (acidic compartments marker) (C1046, Beyotime) for 30 min. After the cells were washed three times with PBS, they were processed to analyze using a confocal microscope.

2.6. Preparation of Ultraviolet (UV)-Inactivated BTV1

To obtain a replication-incompetent BTV1 strain, UV inactivation was performed as described previously [28]. Virus preparations (2 mL) were exposed to 254 nm UV, 10 cm from the source, in a 35-mm diameter dish for 40 min. The loss of infectivity of UV-inactivated BTV1 was experimentally verified in BSR cells.

2.7. Cell Viability and Drug Treatment Assay

Three drugs, 3-methyladenine (3-MA) (M9281, Sigma), chloroquine (CQ) (C6628, Sigma) and rapamycin (Rapa) (S1842, Beyotime), were used in this experiment to explore the effects of autophagy on the replication of BTV. The cytotoxicity of these drugs at optimal concentrations (10 mM 3-MA;

20 μM CQ; 100 nM Rapa) was tested using a WST-1 cell proliferation and cytotoxicity assay kit (C0035, Beyotime) according to the manufacturer's protocol. Next, BSR cells were pretreated with optimal concentrations of Rapa, 3-MA for 2-4 h and were not pretreated with CQ prior to BTV infection. Subsequently, the cells were infected with BTV1 as above and further incubated in fresh media in the absence or presence of these drugs at the same concentrations as for the pretreatments. Corresponding DMSO or ddH2O were used as vehicle controls. At the indicated times, the virus from the supernatants and cells were collected and the viral protein expression were detected as above.

2.8. RNA Interference of Beclin1

To further determine the role of autophagy induced by BTV during its life cycle, small interfering RNAs (siRNA) were synthesized by Shanghai GenePharma Co., Ltd. (Shanghai, China) to target Beclin1. The siRNA sequences for targeting Beclin1 (siBeclin1) were as follows: GAUGGUGUCUCUCGAAGAUTT (sense) and AUCUUCGAGAGACACCAUCTT (antisense). To evaluate the efficiency of the siRNAs, 80 pmol siBeclin1 and negative control siRNA (siNC) were transfected into BSR cells cultured in a 12-well plate using Lipofectamine 2000 following the manufacturer's instructions. At 24 h post-transfection, the knockdown efficiency was determined by examining endogenous Beclin1 expression by Western blot. The cells were infected with BTV1 (MOI = 1, 50 μL per well) at 24 h post-transfection and incubated for 24 h before samples were collected to determine the virus titer and viral protein expression.

2.9. Statistical Analysis

Statistical analyses were performed using GraphPad Prism 6.0 software. Student's t-test was used to analysis significance. p values of <0.05 were considered significantly different.

3. Results

3.1. Notable Autophagy is Triggered by BTV1 Infection in BSR Cells

BTV1, the prevalent serotype of BTV in China, was selected to explore whether BTV can trigger autophagy in infected BSR cells. First, at 12 or 24 hpi, the formation of autophagosomes in BSR cells was examined by transmission electron microscopy (TEM), which is considered the most convincing approach for monitoring autophagosome induction [29]. As shown in Figure 1A, double-membrane vesicles were rarely observed in mock-infected cells. In contrast, in BTV1-infected cells, several double-membrane autophagic vesicles appeared, in which cytosolic components or organelles were sequestered. These double-membrane structures were similar to the autophagosomes in BSR cells induced by rapamycin, a well-known autophagy inducer, suggesting that the autophagosomes were induced in BSR cells by BTV1 infection.

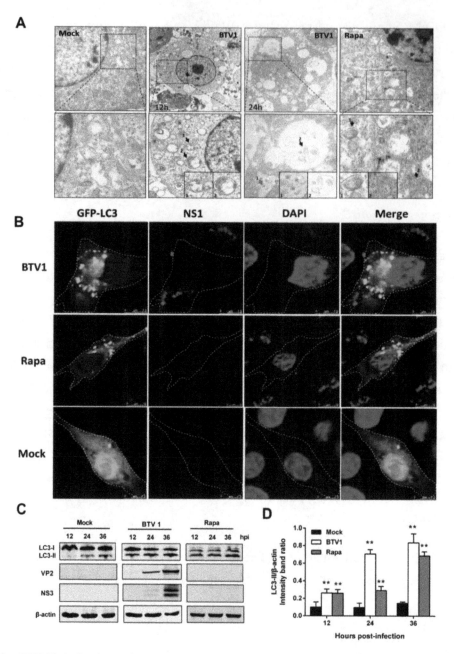

Figure 1. BTV1 infection triggers cellular autophagy in BSR cells. (**A**) TEM analysis. Mock-treated BSR cells as a negative control showed normal distribution of organelles. BSR cells infected with BTV1 at an MOI of 1 were harvested at 12 or 24 hpi and subjected to TEM observation. Cup-shaped membranous structures or autophagosome-like structures in the cytoplasm are indicated by the black arrows and BTV particles were pointed by red arrow. Higher-magnification views were listed below. Scale bars: 500 nm. Rapamycin-treated BSR cells were set as a positive control; (**B**) Confocal microscopy. Aggregation of autophagosomes is shown as green puncta in BSR cells that were transfected with GFP-LC3 followed by treatment with rapamycin (positive control); or infection of BTV1 at an MOI of 1 for 24 h at 24 h post-transfection. Mock treatment was a negative control. In addition, NS1 protein of BTV shown in red was used as indicator for BTV infection. The cell nucleuses were stained with DAPI. Scale bar: 7.5 μm. The outlines of the cells on the image were sketched by white dash line; (**C**) Western blotting. Transformation of LC3-II from LC3-I

was detected in mock or BTV-infected BSR cells using an anti-LC3B antibody, and the expression of VP2 and NS3 of BTV1 was also detected using anti-VP2 and anti-NS3 antibodies, respectively. Rapamycin-treated group was positive control; (**D**) Intensity band ratios of LC3-II/β-actin. The values represent the mean ± SD of three independent experiments. Statistical significance was analyzed with Student's t-test (* $p < 0.05$ and ** $p < 0.01$).

Next, to further verify the observations above, the formation of GFP-LC3 dots was investigated after the pEGFP-LC3-transfected BSR cells were infected or mock-infected with BTV1 for 24 h. As presented in Figure 1B, NS1 expression of BTV1 firstly suggested the cells were indeed infected with BTV1, and a considerable number of punctate GFP-LC3 proteins accumulated in infected cells compared with the uninfected cells. The GFP-LC3 dots induced by rapamycin were used as a positive control.

The lipidated form of LC3 (LC3-II) is a hallmark of autophagy induction and is widely used to determine the presence of autophagy. In our study, the increase of LC3-II was monitored by immunoblotting at the indicated times after BTV infection. Rapamycin treatment was used as a positive control. As shown in Figure 1C, LC3-I-to-LC3-II conversion was significantly greater in BTV-infected cells relative to mock-infected cells. The ratio of LC3-II to β-actin as an indicator was significantly increased about seven times ($p < 0.01$) at 24 and 36 hpi (Figure 1D). Rapamycin-group was also displayed increased trend. Meanwhile, monoclonal antibodies that specifically recognize the BTV1, VP2 and NS3 proteins were used to track infection progression. Based on these experiments, we concluded that BTV1 infection triggered autophagy in BSR cells.

3.2. Autophagy is Induced in Primary Lamb Lingual Epithelial Cells by BTV1 Infection

Although BSR is a good model for the research of the pathogenesis of BTV, a cell line cannot completely replicate the physiology of host cells. Therefore, to explore whether autophagy is triggered by BTV1 infection in host cells, primary lamb lingual epithelial cells were used in this experiment. To firstly confirm BTV could infect and replicate in this cell, viral nonstructural proteins (NS1 and NS3) were detected at 48 h post-infection. In addition, the induced autophagy by BTV infection in the lamb lingual epithelial cells was assessed by both LC3-I-to-LC3-II conversion and GFP-LC3 dots as above. Rapamycin treatment was used as the positive control as above. Similarly, the results shown in Figure 2 indicated that both the LC3-II and the GFP-LC3 dots increased compared to the mock group after BTV1 infection, suggesting that autophagy is induced by BTV1 in lamb lingual epithelial cells.

3.3. BTV1 Infection Increases the Levels of Autophagic Flux

Autophagic flux is a continuous and complete process of autophagy. The fusion of lysosomes and autophagosomes and the degradation of substrates are essential components of autophagic flux, in addition to the formation of autophagosomes. p62, also known as Sequestosome-1 (SQSTM1), is an indicator to assess autophagic flux. When autophagic flux occurs, p62 can target specific cargoes for autophagy and is specifically degraded by the autophagic-lysosomal pathway [30].

To investigate whether BTV infection induces autophagic flux, we measured p62 degradation levels. Immunoblotting revealed significant progressive degradation of p62 in BTV-infected cells compared with the unaltered p62 levels in mock-infected cells (Figure 3A), suggesting that autophagosomes were able to fuse with lysosomes to degrade the cargos. The ratios of p62 to β-actin shown in Figure 3B indicated that p62 reduction was notable at 36 hpi (about one time) and peaked at 48 hpi (about five times).

Additionally, the turnover of LC3-II caused by the inhibition of autophagic degradation further indicated autophagic flux. For this reason, we detected the expression of LC3-II and p62 with or without CQ, an inhibitor of lysosomal acidification and autophagic proteolysis, by Western blotting. As shown in Figure 3C, the accumulation of LC3-II and p62/SQSTM1 was observed with the employment of CQ at 24 h post-infection in BTV-infected cells compared with untreated controls, further indicating enhanced autophagic flux. The confocal image of BTV plus CQ also showed more undegraded and accumulated LC3 puncta in Figure 3D.

Lastly, the fusion of autophagosomes with lysosomes was verified by labeling lysosomes with an acidified compartments marker LysoTracker (red fluorescence) [31,32]. As demonstrated in Figure 3E, the colocalization of GFP-LC3-tagged autophagosomes and LysoTracker-stained lysosomes was detected in BTV1-infected BSR cells, while uninfected cells exhibited almost no overlap. These results indicated that cells underwent a complete autophagic process following BTV infection.

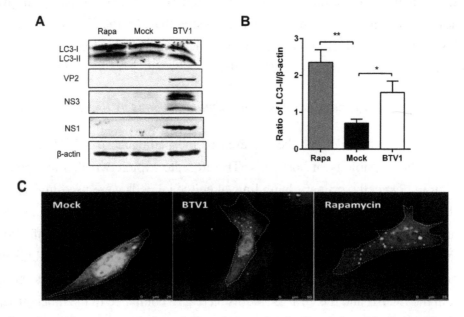

Figure 2. Autophagy was induced in primary lamb lingual epithelial cells by BTV1 infection. (**A**) The change from LC3-I to LC3-II in primary cells after BTV1 infection for 48 h was detected by Western blotting using an anti-LC3B antibody. Rapamycin group was positive control; (**B**) Ratios of LC3-II to β-actin of panel (**A**). The values represent the mean ± SD of three independent experiments. Statistical significance was analyzed with Student's t-test (* $p < 0.05$); (**C**) The primary lamb lingual epithelial cells were transfected with pEGFP-LC3. At 24 h post-transfection, cells were treated with mock-infection (**left**) or BTV1-infection at an MOI of 1 for 48 h (**middle**); or rapamycin treatment for 36 h (**right**). The fluorescence signals were visualized by confocal microscopy. Scale bar: 7.5 μm.

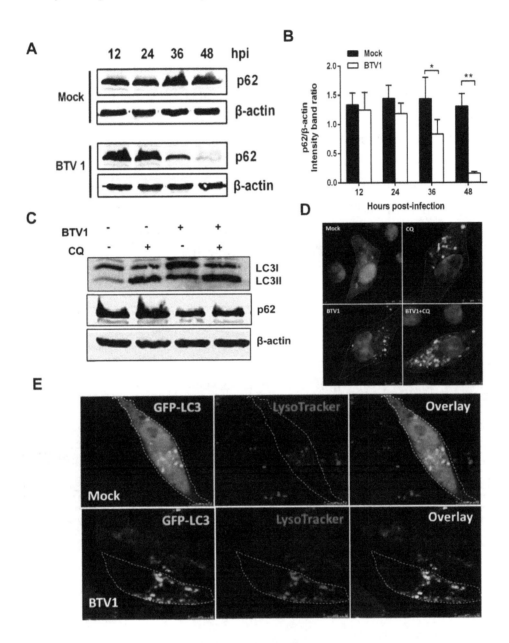

Figure 3. Autophagic flux measurement in BSR cells infected with BTV1. (**A**) p62/SQSTM1 (p62) degradation was tested by Western blotting. p62 in BTV1-infected and mock-infected BSR cells was monitored using an anti-p62 antibody; (**B**) Representative graphs were shown exhibiting the ratios of p62 to β-actin (mean ± SD) from three independent experiments. Statistical significance was analyzed with Student's t test (* $p < 0.05$ and ** $p < 0.01$); (**C**) The expression levels of LC3-II and p62 were detected by Western blotting after mock-infection or BTV1-infection in the presence or absence of 20 μM CQ for 24 h; (**D**) Representative confocal images of BTV with or without CQ treatment for 24 h; (**E**) Representative confocal images of co-localization of the autophagosome marker GFP-LC3 (**green**) with the acidic lysosome marker LysoTracker (**red**) in BSR cells infected with BTV1 for 36 h. Mock-infected cells served as negative control. Scale bar: 7.5 μm.

3.4. The Autophagy is Dependent on BTV-Productive Infection

Based on the results of LC3 lipidation and p62 degradation, we found that autophagy became more and more apparent with BTV replication, and we wondered if the induction of autophagy by BTV required productive infection. First, infective BTV1 was inactivated by UV irradiation. Then, BSR cells were infected with this UV-irradiated BTV1 for the indicated amounts of time, and LC3, p62 and NS3 in the cells were detected by Western blot. Compared with the normal BTV1 infection, neither virus-induced detectable LC3-I-to-LC3-II conversion nor the expression of viral NS3 protein was detected at any time point post-infection of UV-irradiated BTV1. Additionally, p62 degradation did not appear with UV-irradiated BTV1 infection (Figure 4). Taken together, these results suggested that the autophagy induced by BTV was dependent on viral replication.

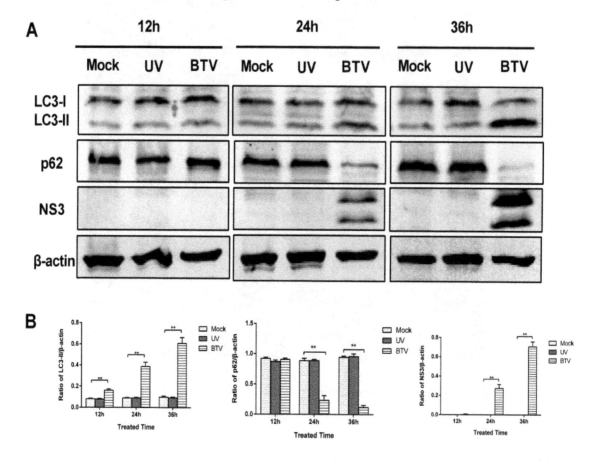

Figure 4. The induction of autophagy by BTV1 was dependent on productive infection. (**A**) BSR cells were untreated (**left**) or incubated with UV-irradiated BTV1 (**middle**) or normal BTV1 (**right**) for the indicated amounts of time. Afterwards, to determine the induction of autophagy, the cells were lysed to detect LC3, p62, β-actin and NS3 of BTV using corresponding antibodies; (**B**) Representative results are shown with graphs representing the intensity of LC3-II/actin or p62/actin or NS3/actin bands. Data represent mean ± SD of three independent experiments (** $p < 0.01$ and * $p < 0.05$).

3.5. Inhibition of Autophagy Reduced BTV1 Replication

Given that the induction of autophagy was closely related to BTV1 replication, we asked the question: What role does autophagy play in the BTV1 life cycle? To answer this question, the effects of autophagy on the replication of BTV were analyzed by inhibiting autophagy. 3-MA is a widely used class III phostphatidylinositol-3-kinase (PI3K) inhibitor that disturbs early autophagosome formation [33]. Thus, 3-MA was first used to investigate the replication of BTV after early autophagosome disturbance. As the results show (Figure 5A,B), the induced-LC3-II was inhibited by 3-MA treatment at the optimal concentration (10 mM) with no cytotoxicity (Figure 7; Lane 2). Meanwhile, expression of the viral proteins VP2 was reduced about seven times in autophagy-inhibited BSR cells compared to control cells (Figure 5C). Similarly, viruses of supernatants or freeze-thaw cells were collected from 3-MA-treated cells at the indicated times. The extracellular virus titers were decreased nearly 1.28 folds at 24 hpi and 1.14 folds at 36 hpi, and the intracellular virus titers were also reduced similarly, (Figure 5D), suggesting that early autophagy was beneficial to BTV growth.

Late autophagy, characterized by autophagosome-lysosome fusion, is also an essential step for the autophagic process. To investigate the impact of late autophagy in virus replication, BSR cells were treated with CQ at an optimal concentration (20 μM) to inhibit autophagy through blocking autophagosome-lysosome fusion. As demonstrated in Figure 5E, F, compared with the control group, there was significant accumulation of LC3-II in CQ-treated cells with no cytotoxicity (Figure 7; Lane 3). Moreover, this figure shows a six-fold decreased expression of VP2 protein in the cells at 36 hpi (Figure 5G). Similarly, extracellular and intracellular virus yields at 24 hpi and 36 hpi decreased nearly 2 folds with CQ treatment (Figure 5H), indicating that autophagy inhibition at late stage also dramatically reduced BTV replication.

To exclude any non-specific effects of these pharmacological experiments, RNA interference was used to suppress the expression of Beclin1, which is at the heart of a regulatory complex during autophagosome formation and maturation. As shown in Figure 5I, compared with negative control siRNA-transfected cells, BSR cells transfected with Beclin1-targeting siRNAs exhibited a notable decrease in endogenous Beclin1 protein with no cytotoxicity (Figure 7; Lane 6). LC3-II formation was also greatly reduced by the suppression of Beclin1 expression. Furthermore, Beclin1 silencing was accompanied by significantly reduced viral protein synthesis, as the ratio of VP2/actin presents nearly nine-fold decrease and extracellular and intracellular virus particle yields also showed more than 1.5-fold reduction at 24 hpi (Figure 5J,K).

These results were consistent with previous pharmacological studies. Taken together, the drug inhibition and Beclin1 interference suggest that autophagy was involved in and required for the effective replication of BTV.

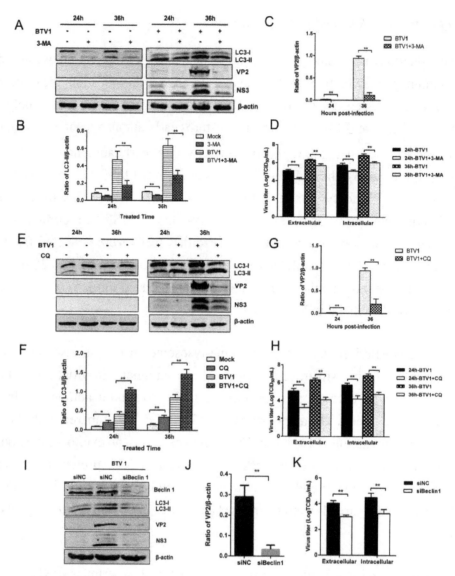

Figure 5. Autophagy inhibition reduced the replication of BTV1. (**A,E**) Effects of 10 mM 3-MA or 20 μM CQ treatment on LC3-II accumulation and VP2 or NS3 expression. The expression of relative proteins were analyzed by immunoblotting with specific antibodies; (**B,F**) Representative results are shown with graphs representing the intensity of LC3-II/actin; (**C,G**) Representative results are shown with graphs representing the intensity of VP2/actin. (**D,H**) Effects of 10 mM 3-MA or 20 μM CQ on the production of infectious particles of BTV1. The extracellular and intracellular virus loads were tested as $TCID_{50}$/mL at 24 hpi and 36 hpi, respectively; (**I**) The effects of Beclin1 silencing on LC3-II and VP2 or NS3 expression. BSR cells were transfected with either specific siRNA targeting Beclin1 or negative control siRNA. At 24 h after transfection, the cells were incubated with BTV1 (MOI = 1). Samples were collected 24 hpi and analyzed by Western blotting with antibodies against Beclin1, LC3, VP2, and NS3; (**J**) Representative results are shown with graphs representing the intensity of VP2/actin; (**K**) The viral titers were determined as $TCID_{50}$/mL at 24 hpi. Data represent the mean ± SD of three independent experiments. * $p < 0.05$ and ** $p < 0.01$, significantly different.

3.6. Induction of Autophagy Promotes BTV1 Replication

Our suppressive results prompted further exploration of the relationship between activated autophagy and virus replication. Thus, we probed for the effects of autophagy induction on viral yield and viral protein expression. Rapamycin, an autophagy inducer that directly inhibits the action of mTOR [34], was used as a positive indictor in autophagy study. LC3-II presented a rising trend in rapamycin-treated cells (Figure 6A) with no cytotoxicity (Figure 7; Lane 4) compared to the control. Viral protein synthesis levels were promoted by enhanced autophagy induction, as reflected in the increased expression of VP2 and NS3 compared with the control group, especially at 36 hpi (Figure 6A,C). Furthermore, enhanced autophagy also markedly increased the yield of BTV progeny at 24 hpi and 36 hpi in extracellular and intracellular virus samples, showing about 1-fold increment (Figure 6D). In brief, these data indicate that enhanced autophagy is favorable for the replication of BTV in BSR cells.

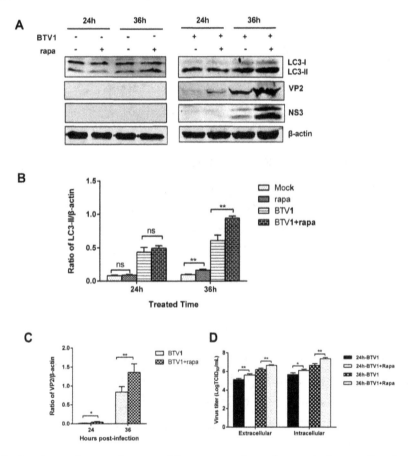

Figure 6. Induction of autophagy with rapamycin enhanced the replication of BTV1. (**A**) Effects of 100 nM rapamycin treatment on LC3-II accumulation and VP2 or NS3 expression. The intensity ratios of proteins relative to β-actin are presented below the blots; (**B,C**) Representative results are shown with graphs representing the intensity of LC3-II/actin or VP2/actin bands; (**D**) The titers of BTV1 produced by rapamycin-treated BSR cells. The virus yields were shown as $TCID_{50}$/mL at 24 hpi and 36 hpi. Results represent the mean ± SD of triplicate experiments. * $p < 0.05$ and ** $p < 0.01$, significantly different; ns, no significant difference.

3.7. Cell Viability Unaffected by Pharmacological Treatment

Considering that high concentrations of drugs or knockdown of Beclin1 might influence cell viability and thus affect our results, the effects of the compounds used in this study on cell viability were analyzed in a WST-1 assay. The data show that the viability of BSR cells was hardly affected by siRNA transfection or drug treatment at the indicated effective doses (Figure 7). These experiments demonstrate the effectiveness of pharmacological treatment and RNA interference on studies of autophagy and BTV replication.

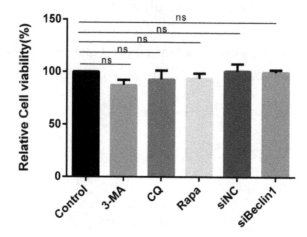

Figure 7. Pharmacological or siRNA treatments had no impact on cell viability. Cell viability was detected by WST-1 assay after treatments with 3-methyladenine (3-MA), chloroquine (CQ), rapamycin (Rapa), or transfection with siBeclin1 for 48 h. Light absorption at 450 nm was recorded and expressed as percentage of relative cell viability, the values represented mean ± SD. "ns" means no significant difference, $p > 0.05$.

4. Discussion

BTV causes a severe disease in ruminant animals and has a major influence on livestock economy. Previously, our understanding of BTV mostly focused on molecular structure, global epidemiology and diagnostic methods; however, the interaction between BTV infection and the host cell response has not been well studied. Recently, autophagy as a cellular adaptive response has been found to be involved in various viral infections [35,36]. However, the relationship between BTV infection and autophagy has not been thoroughly examined.

In this study, we observed that BTV1 infection triggers an increase in autophagosome-like double-membrane vesicles, the accumulation of GFP-LC3 puncta, and the conversion of LC3-I to LC3-II in permissive BSR cells (Figure 1), indicating that autophagosomes were formed [37]. Moreover, we identified similar autophagic phenomena during BTV1 infection in natural target cells: primary lamb lingual epithelial cells (Figure 2). These observations of natural target cells provide a valuable reference for the study of the interactions between BTV infection and the host cell response. Overall, we concluded that BTV infection triggered autophagosomes formation in host cells.

Autophagic flux is dynamic and continuous, refers to the increased autophagosome formation as well as the entire process, including lysosomes fusion and substrate recycle, thus, autophagic flux is a more accurate indicator of autophagy activity than autophagosomes formation [37]. Notably, the autophagic response to different viruses is not always the same. The autophagic flux induced by some viruses is incomplete, such as with porcine reproductive and respiratory syndrome virus and hepatitis C virus (HCV) [38,39], which implies that autophagic flux is virus specific. Our results show that BTV1 infection promoted p62 degradation (Figure 3A), and the lysosome inhibitor CQ increased the levels of LC3-II and p62 in virus-infected cells (Figure 3C). Furthermore, colocalization of GFP-LC3 with LysoTracker in infected cells was observed (Figure 3E). These findings demonstrated that complete autophagic flux was triggered upon BTV infection.

The finding that LC3-II formation and p62 degradation were enhanced by BTV replication implies the autophagy induced by BTV was strictly dependent on productive infection. Indeed, the same phenomena were not observed when BSR cells were incubated with the UV-irradiated BTV1 virus (Figure 4) that lacks replication ability. UV-inactivated BTV could not induce autophagy.

Given that viral replication was required for BTV infection-induced autophagy, what is the effect of autophagy on BTV replication? Autophagy is a balancing mechanism that maintains the homeostasis of cells. Although autophagy commonly serves as a defense mechanism, growing evidence suggests that some RNA viruses utilize or manipulate this cellular process to facilitate their own survival and replication [17,40], for instance influenza A virus [18], epizootic hemorrhagic disease virus [20], and avian reovirus [22]. To define the role of autophagy in BTV1 replication and provide a basis for further study of the relationship between autophagy and BTV, we analyzed the physiological significance of autophagy on BTV1 replication *in vitro*. Pharmacological treatments, including early autophagic inhibitor 3-MA and late autophagic inhibitor CQ, significantly downregulated viral protein expression and virus production. Furthermore, siRNA-mediated silencing of Beclin1 resulted in effects on BTV1 replication similar to those induced by 3-MA and CQ (Figure 5). Correspondingly, the expression of viral proteins and the production of infectious virions were promoted by exposing infected cells to the autophagy inducer rapamycin (Figure 6). Therefore, these results definitively demonstrated that autophagy is a critical process activated by bluetongue virus that benefits its replication.

It is worth mentioning that CQ used in the above assays is a multifunctional drug in biological tests, whose functions vary by dosage and usage. According to reference [41], CQ was also used to block clathrin-mediated endocytosis at a dosage of 200 μM, although we applied CQ as inhibitor of lysosomal acidification and autophagic proteolysis at a concentration of 20 μM during the whole process of virus infection, it is hard to know whether 20 μM CQ pretreatment has effect on endocytosis, which is the entry method used by BTV infection. Therefore, to rule out the possibility that 20 μM CQ pretreatment will affect BTV entry and accurately assess the effect of autophagy on BTV replication, BSR cells in our assays were not pretreated with 20 μM CQ.

Many viruses have evolved to exploit specific components of autophagy to assist in viral replication. Firstly, the autophagosome as a typical membrane vesicle structure could serve as the virus factory or viroplasm for many viruses. The virus factories or viroplasms are special structures that provide a physical scaffold to concentrate viral components and thereby facilitate virus replication and assembly [42]. For example, a study recently found that classical swine fever virus (CSFV) localized at the autophagosome-like vesicle membrane, which was involved in the replication of CSFV [39]. Additionally, a rotavirus pore-forming protein activates a calcium-dependent signaling pathway to initiate autophagy and hijacks this membrane trafficking pathway to transport viral proteins to viroplasms, the site of genome replication and viral assembly [43]. Viroplasms are characteristic of reovirus infections in general [44]. Given that BTV is a member of *Reoviridae* and its encoded proteins serve a similar function to rotavirus, we speculated that the vesicle membrane might also be involved in the replication of BTV. Thus, more work is underway to study how BTV takes advantage of host autophagy (or components of autophagy).

Second, vesicular acidification, which precedes the delivery of cargo to lysosomes during autophagy, may provide an environment that is beneficial to virus production. Autolysosomes formed by the fusion of autophagosomes with lysosomes are the main site for autophagic degradation. The acidic environment in autolysosomes is essential for the degradation or modification of contents trapped in the autophagosome. For some viruses, these events related to vesicular acidification are used to assist in their replication. For example, the production of the infectious poliovirus virion depends on vesicle acidification to modify the capsid protein VP0 [45]. In addition, degraded host components could also be used by the virus. For example, lipid droplets from the degradation of cholesterol in autolysosomes are required for HCV assembly [46]. For BTV, the release of the BTV core is dependent on an acidic pH [47], and accumulating evidence suggests that VP5 is involved in the permeabilization of the endosomal membrane in a pH-dependent manner [48]. An inhibitor of endosomal acidification, CQ, significantly reduced the replication of the virus. Thus, BTV could utilize the acidic environment during late autophagy for its own benefit.

In conclusion, our findings suggest that autophagy is triggered upon BTV infection and is critical for viral replication, implying that the autophagic pathway may be utilized to sustain BTV replication. Clearly, our understanding of the molecular mechanisms driving the interplay between BTV and autophagy is still insufficient, as several aspects of BTV-induced autophagy remain unclear, but the basic data presented here are the first step in our exploration. Our studies provide novel insights into BTV-host interactions and open a new window to examine the pathogenesis of BTV. This knowledge will contribute to the development of antiviral strategies or drugs against BTV infection.

Acknowledgments

This study was supported by grants from the National Natural Science Foundation of China (No. 31402207 and 31302065), Basic Scientific Research Foundation of Central Research Academies and Institutes (0302015008) and National High-Tech Research and Development Program of China (No. 2011AA10A212).

Author Contributions

D.W. and S.L. conceived and designed the experiments. S.L. performed the experiments and wrote the paper. Q.X., E.S., T.Y. contributed reagents/materials/analysis tools and gave valuable advice. J.L., Y.F. were involved in analyzed the data of experiments. Q.Z, H.W. and J.Z. were involved in the interpretation of the results and critically read the manuscript.

References

1. Maclachlan, N.J. Bluetongue: History, global epidemiology, and pathogenesis. *Prev. Vet. Med.* **2011**, *102*, 107–111. [CrossRef] [PubMed]

2. Mertens, P.P.; Diprose, J.; Maan, S.; Singh, K.P.; Attoui, H.; Samuel, A.R. Bluetongue virus replication, molecular and structural biology. *Vet. Ital.* **2004**, *40*, 426–437. [PubMed]

3. Schwartz-Cornil, I.; Mertens, P.P.; Contreras, V.; Hemati, B.; Pascale, F.; Breard, E.; Mellor, P.S.; MacLachlan, N.J.; Zientara, S. Bluetongue virus: Virology, pathogenesis and immunity. *Vet. Res.* **2008**, *39*, e46. [CrossRef] [PubMed]

4. Roy, P. Bluetongue virus proteins and particles and their role in virus entry, assembly, and release. *Adv. Virus Res.* **2005**, *64*, 69–123. [PubMed]

5. Roy, P. Functional mapping of bluetongue virus proteins and their interactions with host proteins during virus replication. *Cell Biochem. Biophys.* **2008**, *50*, 143–157. [CrossRef] [PubMed]

6. Roy, P. Bluetongue virus: Dissection of the polymerase complex. *J. Gen. Virol.* **2008**, *89*, 1789–1804. [CrossRef] [PubMed]

7. Patel, A.; Roy, P. The molecular biology of bluetongue virus replication. *Virus Res.* **2014**, *182*, 5–20. [CrossRef] [PubMed]

8. Lin, L.T.; Dawson, P.W.H.; Richardson, C.D. Viral interactions with macroautophagy: A double-edged sword. *Virology* **2010**, *402*, 1–10. [CrossRef] [PubMed]

9. Mizushima, N.; Komatsu, M. Autophagy: Renovation of cells and tissues. *Cell* **2011**, *147*, 728–741. [CrossRef] [PubMed]

10. Klionsky, D.J. Autophagy: From phenomenology to molecular understanding in less than a decade. *Nat. Rev. Mol. Cell Biol.* **2007**, *8*, 931–937. [CrossRef] [PubMed]

11. Yorimitsu, T.; Klionsky, D.J. Autophagy: Molecular machinery for self-eating. *Cell Death Differ.* **2005**, *12*, 1542–1552. [CrossRef] [PubMed]

12. Fimia, G.M.; Piacentini, M. Toward the understanding of autophagy regulation and its interplay with cell death pathways. *Cell Death Differ.* **2009**, *16*, 933–934. [CrossRef] [PubMed]

13. Chiu, H.C.; Richart, S.; Lin, F.Y.; Hsu, W.L.; Liu, H.J. The interplay of reovirus with autophagy. *Biomed. Res. Int.* **2014**, *2014*. [CrossRef] [PubMed]

14. Lee, H.K.; Lund, J.M.; Ramanathan, B.; Mizushima, N.; Iwasaki, A. Autophagy-dependent viral recognition by plasmacytoid dendritic cells. *Science* **2007**, *315*, 1398–1401. [CrossRef] [PubMed]

15. Shelly, S.; Lukinova, N.; Bambina, S.; Berman, A.; Cherry, S. Autophagy plays an essential anti-viral role in drosophila against vesicular stomatitis virus. *Immunity* **2009**, *30*, 588–598. [CrossRef] [PubMed]

16. Jordan, T.X.; Randall, G. Manipulation or capitulation: Virus interactions with autophagy. *Microbes Infect.* **2012**, *14*, 126–139. [CrossRef] [PubMed]

17. Kim, H.; Lee, S.; Jung, J. When autophagy meets viruses: A double-edged sword with functions in defense and offense. *Semin. Immunopathol.* **2010**, *32*, 323–341. [CrossRef] [PubMed]

18. Zhou, Z.; Jiang, X.; Liu, D.; Fan, Z.; Hu, X.; Yan, J.; Wang, M.; Gao, G.F. Autophagy is involved in influenza a virus replication. *Autophagy* **2009**, *5*, 321–328. [CrossRef] [PubMed]

19. Gu, L.; Musiienko, V.; Bai, Z.; Qin, A.; Schneller, S.W.; Li, Q. Novel virostatic agents against bluetongue virus. *PLoS ONE* **2012**, *7*, e43341. [CrossRef] [PubMed]

20. Shai, B.; Schmukler, E.; Yaniv, R.; Ziv, N.; Horn, G.; Bumbarov, V.; Yadin, H.; Smorodinsky, N.I.;

Bacharach, E.; Pinkas-Kramarski, R.; *et al.* Epizootic hemorrhagic disease virus induces and benefits from cell stress, autophagy, and apoptosis. *J. Virol.* **2013**, *87*, 13397–13408. [CrossRef] [PubMed]

21. Thirukkumaran, C.M.; Shi, Z.Q.; Luider, J.; Kopciuk, K.; Gao, H.; Bahlis, N.; Neri, P.; Pho, M.; Stewart, D.; Mansoor, A.; *et al.* Reovirus modulates autophagy during oncolysis of multiple myeloma. *Autophagy* **2013**, *9*, 413–414. [CrossRef] [PubMed]

22. Meng, S.; Jiang, K.; Zhang, X.; Zhang, M.; Zhou, Z.; Hu, M.; Yang, R.; Sun, C.; Wu, Y. Avian reovirus triggers autophagy in primary chicken fibroblast cells and vero cells to promote virus production. *Arch. Virol.* **2012**, *157*, 661–668. [CrossRef] [PubMed]

23. Chaabane, W.; User, S.D.; El-Gazzah, M.; Jaksik, R.; Sajjadi, E.; Rzeszowska-Wolny, J.; Los, M.J. Autophagy, apoptosis, mitoptosis and necrosis: Interdependence between those pathways and effects on cancer. *Arch. Immunol. Ther. Exp. (Warsz.)* **2013**, *61*, 43–58. [CrossRef] [PubMed]

24. Kang, R.; Zeh, H.J.; Lotze, M.T.; Tang, D. The beclin 1 network regulates autophagy and apoptosis. *Cell Death Differ.* **2011**, *18*, 571–580. [CrossRef] [PubMed]

25. DeMaula, C.D.; Jutila, M.A.; Wilson, D.W.; MacLachlan, N.J. Infection kinetics, prostacyclin release and cytokine-mediated modulation of the mechanism of cell death during bluetongue virus infection of cultured ovine and bovine pulmonary artery and lung microvascular endothelial cells. *J. Gen. Virol.* **2001**, *82*, 787–794. [PubMed]

26. Mortola, E.; Noad, R.; Roy, P. Bluetongue virus outer capsid proteins are sufficient to trigger apoptosis in mammalian cells. *J. Virol.* **2004**, *78*, 2875–2883. [CrossRef] [PubMed]

27. Alexander, D.E.; Ward, S.L.; Mizushima, N.; Levine, B.; Leib, D.A. Analysis of the role of autophagy in replication of herpes simplex virus in cell culture. *J. Virol.* **2007**, *81*, 12128–12134. [CrossRef] [PubMed]

28. Ruscanu, S.; Pascale, F.; Bourge, M.; Hemati, B.; Elhmouzi-Younes, J.; Urien, C.; Bonneau, M.; Takamatsu, H.; Hope, J.; Mertens, P.; *et al.* The double-stranded rna bluetongue virus induces type i interferon in plasmacytoid dendritic cells via a myd88-dependent tlr7/8-independent signaling pathway. *J. Virol.* **2012**, *86*, 5817–5828. [CrossRef] [PubMed]

29. Mizushima, N. Methods for monitoring autophagy. *Int. J. Biochem. Cell Biol.* **2004**, *36*, 2491–2502. [CrossRef] [PubMed]

30. Rusten, T.E.; Stenmark, H. P62, an autophagy hero or culprit? *Nat. Cell Biol.* **2010**, *12*, 207–209. [CrossRef] [PubMed]

31. Juhasz, G.; Neufeld, T.P. Autophagy: A forty-year search for a missing membrane source. *PLoS Biol.* **2006**, *4*, e36. [CrossRef] [PubMed]

32. Gannagé, M.; Schmid, D.; Albrecht, R.; Dengjel, J.; Torossi, T.; Rämer, P.C.; Lee, M.; Strowig, T.; Arrey, F.; Conenello, G.; *et al.* Matrix protein 2 of influenza a virus blocks autophagosome fusion with lysosomes. *Cell Host Microbe* **2009**, *6*, 367–380. [CrossRef] [PubMed]

33. Petiot, A.; Ogier-Denis, E.; Blommaart, E.F.C.; Meijer, A.J.; Codogno, P. Distinct classes of phosphatidylinositol 3′-kinases are involved in signaling pathways that control macroautophagy in ht-29 cells. *J. Biol. Chem.* **2000**, *275*, 992–998. [CrossRef] [PubMed]

34. Klionsky, D.J.; Meijer, A.J.; Codogno, P. Autophagy and p70s6 kinase. *Autophagy* **2005**, *1*, 59–60. [CrossRef] [PubMed]

35. Deretic, V.; Levine, B. Autophagy, immunity, and microbial adaptations. *Cell Host Microbe* **2009**, *5*, 527–549. [CrossRef] [PubMed]

36. Sir, D.; Ou, J.H.J. Autophagy in viral replication and pathogenesis. *Mol. Cells* **2010**, *29*, 1–7. [CrossRef] [PubMed]

37. Mizushima, N.; Yoshimorim, T.; Levine, B. Methods in mammalian autophagy research. *Cell* **2010**, *140*, 313–326. [CrossRef] [PubMed]

38. Sir, D.; Chen, W.L.; Choi, J.; Wakita, T.; Yen, T.S.B.; Ou, J.H.J. Induction of incomplete autophagic response by hepatitis c virus via the unfolded protein response. *Hepatol. (Baltim. Md.)* **2008**, *48*, 1054–1061. [CrossRef] [PubMed]

39. Sun, M.X.; Huang, L.; Wang, R.; Yu, Y.L.; Li, C.; Li, P.P.; Hu, X.C.; Hao, H.P.; Ishag, H.A.; Mao, X. Porcine reproductive and respiratory syndrome virus induces autophagy to promote virus replication. *Autophagy* **2012**, *8*, 1434–1447. [CrossRef] [PubMed]

40. Jain, B.; Chaturvedi, U.C.; Jain, A. Role of intracellular events in the pathogenesis of dengue; an overview. *Microb. Pathog.* **2014**, *69–70*, 45–52. [CrossRef] [PubMed]

41. Chen, C.L.; Hou, W.H.; Liu, I.H.; Hsiao, G.; Huang, S.S.; Huang, J.S. Inhibitors of clathrin-dependent endocytosis enhance tgfbeta signaling and responses. *J. Cell Sci.* **2009**, *122*, 1863–1871. [CrossRef] [PubMed]

42. Wileman, T. Aggresomes and autophagy generate sites for virus replication. *Science* **2006**, *312*, 875–878. [CrossRef] [PubMed]

43. Crawford, S.E.; Hyser, J.M.; Utama, B.; Estes, M.K. Autophagy hijacked through viroporin-activated calcium/calmodulin-dependent kinase kinase-β signaling is required for rotavirus replication. *Proc. Natl. Acad. Sci. USA* **2012**, *109*, E3405–E3413. [CrossRef] [PubMed]

44. Barhoom, S.; Kaur, J.; Cooperman, B.S.; Smorodinsky, N.I.; Smilansky, Z.; Ehrlich, M.; Elroy-Stein, O. Quantitative single cell monitoring of protein synthesis at subcellular resolution using fluorescently labeled trna. *Nucleic Acids Res.* **2011**, *39*, e129. [CrossRef] [PubMed]

45. Richards, A.L.; Jackson, W.T. Intracellular vesicle acidification promotes maturation of infectious poliovirus particles. *PLoS Pathog.* **2012**, *8*, e1003046. [CrossRef] [PubMed]

46. Miyanari, Y.; Atsuzawa, K.; Usuda, N.; Watashi, K.; Hishiki, T.; Zayas, M.; Bartenschlager, R.; Wakita, T.; Hijikata, M.; Shimotohno, K. The lipid droplet is an important organelle for hepatitis c virus production. *Nat. Cell Biol.* **2007**, *9*, 1089–1097. [CrossRef] [PubMed]

47. Eaton, B.T.; Hyatt, A.D.; Brookes, S.M. The replication of bluetongue virus. *Curr. Top. Microbiol. Immunol.* **1990**, *162*, 89–118. [PubMed]

48. Hassan, S.H.; Wirblich, C.; Forzan, M.; Roy, P. Expression and functional characterization of bluetongue virus vp5 protein: Role in cellular permeabilization.*J. Virol.* **2001**, *75*, 8356–8367. [CrossRef] [PubMed]

4

Modes of Human T Cell Leukemia Virus Type 1 Transmission, Replication and Persistence

Alexandre Carpentier [†], **Pierre-Yves Barez** [†], **Malik Hamaidia** [†], **Hélène Gazon, Alix de Brogniez, Srikanth Perike, Nicolas Gillet and Luc Willems** *

Molecular and Cellular Epigenetics (GIGA) and Molecular Biology (Gembloux Agro-Bio Tech), University of Liège (ULg), 4000 Liège, Belgium; E-Mails: a.carpentier@doct.ulg.ac.be (A.C.); py.barez@doct.ulg.ac.be (P.-Y.B.); mhamaidia@ulg.ac.be (M.H.); helene.gazon@ulg.ac.be (H.G.); alix.debrogniez@ulg.ac.be (A.B.); Srikanthperike@gmail.com (S.P.); n.gillet@ulg.ac.be (N.G.)

[†] These authors contributed equally to this work.

* Author to whom correspondence should be addressed; E-Mail: luc.willems@ulg.ac.be

Academic Editor: David Boehr

Abstract: Human T-cell leukemia virus type 1 (HTLV-1) is a retrovirus that causes cancer (Adult T cell Leukemia, ATL) and a spectrum of inflammatory diseases (mainly HTLV-associated myelopathy—tropical spastic paraparesis, HAM/TSP). Since virions are particularly unstable, HTLV-1 transmission primarily occurs by transfer of a cell carrying an integrated provirus. After transcription, the viral genomic RNA undergoes reverse transcription and integration into the chromosomal DNA of a cell from the newly infected host. The virus then replicates by either one of two modes: (i) an infectious cycle by virus budding and infection of new targets and (ii) mitotic division of cells harboring an integrated provirus. HTLV-1 replication initiates a series of mechanisms in the host including antiviral immunity and checkpoint control of cell proliferation. HTLV-1 has elaborated strategies to counteract these defense mechanisms allowing continuous persistence in humans.

Keywords: HTLV-1; viral replication; viral persistence; Tax; HBZ

1. Introduction

HTLV-1 infects approximately 5–10 million people worldwide mainly in subtropical areas [1]. In the vast majority of cases, HTLV-1 infection remains clinically silent. Among asymptomatic carriers, 3% to 5% will develop a leukemia/lymphoma (ATL) or a neurodegenerative disease (HAM/TSP) after long latent periods (40–60 years) [2]. ATL results from proliferation and accumulation of infected cells carrying an integrated proviral genome (here referred to as clones). HAM/TSP is associated with invasion of the central nervous system by infected cells, antiviral immunity, cytokine burst, and inflammation. Main clinical symptoms of HAM/TSP are urinary failures and paralysis of lower legs. Why infected subjects develop either ATL or HAM/TSP is currently unknown. There is no efficient treatment for HAM/TSP, except palliative attenuation of inflammation with corticosteroids. The leukemic form of ATL is initially responsive to general chemotherapy (CHOP) but almost invariably relapses after a few months. An antiviral therapy based on AZT combined with interferon yields 50% survival at five years [3]. Another type of treatment includes hematopoietic stem cell transplantation that yields, if successful, the best long-term survival rates [4–6]. An anti-CCR4 antibody is now in clinical use in Japan [7,8] and other promising approaches such as valproic acid are currently being investigated [8,9].

The HTLV-1 genome contains essential structural and enzymatic genes (Gag, Pro, Pol and Env) shared by all retroviral family members (reviewed by [10]). As a deltaretrovirus, HTLV-1 also encodes a series of accessory and regulatory proteins. Among these, the Tax oncoprotein and HTLV-1 basic leucine zipper factor (HBZ) play pivotal roles in the viral life cycle [11]. Here, we describe how these factors subvert cellular pathways to allow viral transmission, persistence, and replication.

2. Current Model of HTLV-1 Replication

HTLV-1 predominantly infects CD4+ T cells but also targets other cell types such as CD8+ T and B lymphocytes, dendritic cells (DCs), monocytes, and macrophages [12–14]. This pleiotropic pattern is permitted by the presence of membrane-associated receptors that interact with the viral envelope allowing efficient binding and entry. These include heparan sulfate proteoglycans (HSPGs), the glucose transporter 1 (GLUT-1) and neuropilin-1 (NRP-1) [15–18]. The mechanisms of receptor binding and virus entry have been reviewed elsewhere [19–21]. A number of studies have shown that cell-free infection is poorly efficient compared to cell-to-cell virus transfer (about 10,000 fold) [22,23], suggesting that HTLV-1 spread *in vivo* relies more on a cellular intermediate than on the virion itself. Whatever the route of infection used, the initial contact with HTLV-1 mainly occurs via breast feeding, sexual intercourse, and blood transfusion [24]. Except when contamination occurs by blood transfer, initial infection first requires interaction with oral, gastrointestinal, or cervical mucosa. Crossing of the mucosal barrier occurs by different mechanisms as schematized on Figure 1a. Although not formally demonstrated yet, HTLV-1 infected macrophages could transmigrate through an intact epithelium as observed for human immunodeficiency virus (HIV) [25,26]. Viral particles produced by HTLV-1 infected T-cells have been shown to cross the epithelium by transcytosis, *i.e.*, the transit of a virion incorporated into a vesicle from the apical to the basal surface of an epithelial cell [26,27]. Alternatively, HTLV-1 can also infect an epithelial cell and produce new virions that are then released from the basal surface [28]. Finally, HTLV-1 infected cells can directly bypass a disrupted mucosa [28].

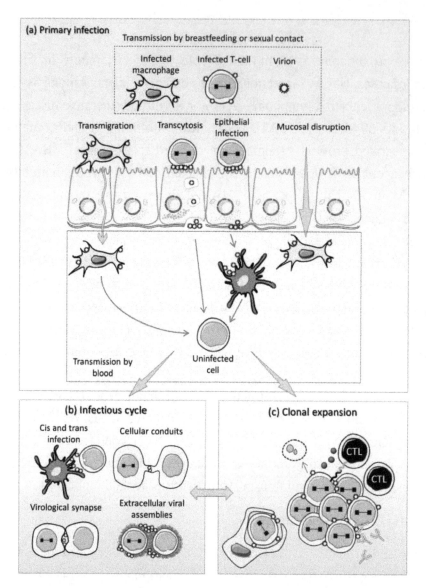

Figure 1. Model of HTLV-1 replication (**a**) HTLV-1 transmission occurs by breastfeeding, sexual intercourse, or blood transfusion. Except for blood transfer, initial infection requires crossing of the mucosal barrier by several mechanisms: (i) transmigration of HTLV-1 infected macrophages, (ii) transcytosis of viral particles, (iii) release of newly produced virions from the basal surface of infected epithelial cell, (iv) bypass of HTLV-1 infected cells through a damaged mucosa. HTLV-1 can then infect mucosal immune cells directly (cis-infection) or via antigen-presenting cells (APCs); (**b**) APCs can either become infected or transfer membrane-bound extracellular virions to T-cells (trans-infection). Cell-to-cell transfer of virions involves different non-exclusive mechanisms: a virological synapse, cellular conduits, or extracellular viral assemblies. Infection of resident cells occurs either in the mucosa or in secondary lymphoid organs. Soon after primary infection, HTLV-1 replicates by cell-to-cell infection (*i.e.*, the infectious cycle) or (**c**) by mitotic division of a cell containing an integrated provirus (clonal expansion). Since an antiviral immune response is quickly initiated, the efficacy of the infectious cycle is severely dampened down soon after infection.

Having crossed the epithelial barrier, HTLV-1 infects mucosal immune cells directly or via APCs such as DCs or macrophages. APCs can either undergo infection or transfer membrane bound extracellular virions to uninfected T-cells (trans-infection) [14]. Cell-to-cell transfer of HTLV-1 virions then potentially involves several non-exclusive mechanisms (reviewed in [28]): a virological synapse [29–31], cellular conduits [32], or extracellular viral assemblies [33,34]. Infection of resident cells occurs either in the mucosa or in secondary lymphoid organs.

Soon after primary infection, HTLV-1 attempts to expand by colonizing new targets by cell-to-cell transfer, reverse transcription of the viral RNA, integration of the provirus into the chromosome, expression of viral proteins and budding of new virions (the infectious cycle; Figure 1b). Another mode of replication involves mitotic division of a cell containing an integrated provirus (clonal expansion; Figure 1c). Recently, host restriction factors such as SAMHD1, APOBEC3 and miR-28-3p have been shown to limit HTLV-1 infection [35–37]. Since an antiviral immune response is also quickly initiated, the efficacy of the infectious cycle is severely attenuated soon after infection, although likely not completely abrogated later on. On the other side, clonal expansion and cell proliferation also require expression of viral factors such as Tax [38,39]. Survival of infected progeny cells therefore requires silencing of viral expression before immune-mediated destruction. This model is consistent with the following observations: (i) to block HTLV-1 infection, reverse transcriptase inhibitors (RTIs) must be administrated simultaneously with viral inoculation [40]; (ii) when used alone, RTIs do not reduce the proviral load in HTLV-1 infected subjects [41,42]; (iii) sustained T-cell proliferation in patients correlates with Tax expression [43], extending previous studies in BLV-infected animal models [44]; (iv) compared to HIV, the HTLV-1 genome undergoes limited variability [45], suggesting a replication mode by cellular DNA polymerase rather than by viral reverse transcriptase; (v) sequential high-throughput sequencing of proviral integration sites reveal a high clonal stability over years [46]. In this context, our recent study in BLV-infected cows also showed that most clones generated during primary infection are destroyed and replaced by others undergoing expansion [47].

Taken together, these data support a model of viral replication by cell-to-cell contact at the early stages of infection, followed by a sustained clonal proliferation counterbalancing the host immune response. Repetitive cycles of viral expression followed by transcriptional silencing continuously challenges the immune response thereby initiating inflammation and ultimately leading to HAM/TSP. By favoring emergence of sporadic mutations in the cell genome, unrestrained proliferation also paves the way to malignant transformation and development of ATL [43].

3. Tax and HBZ Are Two Main Drivers of Viral Replication

According to currently most accepted model, Tax and HBZ are believed to have the highest impact on viral replication and cell transformation, besides other components required to synthesize the viral particle. The modes of action of Tax and HBZ are remarkably pleiotropic and involve a variety of cell signaling pathways (CREB, NFkB and AKT; Figure 2). Tax inhibits tumor suppressors (p53, Bcl11B and TP53INP1 [48–50]) and activates cyclin-dependent kinases (CDKs) [51], both of these mechanisms leading to accelerated cell proliferation. In parallel, Tax attenuates the Mad1 spindle assembly checkpoint protein, induces genomic lesions and interferes with DNA repair thereby promoting aneuploidy [39,52,53]. Experimental evidence also shows that Tax drives tumor formation in transgenic mouse models, supporting its oncogenic potential [54–56]. Tax also induces genomic instability [39,57], generating somatic alterations [58] and promoting cell growth. However, expression of Tax alone fails to systematically immortalize human primary T cells [59], suggesting the involvement of other viral or cellular components. In particular, driver mutations affecting the CCR4 chemokine receptor have been identified in ~25% of ATL cases [60,61]. In about 50% of ATL cases, Tax is either inactivated by genetic mutation or transcriptionally silenced by hyper-methylation or deletion of the 5′-LTR [62–65]. Because of the strong immunogenicity of the Tax protein, it is possible that these mechanisms confer a selective advantage to HTLV-1-transformed T cells [66–69]. In comparison, HBZ triggers a less efficient immunity that is compatible with permanent expression throughout HTLV-1 infection [70,71]. Later in leukemogenesis, cell growth can thereby become independent of Tax and be promoted by HBZ. Indeed, HBZ is constitutively expressed throughout HTLV-1 infection [72,73], counteracts Tax-mediated viral and cellular pathways modulation (such as NF-κB, Akt and CREB) and stimulates cell proliferation [69,74] via apoptosis/senescence inhibition and cell cycle modulation [69,74]. This simplified model thus hypothesizes that Tax initiates transformation while HBZ is required to maintain the transformed phenotype if Tax expression is silenced [75]. Clinical data indicating that Tax mRNA expression allows estimating the risk of HAM/TSP development and that HBZ positively correlates with the severity of symptoms further supports a role of Tax and HBZ in pathogenesis [76,77].

3.1. Tax and HBZ Exert Opposite Functions in Signaling Pathways

Almost systematically, the activities of Tax on a series of cellular pathways are balanced by HBZ.

3.1.1. NF-κB

By controlling T lymphocyte activation and proliferation in response to diverse immune stimuli (such as antigens, cytokines or microbial components), the NF-κB pathway is a key player in regulation of immunity and inflammation [78]. HTLV-1 Tax activates the IKK complex through IKKγ/NEMO binding. Tax requires CADM1/TSLC1 for inactivation of the NF-kappaB inhibitor A20 and constitutive NF-κB signaling [79]. The subsequent translocation of p50/p65 complex into the nucleus activates transcription of NF-κB responsive genes [78,80]. Activation of the canonical NF-κB pathway by Tax requires IL17RB signaling [81]. On the other hand, Tax stimulates IKKα-dependent processing of p100 into p52 [78,80]. Tax also hijacks the cellular ubiquitin machinery to activate ubiquitin-dependent kinases and NF-κB signaling (reviewed in [82]). Tax thereby induces expression of a variety of growth promoting cytokines (such as IL-1, IL-6, TNF, and EGF [83,84]). Tax also upregulates antiapoptotic proteins: caspase-8 inhibitory protein c-FLIP [85,86] and members of the Bcl-2 family (Bcl-2, Bcl-xL, Mcl-1 and Blf-1) [87–90]. By activating the NF-κB pathway, Tax thus favors proliferation and survival of HTLV-1-infected T cells. On the contrary, HBZ suppresses the canonical NF-κB signaling pathway by inhibiting the activity of the RelA/p65 complex and thus mitigates excessive activation of NF-κB by Tax [91]. NF-κB activation by Tax is associated with an upregulation of p21$^{WAF1/CIP1}$ and p27^{KIP1}, leading to cellular senescence [92,93]. In HeLa cells, HBZ prevents Tax-induced senescence through down-regulation of NF-κB [92,94].

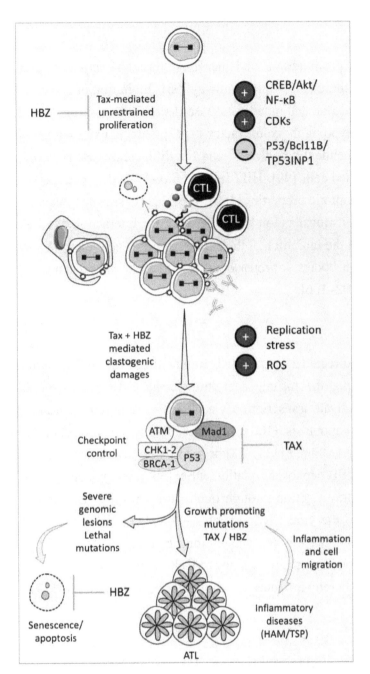

Figure 2. Tax and HBZ promote proliferation and persistence of the infected cell. Tax activates survival pathways (CREB/Akt/NFkB), promotes mitosis (CDKs), and inhibits tumor suppressors (p53, TP53INP1, Bcl11B). Tax-mediated growth-promoting activities are counteracted by HBZ, mitigating unrestrained proliferation. The host immune response further controls infected cell proliferation. Tax-induced proliferation creates replicative stress and generates reactive oxygen species (ROS). Tax interacts with the mitotic checkpoint control protein Mad1 thereby inducing clastogenic damage. Tax attenuates the DNA damage response (DDR) induced by unscheduled cell proliferation. Inhibition of the DDR allows cells to accumulate DNA lesions and stabilize mutations. If uncontrolled by senescence or cell death mechanisms, growth-promoting mutations pave the way to disease development.

3.1.2. Akt

Tax promotes cell proliferation and survival through activator protein-1 (AP-1) and the phosphatidylinositol 3-kinase (PI3K)/Akt pathway [95]. Inhibition of Akt in HTLV-1-transformed cells decreases phosphorylated Bad and induces caspase-dependent apoptosis [96]. Stimulation of PI3K/Akt by Tax activates HiF-1 (hypoxia-inducible factor 1) [97], reduces expression of proapoptotic Bim and Bid and promotes IL-2 independent growth [98] and finally increases Bcl3 whose expression is associated with the growth of infected cells [99]. HBZ inhibits Tax-dependent activation of the PI3K/Akt pathway and downstream anti-apoptotic properties [100]. HBZ suppresses apoptosis by attenuating the function of FOXO3a and altering its localization [101]. Besides, the interaction of HBZ with AP-1 factors (cJun, JunB or MafB) results in the inhibition of their transcriptional activities via several mechanisms, such as sequestration into nuclear bodies or proteasomal degradation, and prevents the subsequent activation of AP-1 regulated genes [102–106].

3.1.3. CREB

Tax activates 5′-LTR-directed transcription by interacting with CREB, modulating its phosphorylation at Ser133 and connecting the histone acetyltransferase CBP (CREB-binding protein/p300) [107]. The ability of Tax to activate transcription via CREB is required to protect murine fibroblasts from serum-depletion-induced apoptosis. [108–110]. Tax modifies the phosphorylation state of CREB (i) by activating the upstream Akt kinase [111,112] or (ii) by decreasing the expression of PTEN phosphatase which is required for CREB dephosphorylation at Ser133 in the nucleus [111,113].

HBZ represses viral transcription through interaction with the bZIP domain of CREB proteins and prevents their binding to the viral CRE elements [114,115]. HBZ interacts with the KIX domain of p300/CBP, competing for Tax binding and inhibiting the association of co-activators with the viral promoter [116]. HBZ modulates the occupancy of KIX domains of p300/CBP by modulating the activity of transcription factors, thereby influencing subsequent gene expression [117].

3.1.4. Wnt

Tax interacts with DAPLE (dishevelled-associating protein with a high frequency of leucine residues) to activate the canonical Wnt pathway. HBZ can suppress this activation by inhibiting DNA binding of TCF-1/LEF-1 transcription factors. On the other side, HBZ promotes transcription of WnT 5a, a key protein of the non-canonical Wnt pathway, by enhancing its promoter activity through transforming growth factor-beta (TGF-beta). Knockdown of Wnt5a represses proliferation and migration of ATL cells, pointing out the role of this pathway in HTLV-1 infected cell growth [118].

3.1.5. TGF-β/Smad

Tax represses TGF-β 1 signaling (i) by blocking the association of Smad proteins with Smad-binding elements, (ii) through its interaction with CREB-binding protein/p300 and (iii) via c-jun activation [119–121]. HBZ counteracts this effect and also interacts with Smad2/3 to enhance TGF-beta/Smad transcriptional responses in a p300-dependent manner, improving transcription of different genes, such as the FOXP3 mediator of regulatory T cells [122].

3.1.6. S Phase Entry and Cell Cycle Progression

Through interaction with cyclins and CDKs, Tax interferes with cell cycle progression by several mechanisms: (i) Tax stabilizes the cyclin D2/CDK4 complex and favors hyperphosphorylation of the retinoblastoma protein (Rb). Phosphorylated Rb frees E2F1 that activates transcription of genes required for G1/S transition; (ii) Tax represses cyclin-dependent kinases inhibitors (CKIs) such as members of INK4 family and KIP1; (iii) Tax interacts with and directs Rb to the proteasome for subsequent degradation; (iv) Tax activates the cyclin D1 transcription by enhancing p300 recruitment to the CRE site of cyclin D1 promoter through interaction with pCREB and TORC2 [123]. As a result, Tax favors S phase entry of HTLV-1 infected cells (reviewed in [51,75]). Tax also accelerates S phase progression by interaction with the replicative helicase (minichromosome maintenance complex, MCM2-7). Tax modulates the spatiotemporal program of replication origins through p300-dependent histone hyperacetylation, resulting in early firing of late replication origins. Tax also fires supplementary origins of replication accelerating S phase progression. This mechanism triggers replicative stress and genomic lesions, such as double strand breaks (DSBs) [39,57]. By modulating replication timing, Tax could also modulate the entire transcriptional landscape of infected cells. Indeed, the level of transcription at replication origins (ORC1 binding sites) correlates with replication timing [124].

In contrast to Tax, HBZ exerts a dual regulatory role in cell cycle progression. Indeed, HBZ interacts with CREB and inhibits transcription of cyclin D1 [125]. HBZ also binds activating transcription factor 3 (ATF3) that modulates expression of cell division cycle 2 (CDC2) and cyclin E2, thereby promoting proliferation of ATL cells [126]. Concomitantly, HBZ suppresses ATF3-induced p53 transcriptional activity. Moreover, the HBZ mRNA increases E2F1 gene transcription and promotes cell proliferation [69].

Together, this series of data on the signaling pathways illustrates the opposite functions of Tax and HBZ in finely tuned regulatory mechanisms of cell proliferation.

3.2. Cellular Checkpoints Control Unscheduled Proliferation

Cellular checkpoints act as a failsafe barrier against unrestrained cellular proliferation. Tax subverts the G1 restriction and the spindle mitotic checkpoints. In G1, the tumor-suppressor protein p53 is the main factor that controls the checkpoint. Although approximately 50% of cancers harbor a mutation in p53, this mechanism only appears in a small percentage of ATL patients. Instead, p53 is functionally inactivated in leukemic and HTLV-1 transformed cells [127]. It remains incompletely understood how Tax inactivates p53: (i) Tax competes with p53 in binding with CBP, thereby repressing p53 trans-activating function [128]; (ii) NF-kappaB p65 subunit is critical for Tax-induced p53 inactivation [129]; (iii) repression of p53 transcriptional activity by Tax is independent of NF-kB and CBP [130]; p53 is invalidated by wild-type p53-induced phosphatase 1 (Wip1) [131,132]. ATL cells are characterized by loss of spindle assembly checkpoint function [133] and aneuploidy [134]. Tax binding to Mad1 perturbs the organization of the spindle assembly and results in multinucleated cells [52]. Moreover, the direct interaction between Tax and the anaphase-promoting complex APC Cdc20 also explains the mitotic abnormalities in HTLV-1 infected cells [135]. Tax promotion of supernumerary centrosomes through recruitment of Ran and Ran-binding protein-1 is another mechanism contributing to leukemia [136].

3.3. Response to DNA Damage

By accelerating the replication-timing program, the Tax protein induces replicative stress and DSBs [39]. Tax expression generates reactive oxygen species (ROS) leading to oxidative and replication-dependent DSBs [53]. Tax-associated DNA damages activate several phosphoproteins of the DDR pathway (H2AX, ATM, CHK1-2, P53, BRCA1), which in turn arrest the cell cycle transiently or lead to apoptosis and senescence. In presence of DNA damaging agents (e.g., UV irradiation), Tax inhibits the DDR machinery by sequestrating key signaling pathway components [137–144]. Induction of genomic lesions and inhibition of the DDR leads to proliferation in presence of DNA mutations, potentially to leukemogenesis.

HBZ induces DNA lesions through activation of miR-17 and miR-21 and downregulation of the DNA damage factor OBFC2A [145]. HBZ association with growth arrest and DNA damage gene 34 (GADD34) also deregulates the cellular responses to DNA damage [146].

3.4. DNA Repair Pathways

Besides modulating the DDR signaling pathway, Tax also directly interferes with the mechanisms of DNA repair. For example, Tax downregulates the expression of β-polymerase [147] and inhibits base excision repair (BER) [148]. Furthermore, Tax activates PCNA and interferes with nucleotide excision repair (NER) [149,150]. Tax decreases Ku80 gene transcription and interacts with Ku80 protein, interfering with non-homologous end joining (NHEJ) [151,152]. In Tax-expressing cells, DSBs are nevertheless preferentially repaired by error-prone NHEJ [153]. Another viral protein, p30, inhibits homologous recombination, shifting repair towards unfaithful pathways [154]. Whether HBZ also interferes with DNA damage repair mechanisms remains to be further clarified.

4. Conclusions

HTLV-1 persists and replicates by means of viral proteins, such as Tax and HBZ that finely tune cellular signaling pathways. Viral replication through the infectious and the mitotic routes requires viral expression and faces destruction by the host immune response. Expression of viral proteins creates genomic stress responsible for DNA lesions that initiate the DDR response. Imperfect repair of these errors stabilizes mutations that potentially drive oncogenesis.

Acknowledgments

This work was supported by the "Fonds National de la Recherche Scientifique" (FNRS), the Télévie, the Interuniversity Attraction Poles (IAP) Program "Virus-host interplay at the early phases of infection" BELVIR initiated by the Belgian Science Policy Office, the Belgian Foundation against Cancer (FBC), the Sixth Research Framework Programme of the European Union (project "The role of infections in cancer" INCA LSHC-CT-2005-018704), the "Neoangio" excellence program and the "Partenariat Public Privé", PPP INCA, of the "Direction générale des Technologies, de la Recherche et de l'Energie/DG06" of the Walloon government, the "Action de Recherche Concertée Glyvir" (ARC) of the "Communauté française de Belgique", the "Centre anticancéreux près ULg" (CAC), the "Subside Fédéral de Soutien à la Recherche Synbiofor and Agricultureislife" projects of Gembloux Agrobiotech (GxABT), the "ULg Fonds Spéciaux pour la Recherche", the "Plan Cancer" of the "Service Public Fédéral". A.C.; S.P. and A.B. are supported by grants of the Télévie. M.H. is a research fellow of the "Agriculture is life" project of GxABT. N.G. is supported by the IAP program. P.-Y. B. (FNRS research fellow), H.G. (post-doctoral researcher) and L.W. (Research Director) are members of the FNRS

Author Contributions

A.C.; P.-Y.B. and M.H. drafted the manuscript. L.W. edited the manuscript. All authors corrected, edited and approved the text.

References

1. Gessain, A.; Cassar, O. Epidemiological Aspects and World Distribution of HTLV-1 Infection. *Front. Microbiol.* **2012**, *3*, e388. [CrossRef] [PubMed]

2. Verdonck, K.; Gonzalez, E.; Van Dooren, S.; Vandamme, A.M.; Vanham, G.; Gotuzzo, E. Human T-lymphotropic virus 1: Recent knowledge about an ancient infection. *Lancet. Infect. Dis.* **2007**, *7*, 266–281. [CrossRef]

3. Bazarbachi, A.; Plumelle, Y.; Carlos Ramos, J.; Tortevoye, P.; Otrock, Z.; Taylor, G.; Gessain, A.; Harrington, W.; Panelatti, G.; Hermine, O. Meta-analysis on the use of zidovudine and interferon-alfa in adult T-cell leukemia/lymphoma showing improved survival in the leukemic subtypes. *J. Clin. Oncol.* **2010**, *28*, 4177–4183. [CrossRef] [PubMed]

4. Shiratori, S.; Yasumoto, A.; Tanaka, J.; Shigematsu, A.; Yamamoto, S.; Nishio, M.; Hashino, S.; Morita, R.; Takahata, M.; Onozawa, M.; *et al.* A retrospective analysis of allogeneic hematopoietic stem cell transplantation for adult T cell leukemia/lymphoma (ATL): Clinical impact of graft-versus-leukemia/lymphoma effect. *Biol. Blood Marrow Transpl.* **2008**, *14*, 817–823. [CrossRef] [PubMed]

5. Ishida, T.; Hishizawa, M.; Kato, K.; Tanosaki, R.; Fukuda, T.; Takatsuka, Y.; Eto, T.; Miyazaki, Y.; Hidaka, M.; Uike, N.; *et al.* Impact of graft-versus-host disease on allogeneic hematopoietic cell transplantation for adult T cell leukemia-lymphoma focusing on preconditioning regimens: Nationwide retrospective study. *Biol. Blood Marrow Transpl.* **2013**, *19*, 1731–1739. [CrossRef] [PubMed]

6. Ishida, T.; Hishizawa, M.; Kato, K.; Tanosaki, R.; Fukuda, T.; Taniguchi, S.; Eto, T.; Takatsuka, Y.; Miyazaki, Y.; Moriuchi, Y.; *et al.* Allogeneic hematopoietic stem cell transplantation for adult T-cell leukemia-lymphoma with special emphasis on preconditioning regimen: A nationwide retrospective study. *Blood* **2012**, *120*, 1734–1741. [CrossRef] [PubMed]

7. Utsunomiya, A.; Choi, I.; Chihara, D.; Seto, M. Recent advances in the treatment of adult T-cell leukemia-lymphomas. *Cancer Sci.* **2015**, *106*, 344–351. [CrossRef] [PubMed]

8. Yamauchi, J.; Coler-Reilly, A.; Sato, T.; Araya, N.; Yagishita, N.; Ando, H.; Kunitomo, Y.; Takahashi, K.; Tanaka, Y.; Shibagaki, Y.; *et al.* Mogamulizumab, an anti-CCR4 antibody,

targets human T-lymphotropic virus type 1-infected CD8+ and CD4+ T cells to treat associated myelopathy. *J. Infect. Dis.* **2015**, *211*, 238–248. [CrossRef] [PubMed]

9. Lezin, A.; Gillet, N.; Olindo, S.; Signate, A.; Grandvaux, N.; Verlaeten, O.; Belrose, G.; de Carvalho Bittencourt, M.; Hiscott, J.; Asquith, B.; *et al.* Histone deacetylase mediated transcriptional activation reduces proviral loads in HTLV-1 associated myelopathy/tropical spastic paraparesis patients. *Blood* **2007**, *110*, 3722–3728. [CrossRef] [PubMed]

10. Katz, R.A.; Skalka, A.M. Generation of Diversity in Retroviruses. *Annu. Rev. Genet.* **1990**, *24*, 409–443. [CrossRef] [PubMed]

11. Matsuoka, M.; Jeang, K.T. Human T-cell leukaemia virus type 1 (HTLV-1) infectivity and cellular transformation. *Nat. Rev. Cancer* **2007**, *7*, 270–280. [CrossRef] [PubMed]

12. Macatonia, S.E.; Cruickshank, J.K.; Rudge, P.; Knight, S.C. Dendritic cells from patients with tropical spastic paraparesis are infected with HTLV-1 and stimulate autologous lymphocyte proliferation. *AIDS Res. Hum. Retrovir.* **1992**, *8*, 1699–1706. [CrossRef] [PubMed]

13. Koyanagi, Y.; Itoyama, Y.; Nakamura, N.; Takamatsu, K.; Kira, J.; Iwamasa, T.; Goto, I.; Yamamoto, N. *In vivo* infection of human T-cell leukemia virus type I in non-T cells. *Virology* **1993**, *196*, 25–33. [CrossRef] [PubMed]

14. Jones, K.S.; Petrow-Sadowski, C.; Huang, Y.K.; Bertolette, D.C.; Ruscetti, F.W. Cell-free HTLV-1 infects dendritic cells leading to transmission and transformation of CD4+ T cells. *Nat. Med.* **2008**, *14*, 429–436. [CrossRef] [PubMed]

15. Manel, N.; Kim, F.J.; Kinet, S.; Taylor, N.; Sitbon, M.; Battini, J.L. The ubiquitous glucose transporter GLUT-1 is a receptor for HTLV. *Cell* **2003**, *115*, 449–459. [CrossRef]

16. Jones, K.S.; Petrow-Sadowski, C.; Bertolette, D.C.; Huang, Y.; Ruscetti, F.W. Heparan sulfate proteoglycans mediate attachment and entry of human T-cell leukemia virus type 1 virions into CD4+ T cells. *J. Virol.* **2005**, *79*, 12692–12702. [CrossRef] [PubMed]

17. Ghez, D.; Lepelletier, Y.; Lambert, S.; Fourneau, J.M.; Blot, V.; Janvier, S.; Arnulf, B.; van Endert, P.M.; Heveker, N.; Pique, C.; *et al.* Neuropilin-1 is involved in human T-cell lymphotropic virus type 1 entry. *J. Virol.* **2006**, *80*, 6844–6854. [CrossRef] [PubMed]

18. Lambert, S.; Bouttier, M.; Vassy, R.; Seigneuret, M.; Petrow-Sadowski, C.; Janvier, S.; Heveker, N.; Ruscetti, F.W.; Perret, G.; Jones, K.S.; *et al.* HTLV-1 uses HSPG and neuropilin-1 for entry by molecular mimicry of VEGF165. *Blood* **2009**, *113*, 5176–5185. [CrossRef] [PubMed]

19. Ghez, D.; Lepelletier, Y.; Jones, K.S.; Pique, C.; Hermine, O. Current concepts regarding the HTLV-1 receptor complex. *Retrovirology* **2010**, *7*, e99. [CrossRef] [PubMed]

20. Jones, K.S.; Lambert, S.; Bouttier, M.; Benit, L.; Ruscetti, F.W.; Hermine, O.; Pique, C. Molecular aspects of HTLV-1 entry: Functional domains of the HTLV-1 surface subunit (SU) and their relationships to the entry receptors. *Viruses* **2011**, *3*, 794–810. [CrossRef] [PubMed]

21. Hoshino, H. Cellular Factors Involved in HTLV-1 Entry and Pathogenicit. *Front. Microbiol.* **2012**, *3*, e222. [CrossRef] [PubMed]

22. Derse, D.; Hill, S.A.; Lloyd, P.A.; Chung, H.; Morse, B.A. Examining human T-lymphotropic virus type 1 infection and replication by cell-free infection with recombinant virus vectors. *J. Virol.* **2001**, *75*, 8461–8468. [CrossRef] [PubMed]

23. Mazurov, D.; Ilinskaya, A.; Heidecker, G.; Lloyd, P.; Derse, D. Quantitative comparison of

HTLV-1 and HIV-1 cell-to-cell infection with new replication dependent vectors. *PLoS Pathog.* **2010**, *6*, e1000788. [CrossRef] [PubMed]

24. Goncalves, D.U.; Proietti, F.A.; Ribas, J.G.; Araujo, M.G.; Pinheiro, S.R.; Guedes, A.C.; Carneiro-Proietti, A.B. Epidemiology, treatment, and prevention of human T-cell leukemia virus type 1-associated diseases. *Clin. Microbiol. Rev.* **2010**, *23*, 577–589. [CrossRef] [PubMed]

25. Takeuchi, H.; Takahashi, M.; Norose, Y.; Takeshita, T.; Fukunaga, Y.; Takahashi, H. Transformation of breast milk macrophages by HTLV-I: Implications for HTLV-I transmission via breastfeeding. *Biomed. Res.* **2010**, *31*, 53–61. [CrossRef] [PubMed]

26. Tugizov, S.M.; Herrera, R.; Veluppillai, P.; Greenspan, D.; Soros, V.; Greene, W.C.; Levy, J.A.; Palefsky, J.M. Differential transmission of HIV traversing fetal oral/intestinal epithelia and adult oral epithelia. *J. Virol.* **2012**, *86*, 2556–2570. [CrossRef] [PubMed]

27. Martin-Latil, S.; Gnadig, N.F.; Mallet, A.; Desdouits, M.; Guivel-Benhassine, F.; Jeannin, P.; Prevost, M.C.; Schwartz, O.; Gessain, A.; Ozden, S.; *et al.* Transcytosis of HTLV-1 across a tight human epithelial barrier and infection of subepithelial dendritic cells. *Blood* **2012**, *120*, 572–580. [CrossRef] [PubMed]

28. Pique, C.; Jones, K.S. Pathways of cell-cell transmission of HTLV-1. *Front. Microbiol.* **2012**, *3*, e378. [CrossRef] [PubMed]

29. Igakura, T.; Stinchcombe, J.C.; Goon, P.K.; Taylor, G.P.; Weber, J.N.; Griffiths, G.M.; Tanaka, Y.; Osame, M.; Bangham, C.R. Spread of HTLV-I between lymphocytes by virus-induced polarization of the cytoskeleton. *Science* **2003**, *299*, 1713–1716. [CrossRef] [PubMed]

30. Majorovits, E.; Nejmeddine, M.; Tanaka, Y.; Taylor, G.P.; Fuller, S.D.; Bangham, C.R. Human T-lymphotropic virus-1 visualized at the virological synapse by electron tomography. *PLoS ONE* **2008**, *3*, e2251. [CrossRef] [PubMed]

31. Nejmeddine, M.; Negi, V.S.; Mukherjee, S.; Tanaka, Y.; Orth, K.; Taylor, G.P.; Bangham, C.R. HTLV-1-Tax and ICAM-1 act on T-cell signal pathways to polarize the microtubule-organizing center at the virological synapse. *Blood* **2009**, *114*, 1016–1025. [CrossRef] [PubMed]

32. Van Prooyen, N.; Gold, H.; Andresen, V.; Schwartz, O.; Jones, K.; Ruscetti, F.; Lockett, S.; Gudla, P.; Venzon, D.; Franchini, G. Human T-cell leukemia virus type 1 p8 protein increases cellular conduits and virus transmission. *Proc. Natl. Acad. Sci. USA* **2010**, *107*, 20738–20743. [CrossRef] [PubMed]

33. Jones, K.S.; Green, P.L. Cloaked virus slips between cells. *Nat. Med.* **2010**, *16*, 25–27. [CrossRef] [PubMed]

34. Pais-Correia, A.M.; Sachse, M.; Guadagnini, S.; Robbiati, V.; Lasserre, R.; Gessain, A.; Gout, O.; Alcover, A.; Thoulouze, M.I. Biofilm-like extracellular viral assemblies mediate HTLV-1 cell-to-cell transmission at virological synapses. *Nat. Med.* **2010**, *16*, 83–89. [CrossRef] [PubMed]

35. Sze, A.; Belgnaoui, S.M.; Olagnier, D.; Lin, R.; Hiscott, J.; van Grevenynghe, J. Host restriction factor SAMHD1 limits human T cell leukemia virus type 1 infection of monocytes via STING-mediated apoptosis. *Cell Host Microbe* **2013**, *14*, 422–434. [CrossRef] [PubMed]

36. Ooms, M.; Krikoni, A.; Kress, A.K.; Simon, V.; Munk, C. APOBEC3A, APOBEC3B, and

APOBEC3H haplotype 2 restrict human T-lymphotropic virus type 1. *J. Virol.* **2012**, *86*, 6097–6108. [CrossRef] [PubMed]

37. Bai, X.T.; Nicot, C. miR-28-3p is a cellular restriction factor that inhibits human T cell leukemia virus, type 1 (HTLV-1) replication and virus infection. *J. Biol. Chem.* **2015**, *290*, 5381–5390. [CrossRef] [PubMed]

38. Twizere, J.C.; Kruys, V.; Lefebvre, L.; Vanderplasschen, A.; Collete, D.; Debacq, C.; Lai, W.S.; Jauniaux, J.C.; Bernstein, L.R.; Semmes, O.J.; *et al.* Interaction of retroviral Tax oncoproteins with tristetraprolin and regulation of tumor necrosis factor-alpha expression. *J. Natl. Cancer Inst.* **2003**, *95*, 1846–1859. [CrossRef] [PubMed]

39. Boxus, M.; Twizere, J.C.; Legros, S.; Kettmann, R.; Willems, L. Interaction of HTLV-1 Tax with minichromosome maintenance proteins accelerates the replication timing program. *Blood* **2012**, *119*, 151–160. [CrossRef] [PubMed]

40. Miyazato, P.; Yasunaga, J.; Taniguchi, Y.; Koyanagi, Y.; Mitsuya, H.; Matsuoka, M. De novo human T-cell leukemia virus type 1 infection of human lymphocytes in NOD-SCID, common gamma-chain knockout mice. *J. Virol.* **2006**, *80*, 10683–10691. [CrossRef] [PubMed]

41. Taylor, G.P.; Goon, P.; Furukawa, Y.; Green, H.; Barfield, A.; Mosley, A.; Nose, H.; Babiker, A.; Rudge, P.; Usuku, K.; *et al.* Zidovudine plus lamivudine in Human T-Lymphotropic Virus type-I-associated myelopathy: A randomised trial. *Retrovirology* **2006**, *3*, e63. [CrossRef] [PubMed]

42. Trevino, A.; Parra, P.; Bar-Magen, T.; Garrido, C.; de Mendoza, C.; Soriano, V. Antiviral effect of raltegravir on HTLV-1 carriers. *J. Antimicrobial Chemother.* **2012**, *67*, 218–221. [CrossRef] [PubMed]

43. Asquith, B.; Zhang, Y.; Mosley, A.J.; de Lara, C.M.; Wallace, D.L.; Worth, A.; Kaftantzi, L.; Meekings, K.; Griffin, G.E.; Tanaka, Y.; *et al.* *In vivo* T lymphocyte dynamics in humans and the impact of human T-lymphotropic virus 1 infection. *Proc. Natl. Acad. Sci. USA* **2007**, *104*, 8035–8040. [CrossRef] [PubMed]

44. Debacq, C.; Asquith, B.; Kerkhofs, P.; Portetelle, D.; Burny, A.; Kettmann, R.; Willems, L. Increased cell proliferation, but not reduced cell death, induces lymphocytosis in bovine leukemia virus-infected sheep. *Proc. Natl. Acad. Sci. USA* **2002**, *99*, 10048–10053. [CrossRef] [PubMed]

45. Ratner, L.; Philpott, T.; Trowbridge, D.B. Nucleotide sequence analysis of isolates of human T-lymphotropic virus type 1 of diverse geographical origins. *AIDS Res. Hum. Retrovir.* **1991**, *7*, 923–941. [CrossRef] [PubMed]

46. Gillet, N.A.; Malani, N.; Melamed, A.; Gormley, N.; Carter, R.; Bentley, D.; Berry, C.; Bushman, F.D.; Taylor, G.P.; Bangham, C.R. The host genomic environment of the provirus determines the abundance of HTLV-1-infected T-cell clones. *Blood* **2011**, *117*, 3113–3122. [CrossRef] [PubMed]

47. Gillet, N.A.; Gutierrez, G.; Rodriguez, S.M.; de Brogniez, A.; Renotte, N.; Alvarez, I.; Trono, K.; Willems, L. Massive depletion of bovine leukemia virus proviral clones located in genomic transcriptionally active sites during primary infection. *PLoS Pathog.* **2013**, *9*, e1003687. [CrossRef] [PubMed]

48. Reid, R.L.; Lindholm, P.F.; Mireskandari, A.; Dittmer, J.; Brady, J.N. Stabilization of wild-type p53 in human T-lymphocytes transformed by HTLV-I. *Oncogene* **1993**, *8*, 3029–3036. [PubMed]

49. Takachi, T.; Takahashi, M.; Takahashi-Yoshita, M.; Higuchi, M.; Obata, M.; Mishima, Y.; Okuda, S.; Tanaka, Y.; Matsuoka, M.; Saitoh, A.; *et al.* Human T-cell leukemia virus type 1 Tax oncoprotein represses the expression of the BCL11B tumor suppressor in T-cells. *Cancer Sci.* **2015**, *106*, 461–465. [CrossRef] [PubMed]

50. Yeung, M.L.; Yasunaga, J.; Bennasser, Y.; Dusetti, N.; Harris, D.; Ahmad, N.; Matsuoka, M.; Jeang, K.T. Roles for microRNAs, miR-93 and miR-130b, and tumor protein 53-induced nuclear protein 1 tumor suppressor in cell growth dysregulation by human T-cell lymphotrophic virus 1. *Cancer Res.* **2008**, *68*, 8976–8985. [CrossRef] [PubMed]

51. Boxus, M.; Twizere, J.C.; Legros, S.; Dewulf, J.F.; Kettmann, R.; Willems, L. The HTLV-1 Tax interactome. *Retrovirology* **2008**, *5*, e76. [CrossRef] [PubMed]

52. Jin, D.Y.; Spencer, F.; Jeang, K.T. Human T cell leukemia virus type 1 oncoprotein Tax targets the human mitotic checkpoint protein MAD1. *Cell* **1998**, *93*, 81–91. [CrossRef]

53. Kinjo, T.; Ham-Terhune, J.; Peloponese, J.M., Jr.; Jeang, K.T. Induction of reactive oxygen species by human T-cell leukemia virus type 1 tax correlates with DNA damage and expression of cellular senescence marker. *J. Virol.* **2010**, *84*, 5431–5437. [CrossRef] [PubMed]

54. Grossman, W.J.; Kimata, J.T.; Wong, F.H.; Zutter, M.; Ley, T.J.; Ratner, L. Development of leukemia in mice transgenic for the tax gene of human T-cell leukemia virus type I. *Proc. Natl. Acad. Sci. USA* **1995**, *92*, 1057–1061. [CrossRef] [PubMed]

55. Hasegawa, H.; Sawa, H.; Lewis, M.J.; Orba, Y.; Sheehy, N.; Yamamoto, Y.; Ichinohe, T.; Tsunetsugu-Yokota, Y.; Katano, H.; Takahashi, H.; *et al.* Thymus-derived leukemia-lymphoma in mice transgenic for the Tax gene of human T-lymphotropic virus type I. *Nat. Med.* **2006**, *12*, 466–472. [CrossRef] [PubMed]

56. Ohsugi, T.; Kumasaka, T.; Okada, S.; Urano, T. The Tax protein of HTLV-1 promotes oncogenesis in not only immature T cells but also mature T cells. *Nat. Med.* **2007**, *13*, 527–528. [CrossRef] [PubMed]

57. Chaib-Mezrag, H.; Lemacon, D.; Fontaine, H.; Bellon, M.; Bai, X.T.; Drac, M.; Coquelle, A.; Nicot, C. Tax impairs DNA replication forks and increases DNA breaks in specific oncogenic genome regions. *Mol. Cancer* **2014**, *13*, 205. [CrossRef] [PubMed]

58. Marriott, S.J.; Semmes, O.J. Impact of HTLV-I Tax on cell cycle progression and the cellular DNA damage repair response. *Oncogene* **2005**, *24*, 5986–5995. [CrossRef] [PubMed]

59. Bellon, M.; Baydoun, H.H.; Yao, Y.; Nicot, C. HTLV-I Tax-dependent and -independent events associated with immortalization of human primary T lymphocytes. *Blood* **2010**, *115*, 2441–2448. [CrossRef] [PubMed]

60. Nakagawa, M.; Schmitz, R.; Xiao, W.; Goldman, C.K.; Xu, W.; Yang, Y.; Yu, X.; Waldmann, T.A.; Staudt, L.M. Gain-of-function CCR4 mutations in adult T cell leukemia/lymphoma. *J. Exp. Med.* **2014**, *211*, 2497–2505. [CrossRef] [PubMed]

61. Shannon, K.M. CCR4 drives ATLL jail break. *J. Exp. Med.* **2014**, *211*, 2485. [CrossRef] [PubMed]

62. Furukawa, Y.; Kubota, R.; Tara, M.; Izumo, S.; Osame, M. Existence of escape mutant in HTLV-I tax during the development of adult T-cell leukemia. *Blood* **2001**, *97*, 987–993. [CrossRef] [PubMed]

63. Koiwa, T.; Hamano-Usami, A.; Ishida, T.; Okayama, A.; Yamaguchi, K.; Kamihira, S.; Watanabe, T. 5′-long terminal repeat-selective CpG methylation of latent human T-cell leukemia virus type 1 provirus *in vitro* and *in vivo*. *J. Virol.* **2002**, *76*, 9389–9397. [CrossRef] [PubMed]

64. Takeda, S.; Maeda, M.; Morikawa, S.; Taniguchi, Y.; Yasunaga, J.; Nosaka, K.; Tanaka, Y.; Matsuoka, M. Genetic and epigenetic inactivation of tax gene in adult T-cell leukemia cells. *Int. J. Cancer* **2004**, *109*, 559–567. [CrossRef] [PubMed]

65. Taniguchi, Y.; Nosaka, K.; Yasunaga, J.; Maeda, M.; Mueller, N.; Okayama, A.; Matsuoka, M. Silencing of human T-cell leukemia virus type I gene transcription by epigenetic mechanisms. *Retrovirology* **2005**, *2*, e64. [CrossRef] [PubMed]

66. Jacobson, S.; Shida, H.; McFarlin, D.E.; Fauci, A.S.; Koenig, S. Circulating CD8+ cytotoxic T lymphocytes specific for HTLV-I pX in patients with HTLV-I associated neurological disease. *Nature* **1990**, *348*, 245–248. [CrossRef] [PubMed]

67. Kannagi, M.; Harada, S.; Maruyama, I.; Inoko, H.; Igarashi, H.; Kuwashima, G.; Sato, S.; Morita, M.; Kidokoro, M.; Sugimoto, M.; *et al.* Predominant recognition of human T cell leukemia virus type I (HTLV-I) pX gene products by human CD8+ cytotoxic T cells directed against HTLV-I-infected cells. *Int. Immunol.* **1991**, *3*, 761–767. [CrossRef] [PubMed]

68. Kannagi, M.; Matsushita, S.; Harada, S. Expression of the target antigen for cytotoxic T lymphocytes on adult T-cell-leukemia cells. *Int. J. Cancer* **1993**, *54*, 582–588. [CrossRef] [PubMed]

69. Satou, Y.; Yasunaga, J.; Yoshida, M.; Matsuoka, M. HTLV-I basic leucine zipper factor gene mRNA supports proliferation of adult T cell leukemia cells. *Proc. Natl. Acad. Sci. USA* **2006**, *103*, 720–725. [CrossRef] [PubMed]

70. Macnamara, A.; Rowan, A.; Hilburn, S.; Kadolsky, U.; Fujiwara, H.; Suemori, K.; Yasukawa, M.; Taylor, G.; Bangham, C.R.; Asquith, B. HLA class I binding of HBZ determines outcome in HTLV-1 infection. *PLoS Pathog.* **2010**, *6*, e1001117. [CrossRef] [PubMed]

71. Hilburn, S.; Rowan, A.; Demontis, M.A.; MacNamara, A.; Asquith, B.; Bangham, C.R.; Taylor, G.P. *In vivo* expression of human T-lymphotropic virus type 1 basic leucine-zipper protein generates specific CD8+ and CD4+ T-lymphocyte responses that correlate with clinical outcome. *J. Infect. Dis.* **2011**, *203*, 529–536. [CrossRef] [PubMed]

72. Usui, T.; Yanagihara, K.; Tsukasaki, K.; Murata, K.; Hasegawa, H.; Yamada, Y.; Kamihira, S. Characteristic expression of HTLV-1 basic zipper factor (HBZ) transcripts in HTLV-1 provirus-positive cells. *Retrovirology* **2008**, *5*, e34. [CrossRef] [PubMed]

73. Matsuoka, M.; Green, P.L. The HBZ gene, a key player in HTLV-1 pathogenesis. *Retrovirology* **2009**, *6*, e71. [CrossRef] [PubMed]

74. Arnold, J.; Zimmerman, B.; Li, M.; Lairmore, M.D.; Green, P.L. Human T-cell leukemia virus type-1 antisense-encoded gene, Hbz, promotes T-lymphocyte proliferation. *Blood* **2008**, *112*, 3788–3797. [CrossRef] [PubMed]

75. Matsuoka, M.; Jeang, K.T. Human T-cell leukemia virus type 1 (HTLV-1) and leukemic transformation: Viral infectivity, Tax, HBZ and therapy. *Oncogene* **2011**, *30*, 1379–1389. [CrossRef] [PubMed]

76. Andrade, R.G.; Goncalves Pde, C.; Ribeiro, M.A.; Romanelli, L.C.; Ribas, J.G.; Torres, E.B.; Carneiro-Proietti, A.B.; Barbosa-Stancioli, E.F.; Martins, M.L. Strong correlation between tax and HBZ mRNA expression in HAM/TSP patients: Distinct markers for the neurologic disease. *J. Clin. Virol.* **2013**, *56*, 135–140. [CrossRef] [PubMed]

77. Saito, M.; Matsuzaki, T.; Satou, Y.; Yasunaga, J.; Saito, K.; Arimura, K.; Matsuoka, M.; Ohara, Y. *In vivo* expression of the HBZ gene of HTLV-1 correlates with proviral load, inflammatory markers and disease severity in HTLV-1 associated myelopathy/tropical spastic paraparesis (HAM/TSP). *Retrovirology* **2009**, *6*, e19. [CrossRef] [PubMed]

78. Sun, S.C.; Yamaoka, S. Activation of NF-kappaB by HTLV-I and implications for cell transformation. *Oncogene* **2005**, *24*, 5952–5964. [CrossRef] [PubMed]

79. Pujari, R.; Hunte, R.; Thomas, R.; van der Weyden, L.; Rauch, D.; Ratner, L.; Nyborg, J.K.; Ramos, J.C.; Takai, Y.; Shembade, N. Human T-cell leukemia virus type 1 (HTLV-1) tax requires CADM1/TSLC1 for inactivation of the NF-kappaB inhibitor A20 and constitutive NF-kappaB signaling. *PLoS Pathog.* **2015**, *11*, e1004721. [CrossRef] [PubMed]

80. Harhaj, E.W.; Harhaj, N.S. Mechanisms of persistent NF-kappaB activation by HTLV-I tax. *IUBMB Life* **2005**, *57*, 83–91. [CrossRef] [PubMed]

81. Lavorgna, A.; Matsuoka, M.; Harhaj, E.W. A critical role for IL-17RB signaling in HTLV-1 tax-induced NF-kappaB activation and T-cell transformation. *PLoS Pathog.* **2014**, *10*, e1004418. [CrossRef] [PubMed]

82. Lavorgna, A.; Harhaj, E.W. Regulation of HTLV-1 tax stability, cellular trafficking and NF-kappaB activation by the ubiquitin-proteasome pathway. *Viruses* **2014**, *6*, 3925–3943. [CrossRef] [PubMed]

83. Karin, M. NF-kappaB as a critical link between inflammation and cancer. *Cold Spring Harb. Perspect. Biol.* **2009**, *1*. [CrossRef] [PubMed]

84. Xiao, G.; Fu, J. NF-kappaB and cancer: A paradigm of Yin-Yang. *Am. J. Cancer Res.* **2011**, *1*, 192–221. [PubMed]

85. Krueger, A.; Fas, S.C.; Giaisi, M.; Bleumink, M.; Merling, A.; Stumpf, C.; Baumann, S.; Holtkotte, D.; Bosch, V.; Krammer, P.H.; *et al.* HTLV-1 Tax protects against CD95-mediated apoptosis by induction of the cellular FLICE-inhibitory protein (c-FLIP). *Blood* **2006**, *107*, 3933–3939. [CrossRef] [PubMed]

86. Okamoto, K.; Fujisawa, J.; Reth, M.; Yonehara, S. Human T-cell leukemia virus type-I oncoprotein Tax inhibits Fas-mediated apoptosis by inducing cellular FLIP through activation of NF-kappaB. *Genes Cells: Devoted Mol. Cell. Mech.* **2006**, *11*, 177–191. [CrossRef] [PubMed]

87. Tsukahara, T.; Kannagi, M.; Ohashi, T.; Kato, H.; Arai, M.; Nunez, G.; Iwanaga, Y.; Yamamoto, N.; Ohtani, K.; Nakamura, M.; *et al.* Induction of Bcl-x(L) expression by human T-cell leukemia virus type 1 Tax through NF-kappaB in apoptosis-resistant T-cell transfectants with Tax. *J. Virol.* **1999**, *73*, 7981–7987. [PubMed]

88. Nicot, C.; Mahieux, R.; Takemoto, S.; Franchini, G. Bcl-X(L) is up-regulated by HTLV-I and HTLV-II *in vitro* and in *ex vivo* ATLL samples. *Blood* **2000**, *96*, 275–281. [PubMed]

89. Swaims, A.Y.; Khani, F.; Zhang, Y.; Roberts, A.I.; Devadas, S.; Shi, Y.; Rabson, A.B. Immune activation induces immortalization of HTLV-1 LTR-Tax transgenic CD4+ T cells. *Blood* **2010**, *116*, 2994–3003. [CrossRef] [PubMed]

90. Macaire, H.; Riquet, A.; Moncollin, V.; Biemont-Trescol, M.C.; Duc Dodon, M.; Hermine, O.; Debaud, A.L.; Mahieux, R.; Mesnard, J.M.; Pierre, M.; *et al.* Tax protein-induced expression of antiapoptotic Bfl-1 protein contributes to survival of human T-cell leukemia virus type 1 (HTLV-1)-infected T-cells. *J. Biol. Chem.* **2012**, *287*, 21357–21370. [CrossRef] [PubMed]

91. Zhao, T.; Yasunaga, J.; Satou, Y.; Nakao, M.; Takahashi, M.; Fujii, M.; Matsuoka, M. Human T-cell leukemia virus type 1 bZIP factor selectively suppresses the classical pathway of NF-kappaB. *Blood* **2009**, *113*, 2755–2764. [CrossRef] [PubMed]

92. Zhi, H.; Yang, L.; Kuo, Y.L.; Ho, Y.K.; Shih, H.M.; Giam, C.Z. NF-kappaB hyper-activation by HTLV-1 tax induces cellular senescence, but can be alleviated by the viral anti-sense protein HBZ. *PLoS Pathog.* **2011**, *7*, e1002025. [CrossRef] [PubMed]

93. Ho, Y.K.; Zhi, H.; DeBiaso, D.; Philip, S.; Shih, H.M.; Giam, C.Z. HTLV-1 tax-induced rapid senescence is driven by the transcriptional activity of NF-kappaB and depends on chronically activated IKKalpha and p65/RelA. *J. Virol.* **2012**, *86*, 9474–9483. [CrossRef] [PubMed]

94. Philip, S.; Zahoor, M.A.; Zhi, H.; Ho, Y.K.; Giam, C.Z. Regulation of human T-lymphotropic virus type I latency and reactivation by HBZ and Rex. *PLoS Pathog.* **2014**, *10*, e1004040. [CrossRef] [PubMed]

95. Peloponese, J.M., Jr.; Jeang, K.T. Role for Akt/protein kinase B and activator protein-1 in cellular proliferation induced by the human T-cell leukemia virus type 1 tax oncoprotein. *J. Biol. Chem.* **2006**, *281*, 8927–8938. [CrossRef] [PubMed]

96. Jeong, S.J.; Dasgupta, A.; Jung, K.J.; Um, J.H.; Burke, A.; Park, H.U.; Brady, J.N. PI3K/AKT inhibition induces caspase-dependent apoptosis in HTLV-1-transformed cells. *Virology* **2008**, *370*, 264–272. [CrossRef] [PubMed]

97. Tomita, M.; Semenza, G.L.; Michiels, C.; Matsuda, T.; Uchihara, J.N.; Okudaira, T.; Tanaka, Y.; Taira, N.; Ohshiro, K.; Mori, N. Activation of hypoxia-inducible factor 1 in human T-cell leukaemia virus type 1-infected cell lines and primary adult T-cell leukaemia cells. *Biochem. J.* **2007**, *406*, 317–323. [PubMed]

98. Higuchi, M.; Takahashi, M.; Tanaka, Y.; Fujii, M. Downregulation of proapoptotic Bim augments IL-2-independent T-cell transformation by human T-cell leukemia virus type-1 Tax. *Cancer Med.* **2014**, *3*, 1605–1614. [CrossRef] [PubMed]

99. Saito, K.; Saito, M.; Taniura, N.; Okuwa, T.; Ohara, Y. Activation of the PI3K-Akt pathway by human T cell leukemia virus type 1 (HTLV-1) oncoprotein Tax increases Bcl3 expression, which is associated with enhanced growth of HTLV-1-infected T cells. *Virology* **2010**, *403*, 173–180. [CrossRef] [PubMed]

100. Sugata, K.; Satou, Y.; Yasunaga, J.; Hara, H.; Ohshima, K.; Utsunomiya, A.; Mitsuyama, M.; Matsuoka, M. HTLV-1 bZIP factor impairs cell-mediated immunity by suppressing production of Th1 cytokines. *Blood* **2012**, *119*, 434–444. [CrossRef] [PubMed]

101. Tanaka-Nakanishi, A.; Yasunaga, J.; Takai, K.; Matsuoka, M. HTLV-1 bZIP factor suppresses apoptosis by attenuating the function of FoxO3a and altering its localization. *Cancer Res.* **2014**, *74*, 188–200. [CrossRef] [PubMed]

102. Matsumoto, J.; Ohshima, T.; Isono, O.; Shimotohno, K. HTLV-1 HBZ suppresses AP-1 activity by impairing both the DNA-binding ability and the stability of c-Jun protein. *Oncogene* **2005**, *24*, 1001–1010. [CrossRef] [PubMed]

103. Hivin, P.; Basbous, J.; Raymond, F.; Henaff, D.; Arpin-Andre, C.; Robert-Hebmann, V.; Barbeau, B.; Mesnard, J.M. The HBZ-SP1 isoform of human T-cell leukemia virus type I represses JunB activity by sequestration into nuclear bodies. *Retrovirology* **2007**, *4*, e14. [CrossRef] [PubMed]

104. Isono, O.; Ohshima, T.; Saeki, Y.; Matsumoto, J.; Hijikata, M.; Tanaka, K.; Shimotohno, K. Human T-cell leukemia virus type 1 HBZ protein bypasses the targeting function of ubiquitination. *J. Biol. Chem.* **2008**, *283*, 34273–34282. [CrossRef] [PubMed]

105. Clerc, I.; Hivin, P.; Rubbo, P.A.; Lemasson, I.; Barbeau, B.; Mesnard, J.M. Propensity for HBZ-SP1 isoform of HTLV-I to inhibit c-Jun activity correlates with sequestration of c-Jun into nuclear bodies rather than inhibition of its DNA-binding activity. *Virology* **2009**, *391*, 195–202. [CrossRef] [PubMed]

106. Ohshima, T.; Mukai, R.; Nakahara, N.; Matsumoto, J.; Isono, O.; Kobayashi, Y.; Takahashi, S.; Shimotohno, K. HTLV-1 basic leucine-zipper factor, HBZ, interacts with MafB and suppresses transcription through a Maf recognition element. *J. Cell. Biochem.* **2010**, *111*, 187–194. [CrossRef] [PubMed]

107. Kashanchi, F.; Brady, J.N. Transcriptional and post-transcriptional gene regulation of HTLV-1. *Oncogene* **2005**, *24*, 5938–5951. [CrossRef] [PubMed]

108. Saggioro, D.; Barp, S.; Chieco-Bianchi, L. Block of a mitochondrial-mediated apoptotic pathway in Tax-expressing murine fibroblasts. *Exp. Cell Res.* **2001**, *269*, 245–255. [CrossRef] [PubMed]

109. Trevisan, R.; Daprai, L.; Acquasaliente, L.; Ciminale, V.; Chieco-Bianchi, L.; Saggioro, D. Relevance of CREB phosphorylation in the anti-apoptotic function of human T-lymphotropic virus type 1 tax protein in serum-deprived murine fibroblasts. *Exp. Cell Res.* **2004**, *299*, 57–67. [CrossRef] [PubMed]

110. Trevisan, R.; Daprai, L.; Paloschi, L.; Vajente, N.; Chieco-Bianchi, L.; Saggioro, D. Antiapoptotic effect of human T-cell leukemia virus type 1 tax protein correlates with its creb transcriptional activity. *Exp. Cell Res.* **2006**, *312*, 1390–1400. [CrossRef] [PubMed]

111. Saggioro, D. Anti-apoptotic effect of Tax: An NF-kappaB path or a CREB way? *Viruses* **2011**, *3*, 1001–1014. [CrossRef] [PubMed]

112. Saggioro, D.; Silic-Benussi, M.; Biasiotto, R.; D'Agostino, D.M.; Ciminale, V. Control of cell death pathways by HTLV-1 proteins. *Front. Biosci.* **2009**, *14*, 3338–3351. [CrossRef]

113. Fukuda, R.I.; Tsuchiya, K.; Suzuki, K.; Itoh, K.; Fujita, J.; Utsunomiya, A.; Tsuji, T. Human T-cell leukemia virus type I tax down-regulates the expression of phosphatidylinositol 3,4,5-trisphosphate inositol phosphatases via the NF-kappaB pathway. *J. Biol. Chem.* **2009**, *284*, 2680–2689. [CrossRef] [PubMed]

114. Gaudray, G.; Gachon, F.; Basbous, J.; Biard-Piechaczyk, M.; Devaux, C.; Mesnard, J.M.

The complementary strand of the human T-cell leukemia virus type 1 RNA genome encodes a bZIP transcription factor that down-regulates viral transcription. *J. Virol.* **2002**, *76*, 12813–12822. [CrossRef] [PubMed]

115. Lemasson, I.; Lewis, M.R.; Polakowski, N.; Hivin, P.; Cavanagh, M.H.; Thebault, S.; Barbeau, B.; Nyborg, J.K.; Mesnard, J.M. Human T-cell leukemia virus type 1 (HTLV-1) bZIP protein interacts with the cellular transcription factor CREB to inhibit HTLV-1 transcription. *J. Virol.* **2007**, *81*, 1543–1553. [CrossRef] [PubMed]

116. Clerc, I.; Polakowski, N.; Andre-Arpin, C.; Cook, P.; Barbeau, B.; Mesnard, J.M.; Lemasson, I. An interaction between the human T cell leukemia virus type 1 basic leucine zipper factor (HBZ) and the KIX domain of p300/CBP contributes to the down-regulation of tax-dependent viral transcription by HBZ. *J. Biol. Chem.* **2008**, *283*, 23903–23913. [CrossRef] [PubMed]

117. Cook, P.R.; Polakowski, N.; Lemasson, I. HTLV-1 HBZ protein deregulates interactions between cellular factors and the KIX domain of p300/CBP. *J. Mol. Biol.* **2011**, *409*, 384–398. [CrossRef] [PubMed]

118. Ma, G.; Yasunaga, J.; Fan, J.; Yanagawa, S.; Matsuoka, M. HTLV-1 bZIP factor dysregulates the Wnt pathways to support proliferation and migration of adult T-cell leukemia cells. *Oncogene* **2013**, *32*, 4222–4230. [CrossRef] [PubMed]

119. Arnulf, B.; Villemain, A.; Nicot, C.; Mordelet, E.; Charneau, P.; Kersual, J.; Zermati, Y.; Mauviel, A.; Bazarbachi, A.; Hermine, O. Human T-cell lymphotropic virus oncoprotein Tax represses TGF-beta 1 signaling in human T cells via c-Jun activation: A potential mechanism of HTLV-I leukemogenesis. *Blood* **2002**, *100*, 4129–4138. [CrossRef] [PubMed]

120. Lee, D.K.; Kim, B.C.; Brady, J.N.; Jeang, K.T.; Kim, S.J. Human T-cell lymphotropic virus type 1 tax inhibits transforming growth factor-beta signaling by blocking the association of Smad proteins with Smad-binding element. *J. Biol. Chem.* **2002**, *277*, 33766–33775. [CrossRef] [PubMed]

121. Mori, N.; Morishita, M.; Tsukazaki, T.; Giam, C.Z.; Kumatori, A.; Tanaka, Y.; Yamamoto, N. Human T-cell leukemia virus type I oncoprotein Tax represses Smad-dependent transforming growth factor beta signaling through interaction with CREB-binding protein/p300. *Blood* **2001**, *97*, 2137–2144. [CrossRef] [PubMed]

122. Zhao, T.; Satou, Y.; Sugata, K.; Miyazato, P.; Green, P.L.; Imamura, T.; Matsuoka, M. HTLV-1 bZIP factor enhances TGF-beta signaling through p300 coactivator. *Blood* **2011**, *118*, 1865–1876. [CrossRef] [PubMed]

123. Kim, Y.M.; Geiger, T.R.; Egan, D.I.; Sharma, N.; Nyborg, J.K. The HTLV-1 tax protein cooperates with phosphorylated CREB, TORC2 and p300 to activate CRE-dependent cyclin D1 transcription. *Oncogene* **2010**, *29*, 2142–2152. [CrossRef] [PubMed]

124. Dellino, G.I.; Cittaro, D.; Piccioni, R.; Luzi, L.; Banfi, S.; Segalla, S.; Cesaroni, M.; Mendoza-Maldonado, R.; Giacca, M.; Pelicci, P.G. Genome-wide mapping of human DNA-replication origins: Levels of transcription at ORC1 sites regulate origin selection and replication timing. *Genome Res.* **2013**, *23*, 1–11. [CrossRef] [PubMed]

125. Ma, Y.; Zheng, S.; Wang, Y.; Zang, W.; Li, M.; Wang, N.; Li, P.; Jin, J.; Dong, Z.; Zhao, G. The HTLV-1 HBZ protein inhibits cyclin D1 expression through interacting with the cellular

transcription factor CREB. *Mol. Biol. Rep.* **2013**, *40*, 5967–5975. [CrossRef] [PubMed]

126. Hagiya, K.; Yasunaga, J.; Satou, Y.; Ohshima, K.; Matsuoka, M. ATF3, an HTLV-1 bZip factor binding protein, promotes proliferation of adult T-cell leukemia cells. *Retrovirology* **2011**, *8*, e19. [CrossRef] [PubMed]

127. Tabakin-Fix, Y.; Azran, I.; Schavinky-Khrapunsky, Y.; Levy, O.; Aboud, M. Functional inactivation of p53 by human T-cell leukemia virus type 1 Tax protein: Mechanisms and clinical implications. *Carcinogenesis* **2006**, *27*, 673–681. [CrossRef] [PubMed]

128. Ariumi, Y.; Kaida, A.; Lin, J.Y.; Hirota, M.; Masui, O.; Yamaoka, S.; Taya, Y.; Shimotohno, K. HTLV-1 tax oncoprotein represses the p53-mediated trans-activation function through coactivator CBP sequestration. *Oncogene* **2000**, *19*, 1491–1499. [CrossRef] [PubMed]

129. Pise-Masison, C.A.; Mahieux, R.; Jiang, H.; Ashcroft, M.; Radonovich, M.; Duvall, J.; Guillerm, C.; Brady, J.N. Inactivation of p53 by human T-cell lymphotropic virus type 1 Tax requires activation of the NF-kappaB pathway and is dependent on p53 phosphorylation. *Mol. Cell. Biol.* **2000**, *20*, 3377–3386. [CrossRef] [PubMed]

130. Miyazato, A.; Sheleg, S.; Iha, H.; Li, Y.; Jeang, K.T. Evidence for NF-kappaB- and CBP-independent repression of p53's transcriptional activity by human T-cell leukemia virus type 1 Tax in mouse embryo and primary human fibroblasts. *J. Virol.* **2005**, *79*, 9346–9350. [CrossRef] [PubMed]

131. Gillet, N.; Carpentier, A.; Barez, P.Y.; Willems, L. WIP1 deficiency inhibits HTLV-1 Tax oncogenesis: Novel therapeutic prospects for treatment of ATL? *Retrovirology* **2012**, *9*, e115. [CrossRef] [PubMed]

132. Zane, L.; Yasunaga, J.; Mitagami, Y.; Yedavalli, V.; Tang, S.W.; Chen, C.Y.; Ratner, L.; Lu, X.; Jeang, K.T. Wip1 and p53 contribute to HTLV-1 Tax-induced tumorigenesis. *Retrovirology* **2012**, *9*, e114. [CrossRef] [PubMed]

133. Kasai, T.; Iwanaga, Y.; Iha, H.; Jeang, K.T. Prevalent loss of mitotic spindle checkpoint in adult T-cell leukemia confers resistance to microtubule inhibitors. *J. Biol. Chem.* **2002**, *277*, 5187–5193. [CrossRef] [PubMed]

134. Yasunaga, J.; Jeang, K.T. Viral transformation and aneuploidy. *Environ. Mol. Mutagen.* **2009**, *50*, 733–740. [CrossRef] [PubMed]

135. Liu, B.; Hong, S.; Tang, Z.; Yu, H.; Giam, C.Z. HTLV-I Tax directly binds the Cdc20-associated anaphase-promoting complex and activates it ahead of schedule. *Proc. Natl. Acad. Sci. USA* **2005**, *102*, 63–68. [CrossRef] [PubMed]

136. Peloponese, J.M., Jr.; Haller, K.; Miyazato, A.; Jeang, K.T. Abnormal centrosome amplification in cells through the targeting of Ran-binding protein-1 by the human T cell leukemia virus type-1 Tax oncoprotein. *Proc. Natl. Acad. Sci. USA* **2005**, *102*, 18974–18979. [CrossRef] [PubMed]

137. Haoudi, A.; Semmes, O.J. The HTLV-1 tax oncoprotein attenuates DNA damage induced G1 arrest and enhances apoptosis in p53 null cells. *Virology* **2003**, *305*, 229–239. [CrossRef] [PubMed]

138. Park, H.U.; Jeong, J.H.; Chung, J.H.; Brady, J.N. Human T-cell leukemia virus type 1 Tax interacts with Chk1 and attenuates DNA-damage induced G2 arrest mediated by Chk1. *Oncogene* **2004**, *23*, 4966–4974. [CrossRef] [PubMed]

139. Park, H.U.; Jeong, S.J.; Jeong, J.H.; Chung, J.H.; Brady, J.N. Human T-cell leukemia virus type 1 Tax attenuates gamma-irradiation-induced apoptosis through physical interaction with Chk2. *Oncogene* **2006**, *25*, 438–447. [PubMed]

140. Gupta, S.K.; Guo, X.; Durkin, S.S.; Fryrear, K.F.; Ward, M.D.; Semmes, O.J. Human T-cell leukemia virus type 1 Tax oncoprotein prevents DNA damage-induced chromatin egress of hyperphosphorylated Chk2. *J. Biol. Chem.* **2007**, *282*, 29431–29440. [CrossRef] [PubMed]

141. Chandhasin, C.; Ducu, R.I.; Berkovich, E.; Kastan, M.B.; Marriott, S.J. Human T-cell leukemia virus type 1 tax attenuates the ATM-mediated cellular DNA damage response. *J. Virol.* **2008**, *82*, 6952–6961. [CrossRef] [PubMed]

142. Durkin, S.S.; Guo, X.; Fryrear, K.A.; Mihaylova, V.T.; Gupta, S.K.; Belgnaoui, S.M.; Haoudi, A.; Kupfer, G.M.; Semmes, O.J. HTLV-1 Tax oncoprotein subverts the cellular DNA damage response via binding to DNA-dependent protein kinase. *J. Biol. Chem.* **2008**, *283*, 36311–36320. [CrossRef] [PubMed]

143. Belgnaoui, S.M.; Fryrear, K.A.; Nyalwidhe, J.O.; Guo, X.; Semmes, O.J. The viral oncoprotein tax sequesters DNA damage response factors by tethering MDC1 to chromatin. *J. Biol. Chem.* **2010**, *285*, 32897–32905. [CrossRef] [PubMed]

144. Boxus, M.; Willems, L. How the DNA damage response determines the fate of HTLV-1 Tax-expressing cells. *Retrovirology* **2012**, *9*, e2. [CrossRef] [PubMed]

145. Vernin, C.; Thenoz, M.; Pinatel, C.; Gessain, A.; Gout, O.; Delfau-Larue, M.H.; Nazaret, N.; Legras-Lachuer, C.; Wattel, E.; Mortreux, F. HTLV-1 bZIP Factor HBZ Promotes Cell Proliferation and Genetic Instability by Activating OncomiRs. *Cancer Res.* **2014**, *74*, 6082–6093. [CrossRef] [PubMed]

146. Mukai, R.; Ohshima, T. HTLV-1 HBZ positively regulates the mTOR signaling pathway via inhibition of GADD34 activity in the cytoplasm. *Oncogene* **2014**, *33*, 2317–2328. [CrossRef] [PubMed]

147. Jeang, K.T.; Widen, S.G.; Semmes, O.J.t.; Wilson, S.H. HTLV-I trans-activator protein, tax, is a trans-repressor of the human beta-polymerase gene. *Science* **1990**, *247*, 1082–1084. [CrossRef] [PubMed]

148. Philpott, S.M.; Buehring, G.C. Defective DNA repair in cells with human T-cell leukemia/bovine leukemia viruses: Role of tax gene. *J. Natl. Cancer Inst.* **1999**, *91*, 933–942. [CrossRef] [PubMed]

149. Kao, S.Y.; Marriott, S.J. Disruption of nucleotide excision repair by the human T-cell leukemia virus type 1 Tax protein. *J. Virol.* **1999**, *73*, 4299–4304. [PubMed]

150. Lemoine, F.J.; Kao, S.Y.; Marriott, S.J. Suppression of DNA repair by HTLV type 1 Tax correlates with Tax trans-activation of proliferating cell nuclear antigen gene expression. *AIDS Res. Hum. Retrovir.* **2000**, *16*, 1623–1627. [CrossRef] [PubMed]

151. Ducu, R.I.; Dayaram, T.; Marriott, S.J. The HTLV-1 Tax oncoprotein represses Ku80 gene expression. *Virology* **2011**, *416*, 1–8. [CrossRef] [PubMed]

152. Majone, F.; Jeang, K.T. Unstabilized DNA breaks in HTLV-1 Tax expressing cells correlate with functional targeting of Ku80, not PKcs, XRCC4, or H2AX. *Cell Biosci.* **2012**, *2*, e15. [CrossRef] [PubMed]

153. Baydoun, H.H.; Bai, X.T.; Shelton, S.; Nicot, C. HTLV-I tax increases genetic instability by inducing DNA double strand breaks during DNA replication and switching repair to NHEJ. *PLoS ONE* **2012**, *7*, e42226. [CrossRef] [PubMed]

154. Baydoun, H.H.; Pancewicz, J.; Nicot, C. Human T-lymphotropic type 1 virus p30 inhibits homologous recombination and favors unfaithful DNA repair. *Blood* **2011**, *117*, 5897–5906. [CrossRef] [PubMed]

5

The Virus-Host Interplay: Biogenesis of +RNA Replication Complexes

Colleen R. Reid [1,†], **Adriana M. Airo** [1,†] **and Tom C. Hobman** [1,2,*]

[1] Department of Medical Microbiology and Immunology, University of Alberta, Edmonton, AB T6G 2E1, Canada; E-Mails: crreid1@ualberta.ca (C.R.R.); airo@ualberta.ca (A.M.A.)

[2] Department of Cell Biology, University of Alberta, Edmonton, AB T6G 2H7, Canada

[†] These authors contributed equally to this work.

[*] Author to whom correspondence should be addressed; E-Mail: tom.hobman@ualberta.ca

Academic Editor: David Boehr

Abstract: Positive-strand RNA (+RNA) viruses are an important group of human and animal pathogens that have significant global health and economic impacts. Notable members include West Nile virus, Dengue virus, Chikungunya, Severe acute respiratory syndrome (SARS) Coronavirus and enteroviruses of the *Picornaviridae* family.Unfortunately, prophylactic and therapeutic treatments against these pathogens are limited. +RNA viruses have limited coding capacity and thus rely extensively on host factors for successful infection and propagation. A common feature among these viruses is their ability to dramatically modify cellular membranes to serve as platforms for genome replication and assembly of new virions. These viral replication complexes (VRCs) serve two main functions: To increase replication efficiency by concentrating critical factors and to protect the viral genome from host anti-viral systems. This review summarizes current knowledge of critical host factors recruited to or demonstrated to be involved in the biogenesis and stabilization of +RNA virus VRCs.

Keywords: +RNA viruses; replication complexes; host factors; membranes

1. Introduction

Positive-sense RNA (+RNA) viruses including the Flaviviruses, enteroviruses of the *Picornaviridae* family, Alphaviruses, and Coronaviruses all dramatically modify cellular membranes to serve as platforms for replication and assembly of new virions. The biogenesis of these replication compartments is a complex interplay of interactions between virus and host proteins. Although considerable progress has been made in identifying host proteins that interact with virus-encoded proteins, much remains to be learned regarding the significance of these interactions. Despite morphological differences in the replication complexes formed by members of each viral family, these viruses have evolved to use common cellular pathways to complete biogenesis. Some of the shared pathways highlighted in this review include lipid metabolism, autophagy, signal transduction and proteins involved in intracellular trafficking (Table 1). Remarkably, even within the higher order of shared pathways, differences within members of specific families (such as *Flaviviridae*) exist, highlighting that the assembly and function of viral replication complexes (VRCs) varies considerably. As such, this review focuses on a broad view of host factors in which there is significant functional evidence linking them to VRCs in effort to highlight commonalities or differences and further advance the understanding of virus-host interactions.

2. *Flaviviridae*

The *Flaviviridae* family includes many significant global pathogens including Hepatitis C virus (HCV), West Nile virus (WNV), and Dengue virus (DENV). This family is comprised of four genera, with the human pathogens belonging to the genera *Flavivirus* and *Hepacivirus*. The *Hepacivirus* genus contains HCV, a prominent blood-borne human pathogen that causes chronic hepatitis and is estimated to have infected 170 million people worldwide. The *Flavivirus* genus includes DENV, WNV, Yellow Fever virus (YFV) and other viruses causing either haemorrhagic or encephalitic disease. Except for YFV and Japanese Encephalitis virus (JEV), vaccines for use in humans are not available against members of this family. Current treatment options are very limited and supportive care is often the only option. Arthropod vectors, mainly mosquitos and ticks are used by most flaviviruses to infect their hosts.

In general, virions are enveloped and contain a single copy of viral genomic RNA ($\sim$11 kilobase (kb)) encoding a single polyprotein that is cleaved by viral and host proteases into three structural and seven non-structural proteins [1]. After binding to cell surface receptors, the virions enter cells through endocytic pathways. Within the acidic environment of endosomes, the virions fuse with endosomal membrane resulting in release of the nucleocapsid into the cytoplasm. After the nucleocapsid disassembles, the viral RNA is translated into a polyprotein, which is then processed into individual viral proteins. VRCs form soon after and serve as platforms for RNA replication. Assembly of nascent virions occurs in close proximity to VRCs on the endoplasmic reticulum (ER). After budding into the ER, virions traverse the secretory pathway before release from the cell.

2.1. Genus Hepacivirus

The biogenesis of the HCV VRCs and the stabilization of these structures have been extensively studied [2,3]. Electron microscopic analysis of infected cells revealed that HCV replicates on altered ER membranes that are closely associated with lipid droplets; termed the "membranous web" [4].

The membrane-associated non-structural protein 4B (NS4B) plays a key role in the formation of this network [5], which consists of double membrane vesicles (DMVs) protruding out of ER. Of note, the DMVs are similar to ER-associated structures induced by members of *Picornaviridae* and *Coronaviridae* [6]. A plethora of host factors involved in lipid metabolism, intracellular signalling, protein folding, and vesicular trafficking are known to be important for HCV VRC activity. Due to the availability of extensive literature on the subject, they will not be discussed here. Instead, we refer readers to the following recent reviews [3,7,8].

2.2. Biogenesis of the Flavivirus Replication Complex

Of the studies investigating the membrane alterations induced by members of this genus, most have focused on DENV and WNV. Infection of mammalian cells with either the Australian attenuated strain WNV$_{KUN}$, or the highly pathogenic WNV$_{NY99}$ strain results in similar phenotypic disruptions of cellular membranes [9,10]. Early studies of cells infected with WNV or DENV revealed dramatic changes in cellular membranes and the formation of single membrane vesicular packets (VPs) and convoluted membranes (CM), which are in close association with smooth membranes and the rough-ER [9,11]. Paracrystalline arrays (PC) were also described in WNV$_{KUN}$-infected cells [9]. Infection of cells derived from the viral vector (mosquito) with DENV or WNV also led to dramatic alterations of membranes resulting in spherules associated with ER membranes [12,13]. These virus-induced structures are thought to segregate viral replication from protein translation [14]. VPs are the sites of viral replication as evidenced by the fact that they contain double-stranded RNA (dsRNA), a replication intermediate, and the viral RNA-dependent RNA polymerase, NS5 [9,11,15–17]. Two other virus-encoded non-structural proteins NS1 and NS3 also associate with these elements. VPs in DENV- and WNV-infected cells are ~85 nm in diameter indicating the conserved nature of these structures. CMs and PCs in WNV$_{KUN}$-infected cells are enriched for NS3/2b, the viral protease and do not contain dsRNA [9,15]. This suggests that CM/PCs may be the sites of viral translation and/or proteolytic processing of the viral proteins. The origins of these membranes vary between viruses. DENV-induced VP membranes contain the ER resident proteins protein disulphide isomerase and calnexin [15], whereas in cells infected with WNV$_{KUN}$, VPs that are positive for dsRNA, contain the *trans*-Golgi network protein, galactosyltransferase, possibly indicating that these structures are derived from Golgi membranes [18]. Moreover, the ER-Golgi intermediate compartment marker ERGIC53, is associated with CMs and PCs. WNV$_{NY99}$ is similar to DENV in that the VRCs colocalize with protein disulphide isomerase, suggesting these structures are ER-derived [10]. Electron tomography was utilized to further characterize the VRCs of DENV [15] (Figure 1B), WNV$_{KUN}$ [9,19], WNV$_{NY99}$ [20], and Tick-borne encephalitis (TBEV)-like virus Langat virus [21]. These "vesicles" in fact appear to be invaginations of the ER membrane with small neck-like openings (~11.2 nm for DENV) that may facilitate trafficking of molecules into and out of these replication sites. In some cases, there were connections between these vesicles within the modified-ER membrane. DENV VPs closely associated with budding sites appear as electron dense invaginations (~60 nm) and can be seen on opposing cisternae [15]. Despite there being good structural information on DENV and WNV VRCs, the exact mechanism by which these membranous organelles form, remains unclear. However, the ER localized non-structural viral proteins, WNV$_{NY99}$ NS4B [20] and DENV-2 NS4A [22] are thought play a role in the initial membrane curvature.

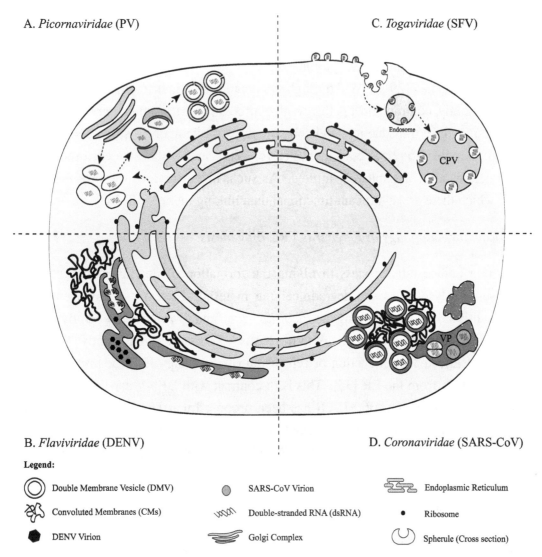

A. *Picornaviridae* (PV)

C. *Togaviridae* (SFV)

Endosome

CPV

B. *Flaviviridae* (DENV)

D. *Coronaviridae* (SARS-CoV)

VP

Legend:

◎ Double Membrane Vesicle (DMV)	● SARS-CoV Virion	⊟ Endoplasmic Reticulum
🕸 Convoluted Membranes (CMs)	⟋⟍ Double-stranded RNA (dsRNA)	• Ribosome
⬢ DENV Virion	⟋ Golgi Complex	∪ Spherule (Cross section)

Figure 1. Biogenesis of Viral Replication Complexes (VRCs): Representative diagram of the structure and biogenesis of the VRCs for each family based on electron microscopy from the following references: Poliovirus (PV) [23] Semliki Forest virus (SFV) [24], Dengue virus (DENV) [15], Severe acute respiratory syndrome coronavirus (SARS-CoV) [25]. Diagram not to scale. (**A**) Formation of the PV VRC: Early in infection single membrane vesicles that contain dsRNA are derived from the ER and Golgi components. Progression of infection results in vesicles wrapping around each other inducing the formation of DMVs; (**B**) Formation of the DENV VRC: Spherule structures containing dsRNA form on modified rough-ER membranes surrounded by convoluted membranes. Assembly sites form on opposing cisternae where newly formed virions are stored; (**C**) Formation of SFV VRC: Spherules that contain dsRNA form at the plasma membrane. Internalization of these structures follows the endo-lysosomal pathway resulting in formation of cytopathic vacuoles (CPV); (**D**) Proposed formation of the SARS-CoV VRC: DMVs containing dsRNA are formed and their outer membranes are continuous with the ER. Convoluted membranes surround the DMVs. Formation of vesicle packets (VPs) is thought to result from DMV fusion. Newly formed virions are associated with these structures.

Following this step, other viral and host factors are likely required for the biogenesis and stabilization of the flavivirus VRC. To date, a large number of host proteins involved in flavivirus replication have been identified by proteomic and transcriptomic studies of infected cells [26–28], mapping the host cell interactome of viral proteins [29–31] and through systematic RNA interference (RNAi) screens [32,33]. Perhaps not surprisingly, common host pathways that affect flavivirus replication include those involved in lipid metabolism, signal transduction, and cell structure. While many host factors that are thought to play a role in virus replication have been identified for WNV and DENV, the corresponding functional and validation studies are comparatively limited. As such, in this review, we have focused mainly on host factors in which there are significant functional data linking them to VRCs.

2.3. Potential Role for Autophagy in Flavivirus VRC Biogenesis

Autophagy is a homeostatic process involving the formation of double membrane vesicles from the ER that fuse with lysosomes and degrade cellular material. Recently, it has been linked to VRC biogenesis for multiple viruses, including the *Picornaviridae* and *Coronaviridae* family (covered later in this review). The requirement for autophagy in flavivirus VRCs varies significantly. WNV propagation for example, is not affected by induction or repression of autophagy [34], nor is autophagy required for biogenesis of VRCs from the ER [35]. This is in contrast with DENV and JEV, which both exploit autophagy for virus propagation [36,37]. It has been proposed that DENV uses autophagy-induction to aid in release of fatty acids from lipid droplets, increase β-oxidation and ATP production [38]. Little is known about the role of autophagy in VRC biogenesis or replication of other members of the *Flavivirus* genus.

2.4. Membrane Remodelling and Lipid Metabolism

Biogenesis of VRCs requires massive expansion of ER-associated membranes and alteration of their lipid compositions. HCV and members of the *Picornaviridae* family are known to recruit phosphatidyl-inositol kinases (PI4Ks) to their VRC networks for the conversion of Phosphatidylinositol (PI) to Phosphatidylinositol 4-phosphate (PI4P) lipids, a process that is essential for viral replication [39,40]. PI4P lipids may serve to recruit host proteins and/or lipid components to these organelles. Interestingly, neither WNV nor DENV seem to require PI4P lipids [41,42] indicating that assembly and function of VRCs varies considerably among the *Flaviviridae* family. However, DENV infection does alter the membrane lipid composition of human cells and two host cell enzymes involved in fatty acid metabolism, fatty acid synthase (FAS) and Acetyl-CoA carboxylase 1 (ACACA) are both important for replication [42]. FAS is recruited to the VRC through interaction with the viral protease NS3, where it upregulates the formation of fatty acids from acetyl-CoA. As the length of fatty acid chains can affect membrane curvature, this process may be important for the formation of the VRC [43].

Newly synthesized fatty acids are incorporated into DENV VRCs and pharmacological inhibition of FAS by cerulenin or C75 negatively affects DENV replication in mammalian [42] as well as mosquito cells [44]. In one current model, the viral protein NS4A initially induces membrane curvature followed by the recruitment of FAS by NS2B/NS3 to the VRC resulting in local production of fatty acids and the expansion of the ER membrane in a more fluid state [42]. Similar to DENV, WNV requires FAS activity for replication [41]. WNV_{NY99} infection also increases the intracellular concentration of sphingolipids and glycerophospholipids, a process that affects the make up of virion envelopes [45]. Virus assembly and release is also dependent on lipid biosynthesis, particularly ceramide. The role of glycerophospholipids such as phosphatidylcholine (PtdCho) in flavivirus replication is less clear, but there is evidence that PtdCho is part of the VRCs and is incorporated into the lipid bilayers of nascent virions [45].

The level of cholesterol also modulates the curvature and plasticity of membranes [46]; a process that is controlled by ER-localized transcription factor sensors, sterol-regulatory element binding proteins (SREBPs). When cholesterol levels are low, SREBP is released from the ER and enters the nucleus where it activates transcription of the genes for FAS, 3-hydroxy-methyglutaryl-CoA reductase (HMGCR), and/or low-density lipoprotein receptor (LDLR). HMGCR catalyzes the rate-limiting step in synthesis of mevalonate, a precursor for cholesterol biosynthesis, whereas LDLR is a cell surface receptor that binds and internalizes cholesterol-containing complexes [47,48]. It is well documented that flaviviruses modulate cholesterol levels in infected cells. During WNV_{KUN} infection, total cholesterol levels rise and this is correlated with upregulation and association of HMGCR with virus-induced membrane structures [49]. This may indicate HMGCR aids VRC formation by producing cholesterol at these sites.

Interestingly, elevated cholesterol (total) levels were not observed in DENV-infected cells [50], even though LDLR transcripts, an indicator of elevated ER cholesterol, were increased. However, inhibition of HMG-CoA reduced replication of DENV replicons indicating cholesterol is important for replication. Another host factor involved in cholesterol biosynthesis, mevalonate diphosphate decarboxylase (MVD), was shown to be important for DENV replication [50]. Thus cholesterol seems to be a key lipid component of the flavivirus VRC and while total cellular cholesterol may not increase in all flavivirus infected cells, this membrane is targeted to the ER membrane where VRCs are produced. Future studies with JEV, YFV, TBEV, and St. Louis Encephalitis Virus (SLEV), are needed to determine how they may alter membrane composition in favour of VRC biogenesis.

2.5. Stabilization and Scaffolding Proteins at the Flavivirus VRC

Microfilaments, microtubules and intermediate filaments are cytoskeletal components essential for cell shape and motility as well as a myriad of other functions including intracellular trafficking, cell division and cell signalling. Identification of host proteins involved in actin polymerization and vesicular trafficking were shown to be important for DENV and WNV replication [32,42], however, comparatively little is known about how this affects VRC formation or function. Conceivably, changes to the cellular structural framework could aid in biogenesis and/or stabilization of newly formed VRCs. Reorganization of the intermediate filament component vimentin occurs after phosphorylation by calcium/calmodulin-dependent protein kinase II; an event that is necessary for productive DENV replication [51]. Moreover, knockdown of vimentin alters the distribution of the VRCs in host cells, indicating that this protein may function in scaffolding/stabilization of these structures. NS4A interactions with vimentin may be the link between intermediate filaments and VRCs [51]. Finally, Stathmin 1 (STMN1), a microtubule destabilizing protein, is another host factor linked to biogenesis of VRCs [52]. DENV infection upregulates STMN1 by reducing levels of miR-223, a microRNA that normally targets the mRNA for STMN1.

3. *Picornaviridae*

The *Picornaviridae* family contains many important human and animal pathogens. Prior to the development of a vaccine, poliovirus (PV) crippled hundreds of thousands of people per year, primarily children. The World Health Organization global PV eradication program started in 1988 has not yet been successful in fully eradicating the virus. Other prominent members of this family include Coxsackie virus (CV), human rhinoviruses (HRV), and the causative agent of hand-foot-and-mouth disease Enterovirus 71 (EV71). Unlike PV, effective vaccines against these pathogens have yet to be developed. Infection by CV, HRV, and EV71 cause a variety of illnesses in humans from self-limiting colds to more serious presentations of encephalitis, myocarditis, and paralysis. Children, elderly, and the immuno-compromised individuals are at highest risk for severe disease.

The majority of research has focused on members of the *Enterovirus* genus including PV, CV, and HRV, and as such our focus will be on host factors linked to formation and stabilization of their VRCs. As with all +RNA viruses, enteroviruses extensively rearrange cellular membranes to facilitate virus replication and assembly. Following entry of the virion, the 5'capped genomic RNA (~7.5 kb) is unpackaged after which translation is initiated from an internal ribosome sequence. The genomic RNA encodes a single large polyprotein that is proteolytically processed into four structural proteins that form the virion, and seven non-structural proteins that function in replication and subverting the host-cellular immune system.

3.1. Biogenesis of the Enterovirus VRC

Early electron microscopy studies of PV-infected cells by Dales and Palade revealed drastic remodelling of the cell cytoplasm [53]. At 5 hours post-infection (hpi), membrane-enclosed bodies were observed in the perinuclear zone. At the peak of viral translation (2.5 hpi), nascent VRCs were not observed but rather, formed later during RNA replication [54]. Isolation of PV VRCs revealed that in addition to dsRNA, a replication intermediate, proteins encoded by the P2 genomic region, specifically 2C containing non-structural proteins, were bound to these membranes [55,56]. More recently, electron tomographic studies were used to examine the biogenesis of these structures in more detail [23]. Early in infection (~2 hpi), 100–200 nm single membrane tubular structures that are involved in RNA synthesis form followed by clustering and bending of these structures at 4 hpi. Later, DMVs, which can be larger (100–300 nm), form through a membrane wrapping process (Figure 1A). These structures evolve from *cis*-Golgi membrane and arise from positive membrane curvature or budding [23]. Despite a tremendous amount of experimental investigation, the origins of the enteroviral VRC remain controversial [57].

3.2. Membrane Remodelling during Enteroviral Infection

Earlier studies suggested that PV VRCs are derived from multiple membrane sources, including lysosome, ER, and Golgi, but do not fully resemble their parent membrane sources [58,59]. Multiple hypotheses exist for how these structures arise during PV infection, including through autophagy [58]. Expression of the PV proteins 2BC and 3A results in the formation of DMVs [59] that colocalize with lysosomal-associated membrane protein 1 (LAMP-1) and LC3-phosphatidylethanolamine conjugate (LC3-II), indicative of autophagic vesicles, early in PV infection [60]. Data consistent with the PV studies were observed with the related enteroviruses HRV-2 and HRV-14 and inhibition of autophagy decreased the amount of intracellular and extracellular virus produced [60–63]. Of note, CVB3 infection induces autophagy in a mouse model *in vivo*, indicated by increased LC3-II [61]. Despite this evidence, the role of autophagy remains controversial. Recently, it was shown that PV vesicles that stained for dsRNA did not colocalize with autophagic marker LC3 early (3 hpi) in infection [64]. However, at 5 hpi LC3 was detected by immuno-electron microscopy in association with dsRNA. In light of the seemingly discrepant data, it has been postulated that autophagy is important for late steps in infection (3 hpi) [64].

COPII-coated vesicles, which are involved in transport of cargo from the ER, have been linked to biogenesis of enterovirus VRCs. During early infection, enteroviruses disrupt anterograde transport and reroute these vesicles to sites of viral replication. PV infection or expression of protein 3A alone has been shown to block ER-Golgi protein transport [65,66]. Moreover, the movement of VRCs is dependent on microtubules leading from the ER to the microtubule organizing center in the Golgi region of the cell [67]. PV proteins 2B and its precursor 2BC colocalize with the COPII component Sec31 [68] and PV infection enhances COPII vesicle budding [69]. However, the effect is transient and is not observed late in infection. In contrast, HRV-1A and -16 have been reported to cause fragmentation of the Golgi without blocking protein secretion [70,71]. Furthermore, recent evidence citing lack of colocalization between dsRNA and the COPII component Sec31 has been interpreted to mean that PV VRC formation is not dependent on COPII [64]. One potential mechanism to account for this discrepancy is that COPII aids in the formation of an intermediary compartment from which nascent VRCs bud. However, attenuation of COPII vesicle formation did not interfere with PV infection suggesting that this budding mechanism is not absolutely required [72]. Clearly, more research is required to fully understand the role of COPII in picornavirus VRC biogenesis and function.

PV infection is sensitive to brefeldin A (BFA), a drug that inhibits activation of ADP-ribosylating factor GTPases (Arfs), which are necessary for formation of COPI vesicles [73,74]. Arfs cycle between a GDP (inactive) and a GTP (active) bound state mediated by guanine-nucleotide exchange factors (GEFs). When bound to GTP, Arfs remodel intracellular membranes to promote COPI dependent budding [75]. COPI mediates budding of vesicles from the Golgi that retrograde traffic to the ER and was identified as a host factor that is required for replication of *Drosophila* C virus, a picorna-like insect virus [72]. Reducing expression of α-COP, a COPI component, was later found to reduce PV infection [72]. Expression of PV 3A or 3CD promotes the association of Arf3 and Arf5 with membranes where viral RNA replication occurs, and this association can be blocked by BFA [76]. Two other GEFs, BIG1 and BIG2 are recruited to VRCs by expression of PV 3CD which then leads to the activation of Arf [77]. This indicates that Arf activation may induce vesicle formation and VRC biogenesis (reviewed in [78]). GBF1, yet another GEF, is also a target of BFA and is the main activator of Arf during PV infection [79]. PV 3A binds GBF1 and recruits it to virus-induced vesicles [77]. While VRCs can still form in the presence of BFA, they are unable to recruit Arf1. This may result in formation of defective VRCs or mislocalization of their contents thereby reducing viral replication and assembly [79]. Although BFA targets BIG1, BIG2, and GBF1, only GBF1 is required for CVB3 replication [80].

More recent studies examined the localization of Arf1 and GBF1 throughout CVB3 and PV infections [39]. Arf and GBF1 colocalize with viral RNA and the viral RNA polymerase, indicating their relocalization to VRCs during infection. Moreover, Arf1 knockdown negatively impacts virus replication [39]. However, since Arf1 interacts strongly with GBF1 and is found in COPI vesicles, it cannot be ruled out that recruitment of Arf1 to the VRC is a consequence of GBF1 recruitment. β-COP, a COPI component, does not colocalize with dsRNA during PV infection and thus, the COPI coat itself may not be involved budding and biogenesis of VRCs [64]. Recruitment of Arf1/GBF1 to VRCs appears to differ among enteroviruses. Expression of CVB3 3A induces the recruitment of GBF1 to membranes, whereas the homologous proteins of HRV-2 or -14 do not [81,82]. These viral proteins also function in recruitment of Phosphatidylinositol-4-OH kinase type III beta (PI4KIIIβ) to VRCs and this is discussed in further detail below.

3.3. Lipid Metabolism

Biogenesis of picornavirus VRCs also requires synthesis and trafficking of specific lipids to membrane organelles. Unlike enveloped viruses such as flaviviruses and togaviruses (also covered in this review) whose replication compartments are formed by invagination into membranes, picornaviruses induce protrusion of cellular membranes to form convoluted tubular-like structures [23,83]. Alterations in the lipid composition of these membranes are needed to allow appropriate curvature and the expansion of membranes that eventually form the VRCs. Early evidence of altered lipid metabolism came from the observation that PV increases PtdCho levels in the cells by upregulating the rate-limiting enzyme phosphocholine cytidylyltransferase [84]. PtdCho is a main component of lipid bilayers at the ER and Golgi network and increased levels of this phospholipid would enable proliferation of membranes (reviewed in [85]). This may indicate that formation of VRCs involves *de novo* lipid synthesis. The role of fatty acids was first reported when the addition of cerulenin, an inhibitor of the enzyme FAS, resulted in a block of PV replication but did not affect viral RNA translation or proteolytic processing [86]. Blocking fatty acid synthesis by inhibiting FAS also reduced proliferation of VRC membranes. Later, the same group reported that specific fatty acids are important for PV replication as evidenced by the observation that incorporation of oleic acid into cellular membranes made them incapable of supporting PV replication [87]. More recently the role of FAS for CVB3 replication has also been demonstrated. FAS upregulation was first observed during a proteomic screen of CVB3 infected cells [88]. FAS protein production is upregulated as early as one hour post CVB3 infection and does not require viral replication, suggesting that FAS gene transcription and translation may be upregulated following signalling cascades induced by CVB3 virions binding [89]. Components of the fatty acid biosynthesis pathway including SREBP and the protein product of the gene it directly regulates, *CG3523* encoding FAS are also required for picorna-like virus *Drosophila C* virus replication, suggesting that FAS may be a common host factor exploited by viruses to alter membrane lipid metabolism [72]. In addition to altering metabolism in cells, picornaviruses may increase uptake of lipids from the extracellular environment. PV infection for example, enhances import of long-chain fatty acids into cells and the viral protein 2A is involved in the initiation of this process [90]. Normally, fatty acids are trafficked and stored in lipid droplets, but in infected cells they colocalize with VRCs. Moreover, activity of long chain acyl-coenzyme A (Acyl-CoA) synthetase Acsl3, involved in the synthesis of PtdCho, was upregulated at 2 hpi, thus further supporting the notion that VRCs are formed from newly synthesized lipids.

Because Arf1 and GBF1 are recruited to VRC membranes, it was thought that Arf1 effectors might also be important for enterovirus replication. Lipid modifying enzymes including (PI4Ks) are downstream effectors of Arf. PI4KIIIβ is normally associated with the Golgi and is involved in the production of (PI4P) [91]. The PI4KIIIβ inhibitor enviroxime exhibits potent antiviral activity against enteroviruses *in vitro*, however in clinical trials, its efficacy was limited [92,93]. Other PI4KIIIβ inhibitors, including GW5074 and BF738735, also efficiently inhibit enteroviral replication *in vitro* and *in vivo* in mice, however in some mice strains inhibition of PI4KIIIβ resulted in harmful side effects thereby limiting the likely therapeutic benefit of this strategy [92,94,95]. PI4KIIIβ colocalizes with Arf1 at replication complexes during CVB3 infection, while other Arf1 effectors, such as COPI components, are lost from these sites [39]. Recruitment of PI4KIIIβ to Arf1-positive membranes can be stimulated by expression of CVB3 3A alone. During infection by PV or CVB3, PI4P levels increase 5-fold and pharmacological inhibition of PI4KIIIβ activity by PIK93, which inhibits PI4P production, also reduces viral replication. Furthermore, PI4KIIIβ interacts with CVB3 3D^{pol}, which strongly interacts with PI4P-containing membranes and thus their production may facilitate the organization and/or association of viral proteins in the VRC [39]. This kinase also interacts with PV 3A and Acyl-CoA binding domain containing 3 (ACBD3) protein [96]. Recruitment of PI4KIIIβ to the VRC seems to be conserved among enteroviruses but the interactions between this kinase and viral as well as other host proteins remain to be fully elucidated.

The production of PI4P lipids, by PI4KIIIβ at replication sites may also be important for recruitment of cholesterol. Both PI4P and cholesterol are enriched at VRCs of CVB3, PV, and HRV-2 [94]. Recently, it was suggested that cholesterol is shuttled from the endosome to the VRC where it colocalizes with CVB3 3A [97] suggesting that the virus re-routes pre-existing pools of cholesterol to VRCs. Moreover, it has also been demonstrated that uptake of extracellular cholesterol by clathrin-mediated endocytosis (CME) is essential for PV and CVB3 replication [98]. After uptake, cholesterol is targeted to recycling endosomes, which then fuse with existing VRCs. Depletion of CME components results in a trafficking of cholesterol from the plasma membrane to lipid droplets, reducing VRC formation [98]. Incorporation of cholesterol into VRCs, which imparts rigidity to membranes, may be important for their curvature and stabilization.

4. *Coronaviridae*

Members of the four genera in the family *Coronaviridae* have enveloped virions that contain very large +RNA, capped and polyadenylated RNA genomes of 26–32 kb. These viruses infect a wide range of mammals and birds causing upper respiratory, gastrointestinal, hepatic, or central nervous system diseases [99]. Members belonging to the genus *Betacoronavirus* include the important human pathogens Severe acute respiratory syndrome coronavirus (SARS-CoV) and Middle East respiratory syndrome coronavirus (MERS-CoV). Prior to the SARS outbreak in 2003, the majority of coronavirus (CoV) research focused on mouse hepatitis virus (MHV) as a model.

Following virion entry into host cells, the +RNA genome is released into the cytoplasm. Coronaviruses employ a rather complex program of gene expression. The *ORF1* encodes the replicase required for transcription of the full-length (genomic) minus-strand template and subgenomic (discontinuous transcription) minus-strand synthesis. Synthesis and processing of the genome results in production of up to 16 nonstructural proteins, of which the predicted multi-spanning membrane proteins nsp3 [100], nsp4 [101] and nsp6 [102] are believed to be involved in biogenesis and stability of the coronavirus replication/transcription complex (RTC). The study of host factors required for biogenesis and stabilization of the coronavirus RTC is a growing field of interest. The involvement of lipid rafts for virus-entry and cell-cell fusion was demonstrated for MHV [103], however, there is a lack of information on how CoVs modulate host lipid composition as previously shown for many +RNA viruses [104]. A recent kinome screen (using small interfering RNAs) has provided a glimpse of the complexity of pro-viral and anti-viral host factors involved at the SARS CoV-cell interplay, including proteins involved in lipid metabolism [105]. Future studies addressing the interplay between CoVs and lipids and the effects on viral replication would also be of considerable interest.

4.1. Coronavirus Replication/Transcription Complex (RTC)

Similar to some of the viruses described above, formation of DMVs is observed during CoV infection in mammalian cells [106]. Early in the SARS-CoV infection process, DMVs, ranging in size from 150 to 300 nm, are distributed throughout the cytoplasm [25]. The ORF1a-encoded multi-spanning transmembrane proteins, nsp3, nsp4 and nsp6 are thought to form the scaffold that facilitates DMV formation and anchors the RTC to intracellular membranes [101,107–109]. The RTC is likely formed through a complex network of interactions involving all 16 CoV non-structural proteins [108]. DsRNA can be detected in the interior of DMVs (Figure 1D) [25], suggesting that these structures serve as sites of RNA replication. Indeed this is supported by the observation that RTC activity correlates with the number of DMVs [110].

Electron tomographic analysis revealed that DMVs are not isolated vesicles but rather, an interconnected network of membranes continuous with the rough-ER [25]. Late in infection, DMVs become concentrated in the perinuclear area and CMs of 0.2 to 2 μm in diameter form in close proximity to the DMV clusters [25]. Compared to DMVs, CMs are highly enriched in SARS-CoV nonstructural proteins that include the replicase proteins [25]; leading to the notion that the active replicase complex is localized to the CM and "dead" dsRNA molecules are found in the DMVs, perhaps as a way to evade immune recognition. Electron microscopy studies revealed that similar membrane formations occur in MERS-CoV-infected cells [111]. The outer membranes of DMVs are thought to fuse together and transition into vesicle packets (VPs) late in infection. VPs are large (1–5 μm) membrane structures where virus budding occurs [25,112] (Figure 1D).

Unexpectedly, electron tomographic analyses failed to reveal connections between the interior of DMVs and the cytosol [25]. This suggests that both membranes of CoV DMVs are sealed raising the question of how the import of metabolites and export of viral RNA occurs from these structures. In contrast, small neck-like openings are clearly discernable in DMVs induced by HCV [6]. In this case, it has been speculated that replication takes place in DMVs as long as the connection to the cytosol is maintained. The question remains as to whether CoVs make use of transport molecules as a means to regulate transport of products in and out of DMVs.

Table 1. Identified cellular interacting proteins with viral replication complexes of +RNA viruses.

	Lipids and Membrane Remodeling	Cellular Trafficking and Signaling Proteins
Flaviviridae *(Flaviviruses)*	Sphingolipids (WNV$_{NY99}$ [45])	Actin polymerization (WNV [32], DENV [42])
	Glycerophospholipids (WNV$_{NY99}$ [45])	Vimentin (DENV [51])
	FAS (DENV [42], WNV$_{NY99}$ [41])	STMN1 (DENV [52])
	ACACA (DENV [42])	
	Cholesterol (DENV [50], WNV$_{KUN}$ [49])	
Picornaviridae	FAS (PV [86,87], CVB3 [88,89])	Arf (CVB3, PV) [39]
	Long chain fatty acids (PV [90])	GBF1 (CVB3, PV) [39]
	PtdCho (PV [84])	
	PI4KIIIβ (CVB3 [39], PV [96])	
	ACBD3 (CVB3, PV) [96]	
	Cholesterol (CVB3, PV) [98]	
Coronaviridae		PDI (SARS-CoV [107])
		Sec61α (SARS-CoV [110])
		EDEM1 (MHV [113])
		OS-9 (MHV [113])
Togaviridae *(Alphaviruses)*	Cholesterol and sphingomyelin lipids (SINV [114])	Vimentin (SINV [115])
		PI3K (SFV [116])
		Amphiphysins (SFV, SINV, CHIKV) [117]

Abbreviations: DENV, Dengue Virus; WNV, West Nile Virus; YFV, Yellow Fever Virus; JEV, Japanese Encephalitis Virus; PV, Poliovirus; CVB3, Coxsackievirus B3; HRV-14, Human Rhinovirus 14; EV71, Enterovirus 71; IBV, Infectious Bronchitis virus; SARS-CoV, Severe Acute Respiratory Syndrome coronavirus; MHV, Mouse Hepatitis Virus; SINV, Sindbis Virus; SFV, Semliki Forest Virus; CHIKV, Chikungunya Virus.

4.2. Potential Role of Autophagy in DMV Formation

Morphological similarities between CoV DMVs and autophagosomes and the co-localization between specific CoV replicase proteins (nsp8, nsp2, nsp3) with microtubule-associated protein Light chain 3 (LC3), a protein marker for autophagic vacuoles [118], are consistent with autophagy playing a role in DMV formation. Moreover, during MHV infection of murine cells lacking the ATG5 gene which functions in the early stages of autophagosome formation [119], no DMVs formed and virus replication was impaired [120]. Replication was restored by expression of the Atg5 protein further supporting the role of autophagy for formation of DMVs, at least in embryonic stem cells. However, another study involving MHV infection of bone marrow-derived macrophages or embryonic fibroblasts concluded that neither Atg5 nor an intact autophagic pathway is required for viral replication [121]. Other morphological studies also found no evidence for autophagy in DMV formation; specifically a lack of co-localization between the autophagy marker LC3 and the SARS-CoV replication complex [107]. Moreover, MHV replication was unaffected in autophagy-deficient cells although depletion of LC3 severely affected CoV replication [113]. Interestingly, MHV replicative structures are decorated with LC3 [106], generally regarded to as the non-functional precursor to the lipidated autophagosome marker LC3-II [122]. The involvement of autophagy and LC3 in DMV formation was further clarified when LC3 was shown to colocalize with MHV proteins nsp2/nsp3, dsRNA, and ER-associated degradation

(ERAD) vesicle markers ER degradation-enhancing alpha-mannosidase-like 1 (EDEM1), Osteosarcoma amplified 9 (OS-9) in embryonic fibroblasts [113]. Down-regulation of LC3 inhibits MHV replication and virion production [113] whereas knocking out autophagy has no effect. Inhibition of MHV replication and virion production was attributed to a defect in DMV biogenesis, which negatively impacts non-structural protein production. These results suggest that LC3 and the ERAD pathway are necessary for DMV formation and the biogenesis of RTCs required for a productive infection. Finally, quantitative proteomics analysis revealed that SARS-CoV infection significantly upregulates BCL2-associated athanogene 3 (BAG3), a protein linked to regulation of the autophagy pathway [123]. Inhibition of BAG3 expression by RNA interference results in significantly reduced replication of SARS-CoV.

Unfortunately, conflicting data make it difficult to derive a definitive conclusion regarding the role of autophagy in CoV RTC formation. Some of these differences may be the result of using different cell lines and different CoVs. What is clear though is that proteins with known roles in autophagy are involved in CoV replication; however, the process of autophagy *per se* may not be functionally relevant to the formation of CoV DMVs.

4.3. The Secretory Pathway and CoV Replication

A number of studies indicate that the ER is involved in the biogenesis of the SARS-Co-V-induced reticulovesicular network (RVN), a membrane compartment involved in virus replication. Specifically, partial co-localization of CoV replicase proteins with the ER resident protein disulfide isomerase (PDI) [107] and the observation that the ER translocon subunit Sec61α redistributes to replicative structures during SARS-CoV infection support this idea [110]. In addition EDEMosome cargo proteins EDEM1 and OS-9, two proteins involved in ER quality control and ER associated degradation (ERAD), associate with CoV replicative structures [113]. However, many protein trafficking and membrane fusion proteins that function downstream of the ER in the early secretory pathway such as Sec13, syntaxin 5, GBF1, and Arf1 have not been detected at the RVN [110]. The involvement of the COPI complex was investigated using the drug BFA, which as mentioned above, blocks COPI-mediated vesicular transport at the ER-Golgi interface. When added to cells early in infection, BFA inhibits RVN formation and decreases, but does not completely abolish viral RNA synthesis [110]. Although the precise role of COPI is unknown, other positive RNA viruses discussed above such as PV and *Drosophila* C virus require the COPI-mediated vesicular transport for replication. This supports the notion of COPI as a common host factor required in viral infection.

5. *Togaviridae*

Togavirus virions are enveloped, spherical particles (50–70 nm in diameter) that contain a single-strand +RNA with a 5-cap and 3′ poly A-tail [124]. The family includes the genera *Rubivirus* and *Alphavirus*. Rubella virus (RUBV) is the sole member in the *Rubivirus* genus and is the causative agent of Rubella (also known as German Measles). The *Alphavirus* genus contains at least 30 members that are separated into New World and Old World viruses. The New World viruses include Venezuelan equine encephalitis (VEEV), Western equine encephalitis (WEEV) and Eastern equine encephalitis virus (EEEV). Old World alphaviruses evolved separately [125] and members include Semliki Forest virus

(SFV), Sindbis virus (SINV), Chikungunya virus (CHIKV) and O'nyong'nyong virus. Transmission occurs mainly through mosquito vectors and human infections are often associated with fever, rash, severe joint pain (arthralgia) and stiffness that can last weeks to months in duration. Some pathogens in this group can cause much more severe illnesses including encephalitis in humans and animals.

Togavirus replication complexes originate from late endosomes and lysosomes and are morphologically similar for RUBV [126,127] and alphavirus-infected cells [128]. Alphavirus VRC biogenesis is comparatively well characterized and as such, we will focus on these structures. Most host factors that interact with alphavirus replication complexes were identified through pull down assays with alphavirus non-structural proteins. In this section we focused on host factors partners that have been speculated or clearly demonstrated to play functional roles in RC biogenesis or function. For a list of additional interacting partners that are not discussed here, readers are referred to the following articles: [129–131].

5.1. Alphavirus Replication

Shortly after alphaviruses infect host cells, small single-membrane bulb-shaped invaginations (~50 nm diameter) called "spherules" form on the external surface of the plasma membrane [132]. The spherules, which are associated with viral nonstructural proteins (nsPs) and dsRNA, each contain a neck-like opening to the cytoplasm (5–10 nm in diameter) that permits exchange of metabolites and export of nascent viral RNA [128]. The fact that dsRNA can be detected inside the spherules and the presence of partially processed non-structural proteins (P123 and nsP4) on the spherule necks [132], suggests that these structures are the sites of viral RNA synthesis [116]. Internalization of spherules by endo-lysosomal membranes gives rise to type 1 cytopathic vacuoles (CPV1), which are 600–2000 nm in diameter [24] (Figure 1C). The endosomal origin of CPVs is confirmed by the observation that these structures are often positive for both endosomal and lysosomal markers [128]. As infection progresses, the non-structural polyprotein precursors are further processed to yield individual non-structural proteins and negative-strand synthesis is inactivated. The fully processed non-structural proteins together form the mature replicase [115,133,134], which is required for efficient synthesis of positive-sense genomic and subgenomic RNA. Spherules are devoid of ribosomes and virus capsid proteins but these structures/proteins are frequently found juxtaposed to the spherule openings [128], suggesting that the site of translation is in close proximity to replication sites.

5.2. Membrane Lipids

The nsP1 protein of alphaviruses, which is required for 5' capping of viral RNAs [135], is involved in attachment of the replication complex to membranes [136] via a highly conserved amphipathic helix [137]. When expressed in the absence of other viral proteins, nsP1 is targeted to the inner surface of the plasma membrane but this not sufficient for cytoplasmic vacuole formation [138]. NsP1 is modified by acylation; however, the significance of this process in its function has yet to be determined [139]. Cholesterol and sphingomyelin in the plasma membrane are important for alphavirus fusion [140–142] and budding [143,144]. Importance of the latter is evidenced by the observation that SINV infection of Niemann-Pick disease-A fibroblasts (NPAF), which cannot degrade sphingomyelin, results in reduced levels of genomic RNA as well as an altered ratio of subgenomic-to-genomic RNA. The authors suggest that due to the build-up of cholesterol and sphingolipids in late endosomes/lysosomes, biogenesis of replication complexes are negatively affected [114]. Interestingly, the alphavirus virions produced in NPAFs were 26 times more infectious than those produced in normal human fibroblasts; resulting in increased titers and cell death. This suggests that cellular production of less infectious virus may be a consequence of host restriction on virus replication.

5.3. Membrane Trafficking Proteins

The study of how alphaviruses enter and exit mammalian cells led to a number of fundamental discoveries about membrane trafficking. Therefore, it is somewhat surprising that comparatively little is known about how membrane trafficking components affect replication. However, it does appear that cytoskeletal elements are involved. Infection of cells with a recombinant SINV encoding GFP-tagged nsP3, followed by anti-GFP pull-downs revealed that this viral protein associates with cytoskeletal proteins, chaperones, elongation factor 1A, and heterogeneous nuclear ribonucleoproteins [115]. Others reported that nsP3 also binds actin, tubulin and myosin [145] [144], vimentin (an intermediate filament protein) [115], and the cytosolic molecular chaperone Heat shock cognate protein 70 (Hsc70) [115,146]. These interaction data suggest that the SINV RCs associate with the cytoskeleton. The concern that the highly abundant cytoskeletal proteins were mere contaminants in the nsP3-binding studies [130] was partially assuaged by imaging studies showing that nsP3 associates with vimentin patches [115]. Imaging studies also provided evidence for Hsc70-ns3p having a role in alphavirus RC formation and/or function [115,146]. Some of the many functions of Hsc70 (reviewed in [147]) are to target proteins for degradation, regulate the translocation of proteins into different cellular organelles such as ER and mitochondria, and regulate apoptosis. Hsc70 has been linked to the replication of many viruses [148] but whether or not this is largely a reflection of its general role as a chaperone has yet to be determined.

NsP3 proteins of SFV, SINV, and CHIKV have also been shown to interact in an SH3 domain-dependent manner with amphiphysin-1 and Bin1/amphiphysin-2, both of which are involved in endocytosis and membrane trafficking [117]. The re-localization of amphiphysins to alphavirus RCs promotes replication, further solidifying the role of these host proteins in formation or stabilization of the replication sites. Finally, recent data suggest a role for phosphatidylinositol 3-kinases (PI3Ks) in alphavirus RC formation. Specifically, the activity of these kinases is required for the initial internalization of spherules at the plasma membrane as well as their subsequent trafficking on microtubule and actin networks [116].

5.4. Ras GTPase-Activating Protein-Binding Proteins

The RNA-binding proteins Ras GTPase-activating protein-binding protein (G3BP)1 and G3BP2 are structural components of stress granules: large cytoplasmic ribonucleoprotein complexes that function in regulating translation. Several studies have identified interactions between G3BP proteins and alphavirus nsP2 and nsP3 [115,129,130,146]. To differentiate between host proteins associated with replication complexes from those that interact with individual nsP2 or nsP3 proteins, Varjak *et al.* used small dextran-covered magnetic beads that incorporated into CPV-1 structures in infected cells, thus permitting the isolation of membranous vesicles. This study confirmed the enrichment of G3BP1 and G3BP2 in CPV-1 vesicles in SFV infected cells [131], however, it was not possible to determine when these host proteins were recruited. Interaction of G3BPs with alphavirus nsPs as evidenced by the observation that the insect G3BP1 homolog Rasputin, was detected in nsP3-containing complexes isolated from mosquito cells infected with SINV [146]. While G3BP proteins may serve an important and conserved function in alphavirus infections, it remains unclear as to how these host proteins function in RC biogenesis. While recruitment of G3PB proteins to RCs is a common feature of alphavirus infection, this was not observed in cells infected with the flavivirus YFV, indicating that G3BP is not a host factor for RCs of all positive strand RNA viruses [130].

SINV may recruit G3BP as a means to block G3BP-dependent export of host mRNAs to the cytoplasm or that host RNAs undergoing nuclear export are sequestered, resulting in translational shutoff. Alternatively, this may reflect a specific host response to counteract infection [130]. The interaction between nsP3 and G3BP1/G3BP2 in CHIKV-infected cells occurs late in the replication cycle where the bulk of G3BP1 and G3BP2 is not associated with the viral RC but rather, is sequestered in nsP3-G3BP aggregates in the cytoplasm [149]. This may indicate that G3BPs play different roles early and late in infection. For example, late in infection interaction between nsP3 and G3BPs, which are nucleating factors for stress granules, could prevent the formation of these RNA granules. Stress granules have been implicated in the antiviral response [150] and have been reported to form late during alphavirus infection, a process that correlates with host translational shutdown [151]. These seemingly contrasting observations may be explained by the temporal dynamics of the nsp-G3BP association (see above for CHIKV). Given the role of G3BPs in stress granule formation, it seems likely that nsPs associate with G3BPs as a means to inhibit their formation; a theory that is supported by recent data showing that translational shut-off in cells infected with the alphavirus VEEV (whose replication mechanisms do not appear to involve G3BPs) is comparatively slower [146].

6. Summary

A common feature of all +RNA viruses is their ability to modulate cellular membranes aiding in the concealment of replication product intermediates from recognition by cellular immune sensors. Despite differences in membrane curvature (DMVs, InVs, *etc.*), all replication complexes are made up of viral proteins, RNA, and cellular factors. In this review, we have focused on identifying host factors that are proposed or validated to play a role in the replication complexes of human pathogens belonging to four different virus families. The cellular proteins at the replication complexes include proteins involved in lipid metabolism, intracellular trafficking, autophagy, secretory pathways, transcription, and translation. Understanding precisely how these host proteins function in virus replication may open avenues for development of novel anti-viral therapeutics.

Acknowledgments

We thank Anil Kumar for his insightful feedback and assistance in editing this review. We apologize to our colleagues whose work was not adequately cited due to space limitations. C.R.R. and A.M.A. hold graduate scholarships from the Canadian Institutes of Health Research (CIHR). T.C.H. holds a Canada Research Chair in RNA virus host interactions. Research in the Hobman laboratory is supported in part by the CIHR and the Li Ka Shing Institute of Virology.

Author Contributions

C.R.R., A.M.A., and T.C.H. wrote this review. C.R.R. provided the illustration for Figure 1.

Abbreviations

Table A1. Commonly used abbreviations in this review article.

Abbreviation	Full Nomenclature
ACACA	Acetyl-CoA carboxylase 1
ACBD3	Acyl-coenzyme A binding domain containing 3 protein
Arf	ADP-Ribosylation factor
ATG5	Autophagy protein 5
BAG3	Bcl-2-associated athanogene 3
BFA	Brefeldin A
ACACA	Acetyl-CoA carboxylase 1
ACBD3	Acyl-coenzyme A binding domain containing 3 protein
Arf	ADP-Ribosylation factor
ATG5	Autophagy protein 5
BIG	Brefeldin A-inhibited guanine nucleotide-exchange factor
CHIKV	Chikungunya virus
CM	Convoluted Membranes
CME	Clathrin-mediate endocytosis
CPV-1	Type 1 cytopathic vacuoles
CV	Coxsackie virus
DENV	Dengue virus
DMV	Double-Membrane Vesicles
dsRNA	Double-stranded RNA
EDEM1	ER degradation-enhancing alpha-mannosidase-like 1
EEEV	Eastern equine encephalitis virus
EV	Enterovirus
FAS	Fatty acid synthase
G3BP	Ras GTPase-activating protein (SH3 domain) binding protein
GBF1	Golgi brefeldin A resistant guanine nucleotide exchange factor 1

Table A1. *Cont.*

HCV	Hepatitis C virus
HMGCR	3-hydroxy-methyglutaryl-CoA reductase
HRV	Human Rhinovirus
Hsc70 (also known as HSPA8)	Heat shock cognate 71 kDa protein
JEV	Japanese Encephalitis virus
LAMP-1	Lysosomal-associated membrane protein 1
LC3 (Cytosolic form; LC-I)	Microtubule-associated protein 1A/1B-light chain 3
LC3-II (Lipidated form)	LC3-phosphatidylethanolamine conjugate
LDLR	Low density lipoprotein receptor
MERS-CoV	Middle East respiratory syndrome coronavirus
MHV	Mouse Hepatitis virus
MVD	Mevalonate diphosphate decarboxylase
NPAF	Niemann-Pick disease-A fibroblasts
OS-9	Osteosarcoma amplified 9, ER lectin
PC	Paracrystalline Array
PDI	Protein disulfide isomerase
PI	Phosphatidylinositol
PI4K	Phosphatidylinositol-4-OH kinase
PI4P	Phosphatidylinositol 4-phosphate
PtdCho	Phosphatidylcholine
PV	Poliovirus
RTC	Replication/Transcription Complex
RUBV	Rubella virus
RVN	Reticulovesicular Network
SARS-CoV	Severe Acute respiratory syndrome coronavirus
Sec-13, -31, -61	ER translocon proteins
SFV	Semliki Forest virus
SINV	Sindbis virus
SLEV	St. Louis Encephalitis virus
SREBP	Sterol-regulatory element binding protein
STMN1	Stathmin 1/oncoprotein 18
TBEV	Tick-borne Encephalitis virus
VEEV	Venezuelan equine encephalitis virus
VP	Vesicular Packet
VRC	Virus Replication Complex
WEEV	Western equine encephalitis virus
WNV	West Nile virus
YFV	Yellow Fever virus

References

1. Mukhopadhyay, S.; Kuhn, R.J.; Rossmann, M.G. A structural perspective of the flavivirus life cycle. *Nat. Rev. Microbiol.* **2005**, *3*, 13–22. [CrossRef] [PubMed]

2. Chatel-Chaix, L.; Bartenschlager, R. Dengue virus- and hepatitis c virus-induced replication and assembly compartments: The enemy inside–caught in the web. *J. Virol.* **2014**, *88*, 5907–5911. [CrossRef] [PubMed]

3. Lohmann, V. Hepatitis c virus rna replication. *Curr. Top. Microbiol. Immunol.* **2013**, *369*, 167–198. [PubMed]

4. Egger, D.; Wolk, B.; Gosert, R.; Bianchi, L.; Blum, H.E.; Moradpour, D.; Bienz, K. Expression of hepatitis c virus proteins induces distinct membrane alterations including a candidate viral replication complex. *J. Virol.* **2002**, *76*, 5974–5984. [CrossRef] [PubMed]

5. Paul, D.; Romero-Brey, I.; Gouttenoire, J.; Stoitsova, S.; Krijnse-Locker, J.; Moradpour, D.; Bartenschlager, R. Ns4b self-interaction through conserved c-terminal elements is required for the establishment of functional hepatitis c virus replication complexes. *J. Virol.* **2011**, *85*, 6963–6976. [CrossRef] [PubMed]

6. Romero-Brey, I.; Merz, A.; Chiramel, A.; Lee, J.-Y.; Chlanda, P.; Haselman, U.; Santarella-Mellwig, R.; Habermann, A.; Hoppe, S.; Kallis, S. Three-dimensional architecture and biogenesis of membrane structures associated with hepatitis c virus replication. *PLoS Pathog.* **2012**, *8*, e1003056. [CrossRef] [PubMed]

7. Paul, D.; Madan, V.; Bartenschlager, R. Hepatitis c virus rna replication and assembly: Living on the fat of the land. *Cell Host Microbe* **2014**, *16*, 569–579. [CrossRef] [PubMed]

8. Romero-Brey, I.; Bartenschlager, R. Membranous replication factories induced by plus-strand rna viruses. *Viruses* **2014**, *6*, 2826–2857. [CrossRef] [PubMed]

9. Westaway, E.G.M.; Mackenzie, J.M.; Kenny, M.T.; Jones, M.K.; Khromykh, A.A. Ultrastructure of kunjin virus-infected cells: Colocalization of ns1 and ns3 with double-stranded rna, and of ns2b with ns3, in virus-induced membrane structures. *J. Virol.* **1997**, *71*, 6650–6661. [PubMed]

10. Whiteman, M.C.; Popov, V.; Sherman, M.B.; Wen, J.; Barrett, A.D.T. Attenuated west nile virus mutant ns1130-132qqa/175a/207a exhibits virus-induced ultrastructural changes and accumulation of protein in the endoplasmic reticulum. *J. Virol.* **2015**, *89*, 1474–1478. [CrossRef] [PubMed]

11. Mackenzie, J.M.; Jones, M.K.; Young, P.R. Immunolocalization of the dengue virus nonstructural glycoprotein ns1 suggests a role in viral rna replication. *Virology* **1996**, *220*, 232–240. [CrossRef] [PubMed]

12. Girard, Y.A.; Popov, V.; Wen, J.; Han, V.; Higgs, S. Ultrastructural study of west nile virus pathogenesis in culex pipiens quinquefasciatus (diptera: Culicidae). *J. Med. Entomol.* **2005**, *42*, 429–444. [CrossRef] [PubMed]

13. Gangodkar, S.; Jain, P.; Dixit, N.; Ghosh, K.; Basu, A. Dengue virus-induced autophagosomes and changes in endomembrane ultrastructure imaged by electron tomography and whole-mount grid-cell culture techniques. *J. Electron. Microsc.* **2010**, *59*, 503–511. [CrossRef] [PubMed]

14. Uchil, P.D.; Satchidanandam, V. Architecture of the flaviviral replication complex. Protease, nuclease, and detergents reveal encasement within double-layered membrane compartments. *J. Biol. Chem.* **2003**, *278*, 24388–24398. [CrossRef] [PubMed]

15. Welsch, S.; Miller, S.; Romero-Brey, I.; Merz, A.; Bleck, C.K.; Walther, P.; Fuller, S.D.; Antony, C.; Krijnse-Locker, J.; Bartenschlager, R. Composition and three-dimensional architecture of the dengue virus replication and assembly sites. *Cell Host Microbe* **2009**, *5*, 365–375. [CrossRef] [PubMed]

16. Mackenzie, J.M.; Kenney, M.T.; Westaway, E.G. West nile virus strain kunjin ns5 polymerase is a phosphoprotein localized at the cytoplasmic site of viral rna synthesis. *J. Gen. Virol.* **2007**, *88*, 1163–1168. [CrossRef] [PubMed]

17. Westaway, E.G.; Khromykh, A.A.; Mackenzie, J.M. Nascent flavivirus rna colocalized *in situ* with double-stranded rna in stable replication complexes. *Virology* **1999**, *258*, 108–117. [CrossRef] [PubMed]

18. Mackenzie, J.M.; Jones, M.K.; Westaway, E.G. Markers for trans-golgi membranes and the intermediate compartment localize to induced membranes with distinct replication functions in flavivirus-infected cells. *J. Virol.* **1999**, *73*, 9555–9567. [PubMed]

19. Gillespie, L.K.; Hoenen, A.; Morgan, G.; Mackenzie, J.M. The endoplasmic reticulum provides the membrane platform for biogenesis of the flavivirus replication complex. *J. Virol.* **2010**, *84*, 10438–10447. [CrossRef] [PubMed]

20. Kaufusi, P.H.; Kelley, J.F.; Yanagihara, R.; Nerurkar, V.R. Induction of endoplasmic reticulum-derived replication-competent membrane structures by west nile virus non-structural protein 4b. *PLoS ONE* **2014**, *9*, e84040. [CrossRef] [PubMed]

21. Offerdahl, D.K.; Dorward, D.W.; Hansen, B.T.; Bloom, M.E. A three-dimensional comparison of tick-borne flavivirus infection in mammalian and tick cell lines. *PLoS ONE* **2012**, *7*, e47912. [CrossRef] [PubMed]

22. Miller, S.; Kastner, S.; Krijnse-Locker, J.; Buhler, S.; Bartenschlager, R. The non-structural protein 4a of dengue virus is an integral membrane protein inducing membrane alterations in a 2k-regulated manner. *J. Biol. Chem.* **2007**, *282*, 8873–8882. [CrossRef] [PubMed]

23. Belov, G.A.; Nair, V.; Hansen, B.T.; Hoyt, F.H.; Fischer, E.R.; Ehrenfeld, E. Complex dynamic development of poliovirus membranous replication complexes. *J. Virol.* **2012**, *86*, 302–312. [CrossRef] [PubMed]

24. Grimley, P.M.; Berezesky, I.K.; Friedman, R.M. Cytoplasmic structures associated with an arbovirus infection: Loci of viral ribonucleic acid synthesis. *J. Virol.* **1968**, *2*, 1326–1338. [PubMed]

25. Knoops, K.; Kikkert, M.; van den Worm, S.H.; Zevenhoven-Dobbe, J.C.; van der Meer, Y.; Koster, A.J.; Mommaas, A.M.; Snijder, E.J. Sars-coronavirus replication is supported by a reticulovesicular network of modified endoplasmic reticulum. *PLoS Biol.* **2008**, *6*, e226. [CrossRef] [PubMed]

26. Zhang, L.K.; Chai, F.; Li, H.Y.; Xiao, G.; Guo, L. Identification of host proteins involved in japanese encephalitis virus infection by quantitative proteomics analysis. *J. Proteome Res.* **2013**

12, 2666–2678. [CrossRef] [PubMed]

27. Mishra, K.P.; Diwaker, D.; Ganju, L. Dengue virus infection induces upregulation of hn rnp-h and pdia3 for its multiplication in the host cell. *Virus Res.* **2012**, *163*, 573–579. [CrossRef] [PubMed]

28. Campbell, C.L.; Harrison, T.; Hess, A.M.; Ebel, G.D. Microrna levels are modulated in aedes aegypti after exposure to dengue-2. *Insect Mol. Biol.* **2014**, *23*, 132–139. [CrossRef] [PubMed]

29. Mairiang, D.; Zhang, H.; Sodja, A.; Murali, T.; Suriyaphol, P.; Malasit, P.; Limjindaporn, T.; Finley, R.L., Jr. Identification of new protein interactions between dengue fever virus and its hosts, human and mosquito. *PLoS ONE* **2013**, *8*, e53535. [CrossRef] [PubMed]

30. Saha, S. Common host genes are activated in mouse brain by japanese encephalitis and rabies viruses. *J. Gen. Virol.* **2003**, *84*, 1729–1735. [CrossRef] [PubMed]

31. Sengupta, N.; Ghosh, S.; Vasaikar, S.V.; Gomes, J.; Basu, A. Modulation of neuronal proteome profile in response to japanese encephalitis virus infection. *PLoS ONE* **2014**, *9*, e90211. [CrossRef] [PubMed]

32. Krishnan, M.N.; Ng, A.; Sukumaran, B.; Gilfoy, F.D.; Uchil, P.D.; Sultana, H.; Brass, A.L.; Adametz, R.; Tsui, M.; Qian, F.; *et al.* Rna interference screen for human genes associated with west nile virus infection. *Nature* **2008**, *455*, 242–245. [CrossRef] [PubMed]

33. Sessions, O.M.; Barrows, N.J.; Souza-Neto, J.A.; Robinson, T.J.; Hershey, C.L.; Rodgers, M.A.; Ramirez, J.L.; Dimopoulos, G.; Yang, P.L.; Pearson, J.L.; *et al.* Discovery of insect and human dengue virus host factors. *Nature* **2009**, *458*, 1047–1050. [CrossRef] [PubMed]

34. Beatman, E.; Oyer, R.; Shives, K.D.; Hedman, K.; Brault, A.C.; Tyler, K.L.; Beckham, J.D. West nile virus growth is independent of autophagy activation. *Virology* **2012**, *433*, 262–272. [CrossRef] [PubMed]

35. Vandergaast, R.; Fredericksen, B.L. West nile virus (wnv) replication is independent of autophagy in mammalian cells. *PLoS ONE* **2012**, *7*, e45800. [CrossRef] [PubMed]

36. Lee, Y.-R.; Lei, H.-Y.; Liu, M.-T.; Wang, J.-R.; Chen, S.-H.; Jiang-Shieh, Y.-F.; Lin, Y.-S.; Yeh, T.-M.; Liu, C.-C.; Liu, H.-S. Autophagic machinery activated by dengue virus enhances virus replication. *Virology* **2008**, *374*, 240–248. [CrossRef] [PubMed]

37. Li, J.-K.; Liang, J.-J.; Liao, C.-L.; Lin, Y.-L. Autophagy is involved in the early step of japanese encephalitis virus infection. *Microbes Infect.* **2012**, *14*, 159–168. [CrossRef] [PubMed]

38. Heaton, N.S.; Randall, G. Dengue virus-induced autophagy regulates lipid metabolism. *Cell Host Microbe* **2010**, *8*, 422–432. [CrossRef] [PubMed]

39. Hsu, N.Y.; Ilnytska, O.; Belov, G.; Santiana, M.; Chen, Y.H.; Takvorian, P.M.; Pau, C.; van der Schaar, H.; Kaushik-Basu, N.; Balla, T.; *et al.* Viral reorganization of the secretory pathway generates distinct organelles for rna replication. *Cell* **2010**, *141*, 799–811. [CrossRef] [PubMed]

40. Berger, K.L.; Cooper, J.D.; Heaton, N.S.; Yoon, R.; Oakland, T.E.; Jordan, T.X.; Mateu, G.; Grakoui, A.; Randall, G. Roles for endocytic trafficking and phosphatidylinositol 4-kinase iii alpha in hepatitis c virus replication. *Proc. Natl. Acad. Sci. USA* **2009**, *106*, 7577–7582. [CrossRef] [PubMed]

41. Martin-Acebes, M.A.; Blazquez, A.B.; Jimenez de Oya, N.; Escribano-Romero, E.;

Saiz, J.C. West nile virus replication requires fatty acid synthesis but is independent on phosphatidylinositol-4-phosphate lipids. *PLoS ONE* **2011**, *6*, e24970. [CrossRef] [PubMed]

42. Heaton, N.S.; Perera, R.; Berger, K.L.; Khadka, S.; Lacount, D.J.; Kuhn, R.J.; Randall, G. Dengue virus nonstructural protein 3 redistributes fatty acid synthase to sites of viral replication and increases cellular fatty acid synthesis. *Proc. Natl. Acad. Sci. USA* **2010**, *107*, 17345–17350. [CrossRef] [PubMed]

43. Seddon, J.M.; Templer, R.H.; Warrender, N.A.; Huang, Z.; Cevc, G.; Marsh, D. Phosphatidylcholine-fatty acid membranes: Effects of headgroup hydration on the phase behaviour and structural parameters of the gel and inverse hexagonal (h(ii)) phases. *Biochim. Biophys. Acta* **1997**, *1327*, 131–147. [CrossRef]

44. Perera, R.; Riley, C.; Isaac, G.; Hopf-Jannasch, A.S.; Moore, R.J.; Weitz, K.W.; Pasa-Tolic, L.; Metz, T.O.; Adamec, J.; Kuhn, R.J. Dengue virus infection perturbs lipid homeostasis in infected mosquito cells. *PLoS Pathog.* **2012**, *8*, e1002584. [CrossRef] [PubMed]

45. Martín-Acebes, M.A.; Merino-Ramos, T.; Blázquez, A.-B.; Casas, J.; Escribano-Romero, E.; Sobrino, F.; Saiz, J.-C. The composition of west nile virus lipid envelope unveils a role of sphingolipid metabolism in flavivirus biogenesis. *J. Virol.* **2014**, *88*, 12041–12054. [CrossRef] [PubMed]

46. Stapleford, K.A.; Miller, D.J. Role of cellular lipids in positive-sense rna virus replication complex assembly and function. *Viruses* **2010**, *2*, 1055–1068. [CrossRef] [PubMed]

47. Bengoechea-Alonso, M.T.; Ericsson, J. Srebp in signal transduction: Cholesterol metabolism and beyond. *Curr. Opin. Cell Biol.* **2007**, *19*, 215–222. [CrossRef] [PubMed]

48. May, P.; Bock, H.H.; Herz, J. Integration of endocytosis and signal transduction by lipoprotein receptors. *Sci. STKE* **2003**, *2003*, PE12. [CrossRef] [PubMed]

49. Mackenzie, J.M.; Khromykh, A.A.; Parton, R.G. Cholesterol manipulation by west nile virus perturbs the cellular immune response. *Cell Host Microbe* **2007**, *2*, 229–239. [CrossRef] [PubMed]

50. Rothwell, C.; Lebreton, A.; Young Ng, C.; Lim, J.Y.; Liu, W.; Vasudevan, S.; Labow, M.; Gu, F.; Gaither, L.A. Cholesterol biosynthesis modulation regulates dengue viral replication. *Virology* **2009**, *389*, 8–19. [CrossRef] [PubMed]

51. Teo, C.S.H.; Chu, J.J.H. Cellular vimentin regulates construction of dengue virus replication complexes through interaction with ns4a protein. *J. Virol.* **2014**, *88*, 1897–1913. [CrossRef] [PubMed]

52. Wu, N.; Gao, N.; Fan, D.; Wei, J.; Zhang, J.; An, J. Mir-223 inhibits dengue virus replication by negatively regulating the microtubule-destabilizing protein stmn1 in eahy926 cells. *Microbes Infect.* **2014**, *16*, 911–922. [CrossRef] [PubMed]

53. Dales, S.; Eggers, H.J.; Tamm, I.; Palade, G.E. Electron microscopic study of the formation of poliovirus. *Virology* **1965**, *26*, 379–389. [CrossRef]

54. Bienz, K.; Egger, D.; Rasser, Y.; Bossart, W. Intracellular distribution of poliovirus proteins and the induction of virus-specific cytoplasmic structures. *Virology* **1983**, *131*, 39–48. [CrossRef]

55. Bienz, K.; Egger, D.; Troxler, M.; Pasamontes, L. Structural organization of poliovirus rna replication is mediated by viral proteins of the p2 genomic region. *J. Virol.* **1990**, *64*, 1156–1163.

[PubMed]

56. Bienz, K.; Egger, D.; Pfister, T.; Troxler, M. Structural and functional characterization of the poliovirus replication complex. *J. Virol.* **1992**, *66*, 2740–2747. [PubMed]

57. Belov, G.A.; Sztul, E. Rewiring of cellular membrane homeostasis by picornaviruses. *J. Virol.* **2014**, *88*, 9478–9489. [CrossRef] [PubMed]

58. Schlegel, A.; Giddings, T.H.; Ladinsky, M.S.; Kirkegaard, K. Cellular origin and ultrastructure of membranes induced during poliovirus infection. *J. Virol.* **1996**, *70*, 6576–6588. [PubMed]

59. Suhy, D.A.; Giddings, T.H.; Kirkegaard, K. Remodeling the endoplasmic reticulum by poliovirus infection and by individual viral proteins: An autophagy-like origin for virus-induced vesicles. *J. Virol.* **2000**, *74*, 8953–8965. [CrossRef] [PubMed]

60. Jackson, W.T.; Giddings, T.H., Jr.; Taylor, M.P.; Mulinyawe, S.; Rabinovitch, M.; Kopito, R.R.; Kirkegaard, K. Subversion of cellular autophagosomal machinery by rna viruses. *PLoS Biol.* **2005**, *3*, e156. [CrossRef] [PubMed]

61. Kemball, C.C.; Alirezaei, M.; Flynn, C.T.; Wood, M.R.; Harkins, S.; Kiosses, W.B.; Whitton, J.L. Coxsackievirus infection induces autophagy-like vesicles and megaphagosomes in pancreatic acinar cells in vivo. *J. Virol.* **2010**, *84*, 12110–12124. [CrossRef] [PubMed]

62. Klein, K.A.; Jackson, W.T. Human rhinovirus 2 induces the autophagic pathway and replicates more efficiently in autophagic cells. *J. Virol.* **2011**, *85*, 9651–9654. [CrossRef] [PubMed]

63. Robinson, S.M.; Tsueng, G.; Sin, J.; Mangale, V.; Rahawi, S.; McIntyre, L.L.; Williams, W.; Kha, N.; Cruz, C.; Hancock, B.M.; *et al.* Coxsackievirus b exits the host cell in shed microvesicles displaying autophagosomal markers. *PLoS Pathog.* **2014**, *10*, e1004045. [CrossRef] [PubMed]

64. Richards, A.L.; Soares-Martins, J.A.P.; Riddell, G.T.; Jackson, W.T. Generation of unique poliovirus rna replication organelles. *MBio* **2014**, *5*, e00833–e00813. [CrossRef] [PubMed]

65. Choe, S.S.; Dodd, D.A.; Kirkegaard, K. Inhibition of cellular protein secretion by picornaviral 3a proteins. *Virology* **2005**, *337*, 18–29. [CrossRef] [PubMed]

66. Beske, O.; Reichelt, M.; Taylor, M.P.; Kirkegaard, K.; Andino, R. Poliovirus infection blocks ergic-to-golgi trafficking and induces microtubule-dependent disruption of the golgi complex. *J. Cell Sci.* **2007**, *120*, 3207–3218. [CrossRef] [PubMed]

67. Egger, D.; Bienz, K. Intracellular location and translocation of silent and active poliovirus replication complexes. *J. Gen. Virol.* **2005**, *86*, 707–718. [CrossRef] [PubMed]

68. Rust, R.C.; Landmann, L.; Gosert, R.; Tang, B.L.; Hong, W.; Hauri, H.P.; Egger, D.; Bienz, K. Cellular copii proteins are involved in production of the vesicles that form the poliovirus replication complex. *J. Virol.* **2001**, *75*, 9808–9818. [CrossRef] [PubMed]

69. Trahey, M.; Oh, H.S.; Cameron, C.E.; Hay, J.C. Poliovirus infection transiently increases copii vesicle budding. *J. Virol.* **2012**, *86*, 9675–9682. [CrossRef] [PubMed]

70. Quiner, C.A.; Jackson, W.T. Fragmentation of the golgi apparatus provides replication membranes for human rhinovirus 1a. *Virology* **2010**, *407*, 185–195. [CrossRef] [PubMed]

71. Mousnier, A.; Swieboda, D.; Pinto, A.; Guedan, A.; Rogers, A.V.; Walton, R.; Johnston, S.L.; Solari, R. Human rhinovirus 16 causes golgi apparatus fragmentation without blocking protein secretion. *J. Virol.* **2014**, *88*, 11671–11685. [CrossRef] [PubMed]

72. Cherry, S.; Kunte, A.; Wang, H.; Coyne, C.; Rawson, R.B.; Perrimon, N. Copi activity coupled with fatty acid biosynthesis is required for viral replication. *PLoS Pathog.* **2006**, *2*, e102. [CrossRef] [PubMed]

73. Irurzun, A.; Perez, L.; Carrasco, L. Involvement of membrane traffic in the replication of poliovirus genomes: Effects of brefeldin a. *Virology* **1992**, *191*, 166–175. [CrossRef]

74. Maynell, L.A.; Kirkegaard, K.; Klymkowsky, M.W. Inhibition of poliovirus rna synthesis by brefeldin a. *J. Virol.* **1992**, *66*, 1985–1994. [PubMed]

75. Behnia, R.; Munro, S. Organelle identity and the signposts for membrane traffic. *Nature* **2005**, *438*, 597–604. [CrossRef] [PubMed]

76. Belov, G.A.; Fogg, M.H.; Ehrenfeld, E. Poliovirus proteins induce membrane association of gtpase adp-ribosylation factor. *J. Virol.* **2005**, *79*, 7207–7216. [CrossRef] [PubMed]

77. Belov, G.A.; Habbersett, C.; Franco, D.; Ehrenfeld, E. Activation of cellular arf gtpases by poliovirus protein 3cd correlates with virus replication. *J. Virol.* **2007**, *81*, 9259–9267. [CrossRef] [PubMed]

78. Belov, G.A.; Ehrenfeld, E. Involvement of cellular membrane traffic proteins in poliovirus replication. *Cell Cycle* **2007**, *6*, 36–38. [CrossRef] [PubMed]

79. Belov, G.A.; Feng, Q.; Nikovics, K.; Jackson, C.L.; Ehrenfeld, E. A critical role of a cellular membrane traffic protein in poliovirus rna replication. *PLoS Pathog.* **2008**, *4*, e1000216. [CrossRef] [PubMed]

80. Lanke, K.H.; van der Schaar, H.M.; Belov, G.A.; Feng, Q.; Duijsings, D.; Jackson, C.L.; Ehrenfeld, E.; van Kuppeveld, F.J. Gbf1, a guanine nucleotide exchange factor for arf, is crucial for coxsackievirus b3 rna replication. *J. Virol.* **2009**, *83*, 11940–11949. [CrossRef] [PubMed]

81. Dorobantu, C.M.; van der Schaar, H.M.; Ford, L.A.; Strating, J.R.; Ulferts, R.; Fang, Y.; Belov, G.; van Kuppeveld, F.J. Recruitment of pi4kiiibeta to coxsackievirus b3 replication organelles is independent of acbd3, gbf1, and arf1. *J. Virol.* **2014**, *88*, 2725–2736. [CrossRef] [PubMed]

82. Dorobantu, C.M.; Ford-Siltz, L.A.; Sittig, S.P.; Lanke, K.H.; Belov, G.A.; van Kuppeveld, F.J.; van der Schaar, H.M. Gbf1- and acbd3-independent recruitment of pi4kiiibeta to replication sites by rhinovirus 3a proteins. *J. Virol.* **2015**, *89*, 1913–1918. [CrossRef] [PubMed]

83. Limpens, R.W.; van der Schaar, H.M.; Kumar, D.; Koster, A.J.; Snijder, E.J.; van Kuppeveld, F.J.; Barcena, M. The transformation of enterovirus replication structures: A three-dimensional study of single- and double-membrane compartments. *MBio* **2011**, *2*. [CrossRef] [PubMed]

84. Vance, D.E.; Trip, E.M.; Paddon, H.B. Poliovirus increases phosphatidylcholine biosynthesis in hela cells by stimulation of the rate-limiting reaction catalyzed by ctp: Phosphocholine cytidylyltransferase. *J. Biol. Chem.* **1980**, *255*, 1064–1069. [PubMed]

85. Van Meer, G.; Voelker, D.R.; Feigenson, G.W. Membrane lipids: Where they are and how they behave. *Nat. Rev. Mol. Cell Biol.* **2008**, *9*, 112–124. [CrossRef] [PubMed]

86. Guinea, R.; Carrasco, L. Phospholipid biosynthesis and poliovirus genome replication, two coupled phenomena. *EMBO J.* **1990**, *9*, 2011–2016. [PubMed]

87. Guinea, R.; Carrasco, L. Effects of fatty acids on lipid synthesis and viral rna replication in poliovirus-infected cells. *Virology* **1991**, *185*, 473–476. [CrossRef]

88. Rassmann, A.; Henke, A.; Zobawa, M.; Carlsohn, M.; Saluz, H.-P.; Grabley, S.; Lottspeich, F.;

Munder, T. Proteome alterations in human host cells infected with coxsackievirus b3. *J. Gen. Virol.* **2006**, *87*, 2631–2638. [CrossRef] [PubMed]

89. Wilsky, S.; Sobotta, K.; Wiesener, N.; Pilas, J.; Althof, N.; Munder, T.; Wutzler, P.; Henke, A. Inhibition of fatty acid synthase by amentoflavone reduces coxsackievirus b3 replication. *Arch. Virol.* **2012**, *157*, 259–269. [CrossRef] [PubMed]

90. Nchoutmboube, J.A.; Viktorova, E.G.; Scott, A.J.; Ford, L.A.; Pei, Z.; Watkins, P.A.; Ernst, R.K.; Belov, G.A. Increased long chain acyl-coa synthetase activity and fatty acid import is linked to membrane synthesis for development of picornavirus replication organelles. *PLoS Pathog.* **2013**, *9*, e1003401. [CrossRef] [PubMed]

91. Godi, A.; Pertile, P.; Meyers, R.; Marra, P.; di Tullio, G.; Iurisci, C.; Luini, A.; Corda, D.; de Matteis, M.A. Arf mediates recruitment of ptdins-4-oh kinase-beta and stimulates synthesis of ptdins(4,5)p2 on the golgi complex. *Nat. Cell Biol.* **1999**, *1*, 280–287. [PubMed]

92. Arita, M.; Kojima, H.; Nagano, T.; Okabe, T.; Wakita, T.; Shimizu, H. Phosphatidylinositol 4-kinase iii beta is a target of enviroxime-like compounds for antipoliovirus activity. *J. Virol.* **2011**, *85*, 2364–2372. [CrossRef] [PubMed]

93. DeLong, D.C.; Reed, S.E. Inhibition of rhinovirus replication in in organ culture by a potential antiviral drug. *J. Infect. Dis.* **1980**, *141*, 87–91. [CrossRef] [PubMed]

94. Van der Schaar, H.M.; Leyssen, P.; Thibaut, H.J.; de Palma, A.; van der Linden, L.; Lanke, K.H.; Lacroix, C.; Verbeken, E.; Conrath, K.; Macleod, A.M.; et al. A novel, broad-spectrum inhibitor of enterovirus replication that targets host cell factor phosphatidylinositol 4-kinase iiibeta. *Antimicrob. Agents Chemother.* **2013**, *57*, 4971–4981. [CrossRef] [PubMed]

95. Spickler, C.; Lippens, J.; Laberge, M.K.; Desmeules, S.; Bellavance, E.; Garneau, M.; Guo, T.; Hucke, O.; Leyssen, P.; Neyts, J.; et al. Phosphatidylinositol 4-kinase iii beta is essential for replication of human rhinovirus and its inhibition causes a lethal phenotype *in vivo*. *Antimicrob. Agents Chemother.* **2013**, *57*, 3358–3368. [CrossRef] [PubMed]

96. Greninger, A.L.; Knudsen, G.M.; Betegon, M.; Burlingame, A.L.; Derisi, J.L. The 3a protein from multiple picornaviruses utilizes the golgi adaptor protein acbd3 to recruit pi4kiiibeta. *J. Virol.* **2012**, *86*, 3605–3616. [CrossRef] [PubMed]

97. Albulescu, L.; Wubbolts, R.; van Kuppeveld, F.J.; Strating, J.R. Cholesterol shuttling is important for rna replication of coxsackievirus b3 and encephalomyocarditis virus. *Cell. Microbiol.* **2015**, *17*, 1144–1156. [CrossRef] [PubMed]

98. Ilnytska, O.; Santiana, M.; Hsu, N.Y.; Du, W.L.; Chen, Y.H.; Viktorova, E.G.; Belov, G.; Brinker, A.; Storch, J.; Moore, C.; et al. Enteroviruses harness the cellular endocytic machinery to remodel the host cell cholesterol landscape for effective viral replication. *Cell Host Microbe* **2013**, *14*, 281–293. [CrossRef] [PubMed]

99. Kuhn, J.H.; Li, W.; Radoshitzky, S.R.; Choe, H.; Farzan, M. Severe acute respiratory syndrome coronavirus entry as a target of antiviral therapies. *Antivir. Ther.* **2006**, *12*, 639–650.

100. Kanjanahaluethai, A.; Chen, Z.; Jukneliene, D.; Baker, S.C. Membrane topology of murine coronavirus replicase nonstructural protein 3. *Virology* **2007**, *361*, 391–401. [CrossRef] [PubMed]

101. Oostra, M.; Te Lintelo, E.; Deijs, M.; Verheije, M.; Rottier, P.; de Haan, C. Localization and

membrane topology of coronavirus nonstructural protein 4: Involvement of the early secretory pathway in replication. *J. Virol.* **2007**, *81*, 12323–12336. [CrossRef] [PubMed]

102. Baliji, S.; Cammer, S.A.; Sobral, B.; Baker, S.C. Detection of nonstructural protein 6 in murine coronavirus-infected cells and analysis of the transmembrane topology by using bioinformatics and molecular approaches. *J. Virol.* **2009**, *83*, 6957–6962. [CrossRef] [PubMed]

103. Choi, K.S.; Aizaki, H.; Lai, M.M. Murine coronavirus requires lipid rafts for virus entry and cell-cell fusion but not for virus release. *J. Virol.* **2005**, *79*, 9862–9871. [CrossRef] [PubMed]

104. Heaton, N.S.; Randall, G. Multifaceted roles for lipids in viral infection. *Trends Microbiol.* **2011**, *19*, 368–375. [CrossRef] [PubMed]

105. De Wilde, A.H.; Wannee, K.F.; Scholte, F.E.; Goeman, J.J.; ten Dijke, P.; Snijder, E.J.; Kikkert, M.; van Hemert, M.J. A Kinome-Wide Small Interfering RNA Screen Identifies Proviral and Antiviral Host Factors in Severe Acute Respiratory Syndrome Coronavirus Replication, Including Double-Stranded RNA-Activated Protein Kinase and Early Secretory Pathway Proteins. *J. Virol.* **2015**, *89*, 8318–8333.

106. Hagemeijer, M.C.; Rottier, P.J.; de Haan, C.A. Biogenesis and dynamics of the coronavirus replicative structures. *Viruses* **2012**, *4*, 3245–3269. [CrossRef] [PubMed]

107. Snijder, E.J.; van der Meer, Y.; Zevenhoven-Dobbe, J.; Onderwater, J.J.; van der Meulen, J.; Koerten, H.K.; Mommaas, A.M. Ultrastructure and origin of membrane vesicles associated with the severe acute respiratory syndrome coronavirus replication complex. *J. Virol.* **2006**, *80*, 5927–5940. [CrossRef] [PubMed]

108. Imbert, I.; Snijder, E.J.; Dimitrova, M.; Guillemot, J.-C.; Lécine, P.; Canard, B. The sars-coronavirus plnc domain of nsp3 as a replication/transcription scaffolding protein. *Virus Res.* **2008**, *133*, 136–148. [CrossRef] [PubMed]

109. Angelini, M.M.; Akhlaghpour, M.; Neuman, B.W.; Buchmeier, M.J. Severe acute respiratory syndrome coronavirus nonstructural proteins 3, 4, and 6 induce double-membrane vesicles. *MBio* **2013**, *4*, e00524–e00513. [CrossRef] [PubMed]

110. Knoops, K.; Swett-Tapia, C.; van den Worm, S.H.; Te Velthuis, A.J.; Koster, A.J.; Mommaas, A.M.; Snijder, E.J.; Kikkert, M. Integrity of the early secretory pathway promotes, but is not required for, severe acute respiratory syndrome coronavirus rna synthesis and virus-induced remodeling of endoplasmic reticulum membranes. *J. Virol.* **2010**, *84*, 833–846. [CrossRef] [PubMed]

111. De Wilde, A.H.; Raj, V.S.; Oudshoorn, D.; Bestebroer, T.M.; van Nieuwkoop, S.; Limpens, R.W.; Posthuma, C.C.; van der Meer, Y.; Bárcena, M.; Haagmans, B.L. Mers-coronavirus replication induces severe in vitro cytopathology and is strongly inhibited by cyclosporin a or interferon-α treatment. *J. Gen. Virol.* **2013**, *94*, 1749–1760. [CrossRef] [PubMed]

112. Goldsmith, C.S.; Tatti, K.M.; Ksiazek, T.G.; Rollin, P.E.; Comer, J.A.; Lee, W.W.; Rota, P.A.; Bankamp, B.; Bellini, W.J.; Zaki, S.R. Ultrastructural characterization of sars coronavirus. *Emerg. Infect. Dis.* **2004**, *10*, 320–326. [CrossRef] [PubMed]

113. Reggiori, F.; Monastyrska, I.; Verheije, M.H.; Calì, T.; Ulasli, M.; Bianchi, S.; Bernasconi, R.; de Haan, C.A.; Molinari, M. Coronaviruses hijack the lc3-i-positive edemosomes, er-derived vesicles exporting short-lived erad regulators, for replication. *Cell Host Microbe* **2010**, *7*,

500–508. [CrossRef] [PubMed]

114. Ng, C.G.; Coppens, I.; Govindarajan, D.; Pisciotta, J.; Shulaev, V.; Griffin, D.E. Effect of host cell lipid metabolism on alphavirus replication, virion morphogenesis, and infectivity. *Proc. Natl. Acad. Sci.* **2008**, *105*, 16326–16331. [CrossRef] [PubMed]

115. Frolova, E.; Gorchakov, R.; Garmashova, N.; Atasheva, S.; Vergara, L.A.; Frolov, I. Formation of nsp3-specific protein complexes during sindbis virus replication. *J. Virol.* **2006**, *80*, 4122–4134. [CrossRef] [PubMed]

116. Spuul, P.; Balistreri, G.; Kääriäinen, L.; Ahola, T. Phosphatidylinositol 3-kinase-, actin-, and microtubule-dependent transport of semliki forest virus replication complexes from the plasma membrane to modified lysosomes. *J. Virol.* **2010**, *84*, 7543–7557. [CrossRef] [PubMed]

117. Neuvonen, M.; Kazlauskas, A.; Martikainen, M.; Hinkkanen, A.; Ahola, T.; Saksela, K. Sh3 domain-mediated recruitment of host cell amphiphysins by alphavirus nsp3 promotes viral rna replication. *PLoS Pathog.* **2011**, *7*, e1002383. [CrossRef] [PubMed]

118. Prentice, E.; McAuliffe, J.; Lu, X.; Subbarao, K.; Denison, M.R. Identification and characterization of severe acute respiratory syndrome coronavirus replicase proteins. *J. Virol.* **2004**, *78*, 9977–9986. [CrossRef] [PubMed]

119. Mizushima, N.; Noda, T.; Yoshimori, T.; Tanaka, Y.; Ishii, T.; George, M.D.; Klionsky, D.J.; Ohsumi, M.; Ohsumi, Y. A protein conjugation system essential for autophagy. *Nature* **1998**, *395*, 395–398. [PubMed]

120. Prentice, E.; Jerome, W.G.; Yoshimori, T.; Mizushima, N.; Denison, M.R. Coronavirus replication complex formation utilizes components of cellular autophagy. *J. Biol. Chem.* **2004**, *279*, 10136–10141. [CrossRef] [PubMed]

121. Zhao, Z.; Thackray, L.B.; Miller, B.C.; Lynn, T.M.; Becker, M.M.; Ward, E.; Mizushima, N.; Denison, M.R.; Virgin, I.; Herbert, W. Coronavirus replication does not require the autophagy gene atg5. *Autophagy* **2007**, *3*, 581–585. [CrossRef] [PubMed]

122. Hayat, M. *Autophagy: Cancer, Other Pathologies, Inflammation, Immunity, Infection, and Aging: Volume 3-Role in Specific Diseases*; Academic Press: Waltham, UK, 2013; Volume 3.

123. Zhang, L.; Zhang, Z.-P.; Zhang, X.-E.; Lin, F.-S.; Ge, F. Quantitative proteomics analysis reveals bag3 as a potential target to suppress severe acute respiratory syndrome coronavirus replication. *J. Virol.* **2010**, *84*, 6050–6059. [CrossRef] [PubMed]

124. Westaway, E.; Brinton, M.; Gaidamovich, S.Y.; Horzinek, M.; Igarashi, A.; Kääriäinen, L.; Lvov, D.; Porterfield, J.; Russell, P.; Trent, D. Togaviridae. *Intervirology* **1985**, *24*, 125–139. [CrossRef] [PubMed]

125. Garmashova, N.; Gorchakov, R.; Volkova, E.; Paessler, S.; Frolova, E.; Frolov, I. The old world and new world alphaviruses use different virus-specific proteins for induction of transcriptional shutoff. *J. Virol.* **2007**, *81*, 2472–2484. [CrossRef] [PubMed]

126. Lee, J.-Y.; Marshall, J.; Bowden, D. Replication complexes associated with the morphogenesis of rubella virus. *Arch. Virol.* **1992**, *122*, 95–106. [CrossRef] [PubMed]

127. Magliano, D.; Marshall, J.A.; Bowden, D.S.; Vardaxis, N.; Meanger, J.; Lee, J.-Y. Rubella virus replication complexes are virus-modified lysosomes. *Virology* **1998**, *240*, 57–63. [CrossRef] [PubMed]

128. Froshauer, S.; Kartenbeck, J.; Helenius, A. Alphavirus rna replicase is located on the cytoplasmic surface of endosomes and lysosomes. *J. Cell Biol.* **1988**, *107*, 2075–2086. [CrossRef] [PubMed]

129. Atasheva, S.; Gorchakov, R.; English, R.; Frolov, I.; Frolova, E. Development of sindbis viruses encoding nsp2/gfp chimeric proteins and their application for studying nsp2 functioning. *J. Virol.* **2007**, *81*, 5046–5057. [CrossRef] [PubMed]

130. Cristea, I.M.; Carroll, J.-W.N.; Rout, M.P.; Rice, C.M.; Chait, B.T.; MacDonald, M.R. Tracking and elucidating alphavirus-host protein interactions. *J. Biol. Chem.* **2006**, *281*, 30269–30278. [CrossRef] [PubMed]

131. Varjak, M.; Saul, S.; Arike, L.; Lulla, A.; Peil, L.; Merits, A. Magnetic fractionation and proteomic dissection of cellular organelles occupied by the late replication complexes of semliki forest virus. *J. Virol.* **2013**, *87*, 10295–10312. [CrossRef] [PubMed]

132. Frolova, E.I.; Gorchakov, R.; Pereboeva, L.; Atasheva, S.; Frolov, I. Functional sindbis virus replicative complexes are formed at the plasma membrane. *J. Virol.* **2010**, *84*, 11679–11695. [CrossRef] [PubMed]

133. Lemm, J.A.; Rice, C.M. Roles of nonstructural polyproteins and cleavage products in regulating sindbis virus rna replication and transcription. *J. Virol.* **1993**, *67*, 1916–1926. [PubMed]

134. Lemm, J.A.; Rümenapf, T.; Strauss, E.G.; Strauss, J.H.; Rice, C. Polypeptide requirements for assembly of functional sindbis virus replication complexes: A model for the temporal regulation of minus-and plus-strand rna synthesis. *EMBO J.* **1994**, *13*, 2925. [PubMed]

135. Leung, J.Y.-S.; Ng, M.M.-L.; Chu, J.J.H. Replication of alphaviruses: A review on the entry process of alphaviruses into cells. *Adv. Virol.* **2011**, *2011*, 249640. [CrossRef] [PubMed]

136. Ahola, T.; Lampio, A.; Auvinen, P.; Kääriäinen, L. Semliki forest virus mrna capping enzyme requires association with anionic membrane phospholipids for activity. *EMBO J.* **1999**, *18*, 3164–3172. [CrossRef] [PubMed]

137. Rozanov, M.N.; Koonin, E.V.; Gorbalenya, A.E. Conservation of the putative methyltransferase domain: A hallmark of the 'sindbis-like'supergroup of positive-strand rna viruses. *J. Gen. Virol.* **1992**, *73*, 2129–2134. [CrossRef] [PubMed]

138. Peränen, J.; Laakkonen, P.; Hyvönen, M.; Kääriäinen, L. The alphavirus replicase protein nsp1 is membrane-associated and has affinity to endocytic organelles. *Virology* **1995**, *208*, 610–620. [CrossRef] [PubMed]

139. Laakkonen, P.; Ahola, T.; Kääriäinen, L. The effects of palmitoylation on membrane association of semliki forest virus rna capping enzyme. *J. Biol. Chem.* **1996**, *271*, 28567–28571. [CrossRef] [PubMed]

140. Smit, J.M.; Bittman, R.; Wilschut, J. Low-ph-dependent fusion of sindbis virus with receptor-free cholesterol-and sphingolipid-containing liposomes. *J. Virol.* **1999**, *73*, 8476–8484. [PubMed]

141. Kielian, M.C.; Helenius, A. Role of cholesterol in fusion of semliki forest virus with membranes. *J. Virol.* **1984**, *52*, 281–283. [PubMed]

142. Kielian, M.; Chatterjee, P.K.; Gibbons, D.L.; Lu, Y.E. Specific roles for lipids in virus fusion and exit examples from the alphaviruses. In *Fusion of Biological Membranes and Related Problems*; Springer: Berlin, Germany, 2002; pp. 409–455.

143. Marquardt, M.T.; Phalen, T.; Kielian, M. Cholesterol is required in the exit pathway of semliki forest virus. *J. Cell Biology* **1993**, *123*, 57–65. [CrossRef]

144. Lu, Y.E.; Cassese, T.; Kielian, M. The cholesterol requirement for sindbis virus entry and exit and characterization of a spike protein region involved in cholesterol dependence. *J. Virol.* **1999**, *73*, 4272–4278. [PubMed]

145. Barton, D.J.; Sawicki, S.G.; Sawicki, D.L. Solubilization and immunoprecipitation of alphavirus replication complexes. *J. Virol.* **1991**, *65*, 1496–1506. [PubMed]

146. Gorchakov, R.; Garmashova, N.; Frolova, E.; Frolov, I. Different types of nsp3-containing protein complexes in sindbis virus-infected cells. *J. Virol.* **2008**, *82*, 10088–10101. [CrossRef] [PubMed]

147. Liu, T.; Daniels, C.K.; Cao, S. Comprehensive review on the hsc70 functions, interactions with related molecules and involvement in clinical diseases and therapeutic potential. *Pharmacol. Ther.* **2012**, *136*, 354–374. [CrossRef] [PubMed]

148. Mayer, M. Recruitment of hsp70 chaperones: A crucial part of viral survival strategies. In *Reviews of Physiology, Biochemistry and Pharmacology*; Springer: Berlin, Germany, 2005; pp. 1–46.

149. Scholte, F.E.; Tas, A.; Albulescu, I.C.; Žusinaite, E.; Merits, A.; Snijder, E.J.; van Hemert, M.J. Stress granule components g3bp1 and g3bp2 play a proviral role early in chikungunya virus replication. *J. Virol.* **2015**, *89*, 4457–4469. [CrossRef] [PubMed]

150. Panas, M.D.; Varjak, M.; Lulla, A.; Eng, K.E.; Merits, A.; Hedestam, G.B.K.; McInerney, G.M. Sequestration of g3bp coupled with efficient translation inhibits stress granules in semliki forest virus infection. *Mol. Biol. Cell* **2012**, *23*, 4701–4712. [CrossRef] [PubMed]

151. McInerney, G.M.; Kedersha, N.L.; Kaufman, R.J.; Anderson, P.; Liljeström, P. Importance of eif2α phosphorylation and stress granule assembly in alphavirus translation regulation. *Mol. Biol. Cell* **2005**, *16*, 3753–3763. [CrossRef] [PubMed]

6

HIV Rev Assembly on the Rev Response Element (RRE): A Structural Perspective

Jason W. Rausch and Stuart F. J. Le Grice *

Reverse Transcriptase Biochemistry Section, Basic Research Program, Frederick National Laboratory for Cancer Research, Frederick, MD 21702, USA; E-Mail: rauschj@mail.nih.gov

* Author to whom correspondence should be addressed; E-Mail: legrices@mail.nih.gov

Academic Editor: David Boehr

Abstract: HIV-1 Rev is an ~13 kD accessory protein expressed during the early stage of virus replication. After translation, Rev enters the nucleus and binds the Rev response element (RRE), a ~350 nucleotide, highly structured element embedded in the *env* gene in unspliced and singly spliced viral RNA transcripts. Rev-RNA assemblies subsequently recruit Crm1 and other cellular proteins to form larger complexes that are exported from the nucleus. Once in the cytoplasm, the complexes dissociate and unspliced and singly-spliced viral RNAs are packaged into nascent virions or translated into viral structural proteins and enzymes, respectively. Rev binding to the RRE is a complex process, as multiple copies of the protein assemble on the RNA in a coordinated fashion *via* a series of Rev-Rev and Rev-RNA interactions. Our understanding of the nature of these interactions has been greatly advanced by recent studies using X-ray crystallography, small angle X-ray scattering (SAXS) and single particle electron microscopy as well as biochemical and genetic methodologies. These advances are discussed in detail in this review, along with perspectives on development of antiviral therapies targeting the HIV-1 RRE.

Keywords: HIV; Rev; Rev response element; RRE; Crm1; nuclear export complex

1. Introduction

Replication of retroviruses and transposition of endogenous retroelements exploits a unique mechanism of post-transcriptional regulation as a means of exporting their full-length and incompletely-spliced mRNAs (which serve as the genomic RNA and the template for protein synthesis, respectively) to the cytoplasm. This is achieved through the concerted interaction of highly structured *cis*-acting regulatory elements in the RNA genome with a variety of obligate viral and host proteins. Examples of such regulatory RNAs include the constitutive transport element (CTE) of the simian retroviruses (MPMV, SRV-1, SRV-2) [1], the musD transport element (MTE) of murine retroelements (intracisternal A particles and musD) [2], the ~1.7 kb posttranscriptional element (PTE) of gammaretroviruses such as murine leukemia virus [3] and the L1-NXF1 element of human LINE-1 transposons [4].

Studies such as these, which have made a significant contribution to our understanding of cellular processes regulating the fate of RNA, resulted from seminal work performed over 30 years ago to understand nucleocytoplasmic RNA transport in human immunodeficiency virus (HIV). In particular, disrupting the HIV-1 genome in the immediate vicinity of the trans-activator of transcription (tat) open reading frame (ORF) had no effect on its expression levels, but inhibited expression of the gag, pol and env genes, resulting in a severe replication defect [5]. This defect could be corrected by inclusion of an overlapping ORF encoding the regulator of expression of viral proteins, or Rev, a small accessory protein containing both nuclear export and localization signals (NES and NLS, respectively).

The notion that Rev was involved in RNA transport rather than modifying splicing arose from observations of Malim *et al.* [5], who elegantly showed that cytoplasmic expression of the non-spliceable HIV-1 *env* gene was also subject to Rev control. At the same time, these authors identified the Rev response element (RRE), an ~350 nt highly-structured *cis*-acting RNA within the *env* gene (nucleotides 7709–8063), as the target of Rev.

The principles of the Rev/RRE axis are outlined schematically in Figure 1. Early in the virus life cycle, Rev and additional HIV regulatory proteins are translated from completely spliced RNAs that are exported from the nucleus in a manner analogous to cellular mRNAs. Subsequently, the arginine-rich NLS facilitates entry of Rev into the nucleus, where it interacts with the RRE in unspliced and incompletely-spliced HIV RNAs. Rev initially binds to stem-loop IIB, a purine rich RNA secondary structure motif (*vide infra*), after which several additional Rev molecules assemble along the RRE to generate the Rev-RRE complex.

This complex then recruits Crm1 and other host proteins into a larger complex that is exported from the nucleus into the cytoplasm. The goal of this review is to provide an updated account of our understanding of the HIV Rev-RRE complex with respect to the recently-elucidated structures of the protein and RNA components. For a broader perspective, the reader is referred to excellent reviews from the Cullen [6], Hope [7], Malim [8], Daelemans [9] and Frankel [10] groups.

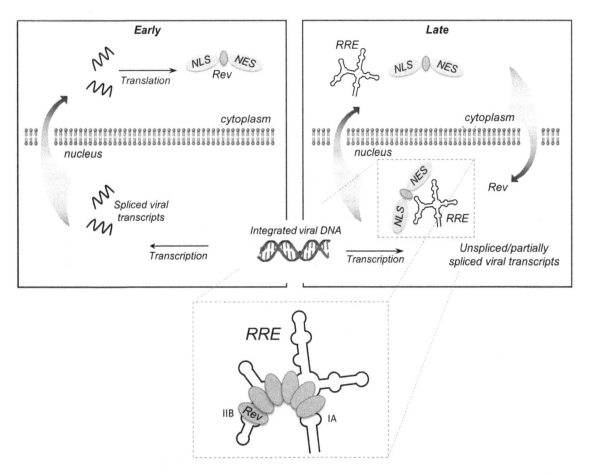

Figure 1. Rev-mediated nucleocytoplasmic transport of HIV-1 RNA containing the Rev RRE. RNA transport: In the early phase of the HIV life cycle, the genomic RNA transcript is completely spliced, generating RRE-free messages, which are transported to the cytoplasm *via* standard nuclear export pathways. One of these messages encodes Rev, which is imported into the nucleus *via* its nuclear localization sequence (NLS). The late phase of the viral life cycle is characterized by the expression of viral proteins encoded by the unspliced (9 kb) or partially spliced (4 kb) RRE-containing mRNAs. These large intron-containing RNAs are retained in the nucleus for splicing/degradation until a sufficient level of Rev accumulates, after which they are exported to the cytoplasm *via* a Rev-dependent export pathway. This involves assembly of the Rev-RRE complex and recruitment of host proteins CRM1 and Ran-GTP *via* the Rev nuclear export sequence (NES). Export of the Rev-RRE -CRM1/RanGTP complex to the cytoplasm provides mRNAs that are translated to produce the remaining viral proteins and full-length genomes that are packaged into the budding virion. Inset: Rev initially binds the IIB secondary structure motif, then cooperatively assembles along the RRE *via* a series of Rev-Rev and Rev-RNA interactions. Rev has also been reported to bind motif IA with high affinity.

2. Rev Structural Organization and RNA Binding

HIV-1 Rev is a 116 amino acid, ~13 kD protein generated by translation of a fully spliced 2 kb mRNA during the early phase of viral replication and organized as shown in Figure 2A [8]. The *N*-terminus of the protein assumes a helix-turn-helix configuration containing two functional domains: the nuclear localization signal and RNA binding domain (NLS/RBD) and the Rev multimerization domain. The NLS/RBD is housed within the distal portion of the helix-turn-helix motif. A stretch of amino acids within α-helix 2 in this domain contains several functionally important arginines and has thus been dubbed the arginine rich motif (ARM). The multimerization domain houses a number of hydrophobic amino acids located at opposite ends of the helix-turn-helix primary sequence, but proximal to each other in three dimensions (3D). Contacts among these hydrophobic residues stabilize the overall Rev structure, and their arrangement generates the two-sided interface for Rev multimerization. Outside of the helix-turn-helix motif, the *C*-terminus of Rev is intrinsically disordered; however, this segment of the protein contains a third, leucine-rich functional domain known as the nuclear export signal (NES). This is the effector domain of Rev, and is required for recruitment of Crm1 and other host proteins that facilitate nuclear export of the Rev-RRE complex.

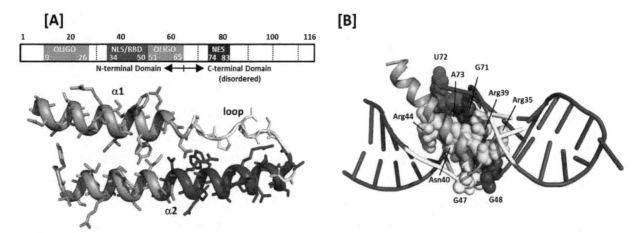

Figure 2. Rev structural organization and arginine-rich motif (ARM)/stem loop IIB complex: (**A**) Rev organization according to primary sequence and 3D structure. The bipartite oligomerization domain and the nuclear localization signal/RNA-binding domain (NLS/RBD) are depicted in green and blue, respectively. The C-terminal domain of Rev, which houses the nuclear export signal (NES), is intrinsically disordered; (**B**) ARM peptide in complex with stem-loop IIB model RNA. The ARM peptide binds in the RNA major groove widened by non-canonical base pairs (G47-A73, G48-G71) and an unpaired, unstacked uridine (U72). These nucleotides are space-filled in the model. Arginines that make specific contacts with nucleic acid bases, and the contacted nucleotides, are highlighted by space-filling and yellow coloration, respectively.

As its designation suggests, the NLS/RBD of HIV-1 Rev is important for both nuclear import of cytoplasmic Rev and binding to the RRE. Multiple reports indicate that initial Rev binding occurs at a purine rich segment in stem loop IIB, a substructure of the HIV-1 RRE for which Rev has been shown to bind with high affinity (*vide infra*) [11–15]. The first high-resolution picture of the Rev–stem loop

IIB interaction was obtained from NMR structures of a Rev ARM peptide in complex with a short synthetic RNA engineered to stably recapitulate the IIB RNA substructure [16]. The averaged structure of the ARM-IIB complex is depicted in Figure 2B, wherein nucleotide numbering reflects that used for the artificial IIB construct in the file deposited in the Protein Data Bank (PDB). The IIB binding site is notable in that it contains two non-canonical base-pairs (G47-A73 and G48-G71) separated by an unstacked, bulged uridine (U72) [17,18]. This arrangement serves to widen the helical major groove at the site where the Rev peptide contacts the RNA [16,19,20]. Upon binding, the Rev ARM penetrates deeply into the stem loop IIB major groove [16,21–23], thereby inducing a conformational change that widens the groove even further [24–26]. Four ARM residues make base-specific contacts with nucleotides in stem loop IIB: Arg35 and Arg39 contact nucleotides U66, G67 and G70 on one side of the RNA major groove, while Asn40 and Arg44 contact U45, G46, G47 and A73 on the other. In addition, Thr34, located at the non-helical turn segment of the Rev helix-turn-helix motif, and six arginine residues (Arg38, Arg41–Arg43, Arg46 and Arg48) within the ARM, interact nonspecifically with the RNA sugar-phosphate backbone.

More recently, a crystallographic structure of a rev dimer in complex with an engineered RRE-like RNA was resolved that has greatly enhanced our understanding of both Rev interactions and how Rev binds RNA [27]. Unlike the NMR structure, the 47-nt RNA in the Rev dimer complex contains both the IIB binding site and an adjacent "junction site" engineered to resemble the three-way junction within stem loop II. This secondary site contains a non-canonical G-A base pair like that found in IIB, as well as an unpaired U. The unpaired U and adjacent A were shown to be essential for binding of the second Rev, as their removal results in a complex in which only a single Rev is bound. In the crystal structure, the ARMs of the dimerized Rev molecules bind along the RNA major grooves of the adjacent binding sites. Contacts at the IIB site closely resemble those observed in the NMR structure, with the notable exception of those involving Rev amino acid Asn40. At the junction site, all contacts except those involving Arg43 and Arg44 are with the RNA phosphate backbone, consistent with finding that binding at the second site is relatively sequence non-specific and requires only that the site contain an RNA bulge [28]. While the rotational positioning of the two Rev ARMs in the major grooves of the IIB and junctional binding sites appears to be similar, their linear positioning is displaced by approximately the length of an alpha helical turn. The structure also addresses how binding of the Rev dimer to adjacent sites on the RRE RNA affects the Rev dimer interface, which will be discussed below.

Another Rev binding site in RRE stem loop I has been identified using structural and biochemical techniques [29]. Like the IIB and junctional sites, the stem I site (designated site IA) is also comprised of a purine-rich bulge in which the major groove might be expected to be widened, flexible, and/or defined by non-canonical G-A base pairs. However, mutational analysis suggests that residues Arg38, Arg 41 and Arg 46 are crucial for binding at site IA, indicating a different rotational positioning of Rev in the RNA major groove and suggesting that the Rev-RRE interface is flexible and may be substantially different at distinct binding sites. Other segments of stem I have been implicated in Rev binding [30], but the Rev-RNA interactions at these sites have not been characterized structurally.

3. Structural Basis for Rev Oligomerization

Although oligomerization has been shown to be an important aspect of Rev function in virus replication [29,31,32], this property was first observed with recombinant protein. At low concentrations, Rev exists as monomer [11], dimer [33] or tetramer [11,34]. Above a critical concentration of ~80 ng/mL (~6 μM), however, Rev forms regular, unbranched filaments of indeterminate length [11,33,35,36]. The structural basis for filament formation resides in the two-sided multimerization interface in which a given Rev molecule can be flanked by additional Rev on either side.

The oligomerization domain of HIV-1 Rev is a flat, two-sided structure formed by juxtaposition of two α helical regions located at opposite ends of the Rev helix-turn-helix motif primary sequence. The domain has a polarity named for the sides of a coin, with the "heads" (H) and "tails" (T) faces interacting specifically with their equivalent counterparts in adjacent Rev molecules; i.e., "H/H" or "T/T" interfaces are preferentially formed in Rev oligomers [37]. Although the T/T Rev dimer is more stable than the H/H and is the form assumed by Rev dimers in solution and after initial binding to the RRE IIB substructure, a high resolution structure of a Rev dimer containing an H/H interface was the first to be resolved by X-ray crystallography [38]. Crystal formation was promoted and filament formation inhibited in this case by blocking the outer T faces of the Rev dimer with monoclonal F_{ab} fragments. Although present in the crystalized protein, the disordered C-terminal portion of Rev, including the NES, was unresolved.

In support of prior genetic and NMR structural data [16,37], a hydrophobic core comprising residues Leu12, Ile19, Leu22, Tyr23 of alpha-helix 1 and Trp45, Ile52, Ile59, Leu60 and Tyr63 of α-helix 2 were shown to be major contributors to Rev helical hairpin stability. Intermolecular interactions among an overlapping set of multimerization domain hydrophobic residues (Leu12, Leu13, Val16, Ile19, Leu60 and Leu64) were shown to comprise much of the H/H interface, of which the roles of Leu12, Val16 and Leu60 were previously established genetically [37]. Because the H/H junction involves residues located at the extreme pronged end of the Rev helix-turn-helix, viewing the structured portion of the dimer from the intermolecular axis orthogonal to the plane of the interface demonstrates that the two molecules assume a "V-like" configuration relative to each other (i.e., rather than one resembling an "X") (Figure 3A). Moreover, the helix-turn-helix motifs are arranged such that the two ARMs form an angle of approximately 140°. The spacing between NLS/RBDs and the opposing trajectories of the two motifs place their extreme termini approximately 8 nm apart, suggesting that the ARMs in such a dimer would be unlikely to bind the RRE in adjacent regions of the major groove on the same helix. Instead, a model based on superimposition of H/H Rev dimer crystal structure and the ARM/IIB NMR structure that suggests the H/H dimer may be best suited for linking either two separate RNAs or two RNA helices located in distant regions of the same RNA [38].

Details of the T/T Rev dimer interface were first resolved in a crystal structure in which H-surface residues Leu12 and Leu60 were mutated to suppress higher order multimerization [39] (Figure 3B). Despite these modifications, asymmetric units in these crystals contained four Rev molecules linked sequentially by T/T, H/H and T/T interfaces, where the H/H interaction matched that observed in the original dimer crystal structure almost exactly. Packing of hydrophobic residues on the T surfaces of adjacent Rev molecules formed an interface that buries over 1500 Å² of surface area. Leu18 and Ile55 form symmetric contacts between monomers at the T/T interface, are highly conserved, and essential for cooperative RNA binding and export [29,37,40]. Phe21, Leu22 and Ile59 are likewise present at the dimer interface, although these residues are also important for stabilizing monomeric Rev. From the vantage point of the oligomerization axis, the ordered portion of the T/T Rev dimer assumes a configuration resembling an asymmetric "X", with long and short NLS/RBD and multimerization domain protrusions flanking the Rev-Rev interface on either side, respectively. Moreover, at 120°, the ARM angle formed in the T/T dimer is narrower than observed in the H/H dimer structure. A molecular model of the T/T dimer in complex with a short IIB-like RNA suggests that both opposing ARMs could bind in the major groove on the same face of an RNA helix with contacts separated by approximately one A-form helical turn. It is further suggested that Rev-Rev and Rev-RNA interactions may propagate from the IIB initiation site along a contiguous region of the RRE, placing the Rev molecules—each with a disordered NES—projecting away from the RNA in a common direction like "tentacles of a jellyfish" [39]. The jellyfish model provides important insight into how multiple Revs may assemble on a comparatively large RRE, and will be discussed in more detail below. However, it is worth noting that in this initial manifestation of the model, the helical axis of the RNA was nearly perpendicular to the axis of Rev oligomerization—an arrangement that would likely be incompatible with the proposed coordination of multiple Revs binding at adjacent sites on the RRE.

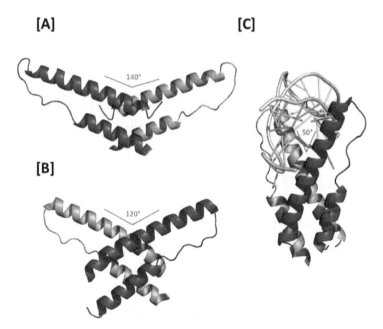

Figure 3. Crystal structures of Rev dimers: (**A**) Ribbon representation of the H/H Rev dimer. The angle formed by the two ARMs is 140°, which separates the apices of the two NLS/RBDs by approximately 8 nm and precludes binding at adjacent sites on an RNA helix; (**B**) T/T Rev dimer structure. In the absence of RNA, the ARM angle is 120°. Binding at adjacent RNA sites is conceivable, but higher order Rev multimerization would likely require a reduced ARM angle at H/H interfaces; (**C**) Structure of the T/T Rev dimer in complex with RNA. The two ARMs bind RNA at adjacent sites on the helix and are oriented at an angle of 50° relative to each other. The Rev oligomerization and RNA helical axes in this structure are roughly parallel, thereby facilitating consecutive binding and the higher order structures proposed in the jellyfish model of Rev assembly.

How Rev-Rev and Rev-RNA interactions can occur concomitantly was largely explained by the aforementioned co-crystal structure in which a T/T Rev dimer binds a truncated RRE RNA at adjacent IIB and junctional sites [27]. ARM binding in the RNA major groove at the two sites appears to change the organization of the T/T interface relative to the naked Rev dimer, and a reciprocal effect can be observed in the distorted binding site major grooves. Among of the more important consequences of this reorganization are that the angle formed by the two Rev ARMs is reduced to $\sim$50° and the Rev multimerization and RNA helical axes are nearly parallel (Figure 3C). Both of these observations are compatible with the jellyfish model of Rev assembly.

Relative to the naked dimer, the T/T interface of the Rev dimer-RNA complex is rotated around Ile55, substantially altering contacts among the hydrophobic residues of both T-surfaces. Although the interacting residues at the T/T interface are largely the same regardless of whether RNA is present, specific points of contact are altered considerably. Collectively, surface area buried at the T/T interface in the Rev dimer-RNA complex is reduced by $\sim$33% (to $\sim$1000 Å^2), suggesting that the dimer may be less stable in the presence of RNA. However, the energetic favorability of RNA binding likely compensates for reduced hydrophobic interactions, thereby rendering the Rev-dimer-RNA complex more stable than the naked dimer. The Phe21 residues that facilitated dimerization in the naked dimer are excluded from the T/T interface in the Rev-dimer RNA complex; moreover, reciprocal Gln51-Gln51 hydrogen bonding is observed only in the presence of RNA. Interestingly, introducing a Gln51Ala mutation into Rev appears to both impair Rev dimerization and reduce the affinity of Rev for the truncated RRE RNA approximately 30-fold. However, the effects of this mutation on the fully assembled Rev-RRE complex are relatively modest, suggesting that Gln51 hydrogen bonding may not be as important during later stages of Rev assembly.

4. Secondary and 3D Structures of the RRE

Our structural understanding of Rev, Rev dimer and Rev–IIB interactions has been greatly facilitated by X-ray crystallography and NMR. Using the same approaches to study the structure of the HIV-1 RRE is problematic, however, given the size and flexibility of this highly structured RNA element. For the most part, experimental approaches have been restricted to using a number of enzymatic and chemical probing techniques to characterize the RRE secondary structure. Although these efforts sometimes produced differing secondary structural models for similar RRE variants, common structural features have been identified that help both in developing 3D models of the motif and for understanding how Rev assembles along this conserved segment of RNA.

Although the HIV-1 RRE is approximately 350 nt in length, the secondary structure of this RNA element was initially characterized using RNA folding prediction software together with enzymatic and chemical RNA probing experiments conducted using a truncated ($\sim$235 nt), *in vitro* transcribed version of the RNA [41]. The sub-structure designations defined in this seminal work will also be used here. In the original model, the HIV-1 RRE RNA assumes a secondary structure comprised of five stems, stem loops or bifurcated stem-loops (I-V) arranged around a central 5-way junction (Figure 4A). Stem I is the longest of the stems/stem-loops, and is interrupted by a number of internal loops and bulges of varying size. Some of these internal loops are purine rich, and that most proximal to the central junction (stem IA) has been identified as a high-affinity Rev binding site [29]. Stem-loop II is bifurcated to form smaller substructures designated IIA, IIB and IIC. As previously noted, IIB contains a purine-rich internal segment characterized by non-canonical G-A and G-G base pairing and a widened major groove, and has been identified as the site at which Rev initially binds to the RRE to initiate assembly [31,32,42]. Moreover, according to the Rev dimer-RNA co-crystal structure [27], the adjacent stem-loop II junctional region serves as the binding site for the second Rev. Proceeding clockwise around the central junction in Figure 4A, stem loops III-V complete this 5-stem secondary structural model of the RRE.

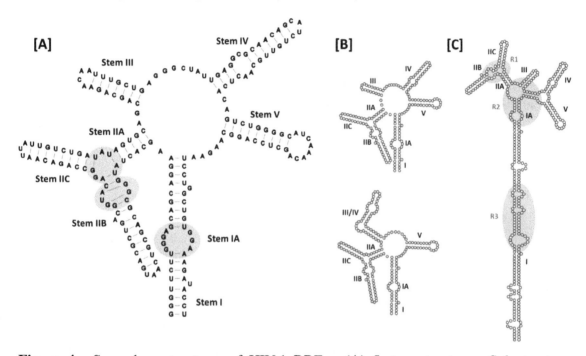

Figure 4. Secondary structures of HIV-1 RREs: (**A**) 5-stem structure. Sub-structure designations and base-pairing patterns, including non-canonical G-A and G-G base pairs in stem IIB, are indicated. Established Rev binding sites at IIB, the stem loop II junction and stem IA are also shown (gray ovals); (**B**) Comparison of the 5- and 4-stem structures. Differences between the two models are limited to the base pairing patterns of nucleotides comprising stem loops III and IV, and stem loop III/IV, in the respective structures. Due to space limitations, stem I has been truncated in the models presented in panels (**A–C**). Alternative 5-stem RRE structure proposed for an ARV-2/SF2 HIV-1 isolate. The entire RRE is shown. Stem I is truncated relative to the original 5-stem model, and there are more single-stranded nucleotides separating stem I and stem IIA. In addition, base pairing across the central junction separates stem loops IV and V from the rest of the alternative 5-stem structure. R1, R2 and R3 refer to Rev binding regions 1–3 identified for this RRE variant by time-resolved selective 2′-hydroxyl acylation analyzed by primer extension (SHAPE).

A later 4-stem model obtained using enzymatic and chemical probing techniques suggests that stem loops III and IV, together with the intervening loop region, are organized to form a hybrid III/IV stem loop [32] (Figure 4B). While the central junction is compacted somewhat in this alternative folding, no additional differences between the 4- and 5-stem RRE secondary structures were proposed. Which of the two structures is assumed by the RRE in the context of HIV replication has been a subject of considerable debate, with multiple studies supporting both forms [2,43,44]. Recently, however, it has been demonstrated using native gel electrophoresis and in-gel SHAPE [45] that the HIV-1 RRE can assume either the 5- or the 4-stem conformation [46]. Moreover, supporting virological data suggest that the two forms may not be equivalent, *i.e.*, HIV-1 housing a mutant RRE that exclusively assumes the 5-stem conformation outgrows virus with a 4-stem RRE in growth competition experiments. This finding is in agreement with the observation that RRE61, an RRE mutant shown to assume a conformation resembling the 5-stem structure, confers resistance to the *trans*-dominant Rev mutant RevM10 [2].

Recently, another secondary structural model was proposed for the RRE of an ARV-2/SF2 HIV-1 isolate (Figure 4C) that closely resembles the RRE configurations proposed for HIV-2 and SIVmac [30,43,47]. This model resembles the original 5-stem HIV-1 RRE structure in that stem-loops II, III, IV and V are identical in the two versions. However, in the more recent model, the loop region between I and IIA is larger than in original model, with a 6-basepair "bridge" across the central junction separating stem loops IV and V from the rest of the structure. Since nucleotides contributing to these alternative arrangements are base-paired at the proximal terminus of stem I in the original structure, the stem I motif is four base pairs shorter in the newer model.

The modified 5-stem RRE structure is arguably the most compatible with the jellyfish model of Rev assembly, as single-stranded, purine rich, Rev binding sites at the IIB/stem II junction, central junction and proximal stem I internal loop are more evenly spaced than in other secondary structural models. However, the authors of this study also provide evidence for a heretofore unrecognized Rev binding region among the purine-rich internal loops located near the center of stem I, and propose a novel mechanism of Rev assembly that differs somewhat from the jellyfish model to explain binding at that site [30]. This model of assembly will be discussed in more detail in the following section.

While varying RRE sequence and secondary structure has been shown to affect RRE function, the nature and degree of these effects are not entirely predictable. For example, whereas large engineered changes such as deletion of stem loop II almost completely abolishes Rev binding in *in vitro* assays [14], disruption of stem loops III and IV (or III/IV) does not. The latter changes do, however, substantially impair Rev/RRE-mediated nuclear export function in cell culture [46]. Nuclear export of HIV-1 RNA is similarly reduced upon serial truncation of RRE stem I, although this process is gradual, and considerable function is retained in RRE variants as small as ~230 nt [32].

The effects of natural RRE sequence and structural variations in the context of viral replication can be more difficult to interpret. Since the RRE is embedded in the *env* gene, changes in RRE sequence may affect the functionality of Env as well as the nuclear export of HIV-1 RNA. Another consequence of this dual functionality may be that genetic flexibility is limited, making the RRE sequence a potentially promising target for antiviral therapies. Despite these seeming limitations, RRE sequences derived from clinical samples do exhibit a degree of genetic variation, and the effects of these sequence differences are not always easy to explain. For example, in one study, the function of patient-derived RRE variants appeared to be more affected by select single nucleotide polymorphisms than by more pronounced changes predicted to substantially alter RRE secondary structure [48]. In other work, select polymorphic RREs obtained from clinical isolates were shown to decrease Rev-dependent nuclear export 2–3-fold, although a corresponding effect on RRE structure was not established [49]. Although it is not clear how specific RRE mutations correlate with RRE structure and function, it has been suggested that mutational attenuation of RRE function could potentially serve as a natural means of down-regulating HIV-1 replication during the course of infection [46].

Thus far, molecular modeling and small angle X-ray scattering (SAXS) have been the only means of generating 3D models of the RRE. One option for the former approach is to use RNA Composer, a web-based RNA folding application that uses homology modeling and energy minimization to assemble a 3D RNA structure from primary sequence and an associated secondary structure map [50]. This software was used to generate the 3D model of the HIV-2 RRE depicted in Figure 5A [47]. While less is known about HIV-2 Rev assembly on its cognate RRE, alignment of the homologs of IIB, the central junction and stem loop I in this model suggest that the RRE 3D structures and mechanisms of Rev assembly may be similar between HIV-2 and HIV-1.

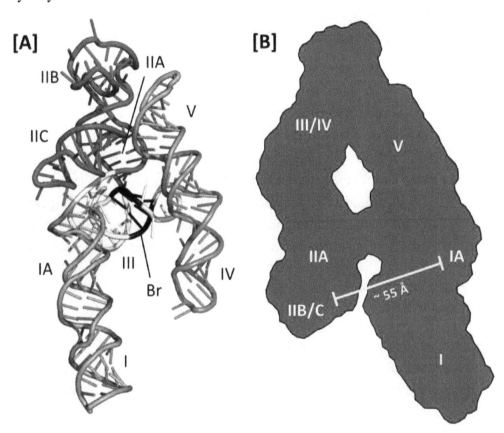

Figure 5. Three-dimensional models of the HIV-2 and HIV-1 RREs: (**A**) HIV-2 RRE model structure obtained using SHAPE and molecular modeling with RNA Composer. Substructures homologous to those reported for the HIV-1 RRE are indicated. The region of base pairing that bridges the gap between stem loops IV and V and the rest of the structure is also shown (Br, black ribbon); (**B**) A-like SAXS envelope obtained for a truncated HIV-1 RRE (233 nt). Sub-structure designations and positioning are determined by fitting an RRE molecular model into the SAXS envelope. The molecular model was generated using the RRE 4-stem secondary structure as a template. High-affinity Rev binding sites at IIB and IA are separated by approximately 55 Å.

A combination of SAXS and molecular modeling was used to generate a 3D model of the HIV-1 RRE [51]. Using the 4-stem secondary structure map as a model, individual sub-structures were constructed (e.g., stem loop III/IV) and analyzed by SAXS. The molecular boundaries of the sub-structures were then mapped onto the global envelope of a truncated (233 nt) HIV-1 RRE, which was itself used to constrain and define the 3D organization of the RNA predicted by molecular modeling (Figure 5B). In the resulting structure, the HIV-1 RRE assumes an A-like configuration in which stem IIA and stem-loop III/IV are collinear, stem I is collinear with stem-loop V and the two extended motifs are centrally linked by a rigid, elongated 4-way junction. Moreover, whereas the loop regions of III/IV and V are proximally located, and perhaps contacting each other at the apex of the A-like configuration, IIB and the putative high affinity Rev binding site in stem I are opposite each other and separated by ~55 Å. Because this distance approximates the separation between the ARMs of the Rev dimer in the absence of RNA [39], it was suggested that dimeric Rev spans from IIB to the high affinity site in I in the early stages of assembly [51]. In this model, multimerization then proceeds outward from the initiating dimer in both directions along the multimerization axis on one face of the A-like RRE RNA.

Subsequent SAXS analysis of the full length RRE extends the preceding model to include all of stem I [30]. These data suggest that an intact stem I folds back on itself so that the distal portion of the motif interacts with the proximal portion, and perhaps also with regions near the central junction. Interestingly, although A-like RRE SAXS envelopes are reported in both studies, probably in part because the former model was used to derive the latter, the 4-stem secondary structure used to model the former structure was not used in the more recent study. Instead, SHAPE analysis [52,53] determined that the RRE variant used in the latter work assumed the modified 5-stem structure with base-pairing across the central junction. While this secondary structural variation was used for subsequent SHAPE-based mapping of Rev binding sites on the RRE, the modified 5-stem RRE was not modeled into the A-like SAXS envelope in this study.

5. Rev Assembly on the RRE

It is well established that multiple copies of HIV-1 Rev bind the RRE, and when this capacity is mutationally abolished, Rev-mediated nuclear export is adversely affected [32,42,54,55]. The stoichiometry of the saturated Rev-RRE complex, however, remains a subject of some debate. Reported binding ratios range from of 5:1 to 13:1 Rev/RRE, with an 8:1 ratio most often represented in early work in this area [11,32,33,39,56–63]. A study of particular note utilized surface plasmon resonance and a 244-nt RRE to measure Rev binding kinetics, with results indicating that up to 10 Rev molecules could bind the RRE before the complex became saturated [63]. Moreover, while the kinetics of the first four binding events suggest that Rev binding is sequential and specific, the subsequent Rev binding appeared to occur non-specifically. Most recently, a Rev/RRE ratio of 6:1 was determined for a 242-nt RRE using size exclusion chromatography [39]. The hexameric-Rev-RRE in this study was also shown to migrate as a discrete species by native polyacrylamide gel electrophoresis, suggesting structural uniformity, while complexes containing the Rev Leu18Gln/Leu60Arg multimerization mutant migrated as a broad, diffuse band to a position consistent with a Rev/RRE ratio of approximately 3–4:1.

Biochemical and biophysical experiments, together with single complex FRET measurements, demonstrate that Rev assembles on the RRE one molecule at a time [2,64], binding initiates at stem-loop IIB [11,12,14,15,31] and the binding of successive Rev molecules is cooperative [29]. Nuclease protection analysis further indicates that Rev binding occurs principally on stem loop IIB and stem I of the RRE, as other structural motifs remained susceptible to nuclease cleavage in the presence of Rev [32]. Although a great deal is now known about how Rev interacts with itself and with RNA, the precise sequence of events required for Rev assembly, the positioning of individual Rev molecules on the RNA and the 3D structure of the saturated Rev-RRE complex remain unknown.

From the 3D structure of a 232-nt RRE obtained using molecular modeling in conjunction with SAXS, a model for Rev assembly was proposed in which 8 Rev molecules are coaxially arranged along one face of the A-like structure of the truncated RRE (Figure 6A) [51]. Support from this proposal comes primarily from observations that the high affinity Rev binding sites on IIB and stem-loop I are separated by ∼55 Å, matching the approximate separation between distal portions of the ARMs in the T/T Rev dimer crystal structure [39].

However, the model also suggests direct binding of Rev to stem loops III/IV and V in higher order complexes and does not involve distal regions of stem I in Rev assembly. Both of these suppositions are inconsistent with prior nuclease protection analysis [32]. The high resolution Rev-dimer-RNA co-crystal structure [27], in which tandem NLS/RBSs bind at adjacent sites on the same helix likewise does not support the IIB-I bridging model of Rev assembly suggested by Wang and colleagues.

As noted previously, SAXS analysis of the full-length RRE suggests that the distal segment of stem I folds back on itself over one face of the A-like RRE structure [30]. Probing experiments further show highly variable sensitivity to chemical acylation at select regions of the RRE in the presence and absence of Rev. These area are designated Regions 1–3, and correspond, respectively, to (i) IIB and the stem loop II junction; (ii) the central junction and proximal purine rich internal loop of stem I; and (iii) a sequence of three adjacent, purine rich bulges located near the center of stem I. These regions of variable acylation sensitivity reportedly mark Rev binding sites, with Region 1 containing sites included in the Rev-dimer-RNA crystal structure [27].

Based on these data, a model for Rev assembly was presented in which Rev dimers bind sequentially at Regions 1–3 (Figure 6B). As in both previous models, assembly is proposed to initiate at the IIB nucleation site of Region 1. Moreover, as would be predicted by the jellyfish model, Rev binding at Regions 1 and 2 is temporally coupled, likely due to their spatial proximity. Unlike in the jellyfish model, however, this intriguing new model suggests that the 4-Rev complex subsequently undergoes an induced-fit conformational change to accommodate a third Rev dimer binding at Region 3, which is not adjacent to Regions 1 and 2 in the RNA secondary structure but is brought into proximity by tertiary RNA interactions.

Specific Rev-Rev associations within the hexameric complex are not otherwise detailed in this model, and the relative positioning and orientation of individual Rev molecules in the fully assembled complex is likewise not predicted. Moreover, it is worth noting that in seeming contradiction to this model, a hexameric Rev-RRE complex has been reported to assemble on a 242-nt truncated RRE that lacks almost all of the Domain 3 Rev binding site(s) [29].

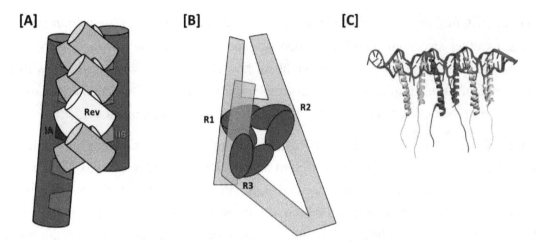

Figure 6. Models for assembly of HIV-1 Rev on the RRE: (**A**) "Bridging" model. The first two Rev in the complex bind to IIB and IA high affinity sites on the A-like RRE, as well as to each other, thereby forming a Rev dimer that "bridges" the distance separating two sub-structures. This proposal is based largely upon the observations that the distances separating the apices of the Rev arms in the T/T dimer crystal structure and the IIB and IA motifs modeled to fit the HIV-1 RRE SAXS envelope are both approximately 55 Å. After this initial Rev dimer-RRE complex is formed, assembly is proposed to propagate in both directions along the A-like RRE structure and involve stem loops III/IV, V and more distal portions of stem I; (**B**) Stem I "loop-back" model. Based on analyses using SAXS and time resolved SHAPE, it has been proposed that following relatively rapid assembly of Rev dimers at R1 and R2 (R1, Region 1—IIB and stem II junction; R2, Region 2—central junction and IA), the remaining dimer in the hexameric Rev complex assembles at R3 (Region 3—a series of purine-rich bulges in a more distant segment of stem I). RNA tertiary interactions bring the stem I terminus into proximity with the RRE central junction (even in the absence of Rev), and the last Rev dimer is accommodated in the fully assembled complex by an induced fit mechanism of conformational sampling; (**C**) Jellyfish model. Rev initially binds to the IIB high-affinity site, after which five additional Rev molecules assemble on the RRE *via* a series of consecutive T/T and H/H multimerization domain interactions. These Rev-Rev interactions are facilitated by concomitant Rev ARM binding at adjacent regions on the RRE RNA, such as is observed in the T/T Rev dimer-RNA crystal structure. The six *C*-terminal domains in this putative Rev-RRE complex would be expected to project in a common direction, which would in turn facilitate NES binding at the Crm1 dimer interface and promote nuclear export.

In the jellyfish model, Rev assembly extends from the IIB nucleation site through the RRE central junction and into the proximal segment of stem I, including stem IA (Figure 6C). Size exclusion chromatography suggests that coordinated assembly stops after six Rev molecules have been bound in the complex [29]. Rev-Rev and Rev-RNA interactions in such a complex would be consistent with the recent Rev-dimer RNA crystal structure [27], but a stable Rev hexamer would likely require similarly flexible H/H interface. More specifically, it would be necessary that the angles formed by the ARMs in adjacent Rev molecules are relatively narrow at both T/T and H/H interfaces, allowing for sequential binding at widened major grooves in adjacent segments of a contiguous RNA helix. Such an arrangement would be consistent with nuclease protection studies that suggest that Rev binds at IIB and along stem I [32], and also predict that the Rev effector domains align along the same face of the RRE.

Some RRE mutations have been shown to alter both RNA secondary structure and the capacity of the RRE to mediate nuclear export of HIV-1 RNA, yet do not appear to affect the binding kinetics or stoichiometry of Rev binding [2,46]. Such observations may be explained by aberrant Rev-RRE complexes that assemble with normal kinetics but where the relative orientations of Rev molecules are affected. Such subtle distinctions among complexes would be invisible in most biochemical assays, yet may substantially affect Rev-RRE function in a cellular context. In a recent study, the structure of a nuclear export complex containing Rev, the RRE and a Crm1 dimer was elucidated using single particle electron microscopy [65]. In this elegant work, the authors demonstrate that the Rev-RRE complex binds at the Crm1 dimer interface. It is suggested that each Crm1 subunit is bound by three NES domains of the hexameric Rev, although there is sufficient diversity among the complexes to allow for other binding stoichiometries. Crm1 binding appears to require a functional NES, as Rev bearing a dominant-negative M10 mutation [66] in the NES did not bind Crm1 under any condition. Although the resolution of these images is insufficient to definitively verify the jellyfish arrangement of Rev on the RRE, the observed interactions are consistent with a hexameric Rev-RRE in which the NES domains are concentrated in a small area and oriented in a common direction.

6. Therapeutic Targeting of the HIV-1 RRE

Based on its crucial role in HIV replication, interrupting RRE function clearly offers an attractive avenue for therapeutic intervention. The notion of using a transdominant negative version of the HIV-1 Rev protein harboring mutations in its nuclear export signal (RevM10), while disrupting export of Rev-dependent viral transcripts [67], was later shown to promote acquisition of resistance-conferring mutations [2]. Early attempts to target the HIV-1 RRE with small molecules have demonstrated the usefulness of neomycin B and related aminoglycosides as structural probes [68,69]. However, the therapeutic potential of this class of antagonist faces the challenges of specificity and poor cellular uptake [70]. Thus, alternative approaches to target the Rev/RRE interaction are warranted.

Both structural and mutagenic analyses have shown that the α-helical arginine-rich motif (ARM) of Rev mediates a critical interaction with the RRE by inserting into the major groove of an asymmetric bulge in IIB. Based on these observations, Mills *et al.* [71] synthesized a series of ARM-like conformationally-constrained (*i.e.*, α-helical) peptidomimetics to antagonize the Rev/RRE interaction. Of 15 candidate peptidomimetics examined, one bound the RRE with an affinity equivalent to that of the Rev ARM (Kd $\sim$50 nM), while control experiments indicated that its unconstrained counterpart showed little specificity. An alternative approach has suggested targeting the RRE with an ARM peptide modified to incorporate reactive metal chelates [72,73]. This strategy envisions "catalytic" metallo-inhibitors, which, following destruction of their target biomolecule, would dissociate and circulate through the viral RNA population, and thus would not be required in saturating amounts to achieve maximum potency. Bifunctional metallo-inhibitors linking a high affinity targeting motif (the Rev ARM) with a metal chelate complex capable of damaging RNA in its vicinity *via* a variety of oxidative chemistries have been synthesized, and proof of concept was demonstrated by the ability of

Cu^{2+}-Gly-Gly-His-ARM complexes to selectively cleave RRE RNA *in vitro* [74] and *in vivo* [72]. An extension of this strategy has investigated the oxidative properties of Rev ARM peptides linked to metal chelators such as tetraazacyclododecane-tetraacetic acid (DOTA), diethylenetriaminepentaacetic acid (DTPA), ethylenediaminetetraacetic acid (EDTA) and nitrilotriacetic acid (NTA) [75]. All complexes retained high affinity binding to RRE IIB (0.2–16 nM) and their Cu^{2+}-bound chelates efficiently induced RRE cleavage, with activity varying in the order Cu-NTA-Rev > Cu-DOTA-Rev > Cu-DTPA-Rev > Cu-EDTA-Rev. From a therapeutic perspective, oxidative damage of the RRE would result in dissociation of the metal chelate-ARM complex, which can be "reactivated" by reducing agents such as ascorbic acid or glutathione, whose concentrations are sufficiently high *in vivo*. Also, since the metal is extremely tightly chelated (Kd $\sim 10^{-15}$ M), the risk of toxicity due to metal ion leaching from the complex would be negligible.

Following initial reports that promoted multiple antigenic peptides for vaccine design [76], branched peptides comprising natural or unnatural amino acids, or combinations thereof, have been gaining attention as therapeutic agents, based on their potential for multivalent targeting of RNA and their resistance to proteolysis [77]. In screening a branched peptide boronic acid (BPBA) library whose boronic acid substituent was designed to mimic an acceptor for RNA 2′ OH groups and improve selectivity for RNA over DNA, Zhang *et al.* [78] identified BPBA1, a compound that bound RRE IIB with micromolar affinity and a 1:1 stoichiometry. Mutational studies supported the notion that the IIB tertiary structure was necessary for high affinity binding of BPBA1, while enzymatic footprinting highlighted several nuclease-insensitive regions in the presence of the branched peptide. Finally, studies with a fluorescent BPBA1 derivative suggest cellular uptake can be achieved, although suppression of HIV-1 replication remains to be established.

Of the currently available approaches, small molecules have superior delivery properties and can be "fine-tuned" through medicinal chemistry. With respect to the Rev/RRE interaction, several recent studies support their further development. By combining NMR spectroscopy and computational molecular dynamics, Steizer *et al.* have taken advantage of virtual screening to identify several compounds that antagonize the Tat/TAR interaction by interacting with nucleotides of both the apical loop and trinucleotide bulge [79]. Our laboratory has exploited small molecule microarrays (SMMs) with fluorescently-labeled, structured RNA motifs, where we identified a novel chemotype that binds the TAR hairpin with micromolar affinity, inhibits virus replication in culture and is not cytotoxic [80]. Finally, Informa, a computational approach to designing lead small molecules targeting RNA motifs based on sequence alone, has been used to identify a bioactive benzimidazole that induces apoptosis in cancer cells by sequestering the nuclease processing site of pre-micro RNA-96 [81]. Extending this approach to target the RRE would seem a logical next step.

Although this section has concentrated on small molecule inhibition of the Rev/RRE axis by targeting the viral *cis*-acting RNA, we should not rule out the possibility of targeting the HIV Rev protein to antagonize its interaction with host factors, thereby interrupting nucleocytoplasmic transport of unspliced and singly-spliced viral RNAs. In this respect, Campos *et al.* [82] have recently made the exciting discovery that the quinolin-2-amine ABX464 specifically targets Rev-dependent nuclear transport of HIV RNA without compromising cellular function. The lack of toxicity displayed by ABX464, combined with its ability to suppress viral load sustainably after treatment arrest, represent another powerful addition to the armament of HIV antivirals that will be needed until either a cross-clade vaccine or functional cure is achieved.

7. Summary and Perspective

The Rev-RRE interaction is vital to HIV replication and therefore constitutes an important axis for antiviral therapy. Although recent SAXS and crystallographic studies have greatly enhanced our understanding of these two viral components on a structural level, much remains to be learned—particularly regarding the details of Rev assembly and the overall structure of the nuclear export complex. Biochemical studies tell us that six Rev molecules cooperatively assemble on each RRE, and the resulting complex binds Crm1 at the dimer interface according to single particle electron microscopy. The latter study also suggests that while they share many common features, the structures of Rev_6-RRE-$Crm1_2$ complexes may not be completely uniform. Other factors, including the observed flexibility in the Rev T/T interface, potential flexibility in the H/H interface, and variability in RRE RNA sequence and secondary structure suggest this as well. It is conceivable, therefore, that select components of the "bridging", "loop-back" and jellyfish models of Rev assembly may all contribute to formation of the nuclear export complex in the context of the HIV-infected cell.

Acknowledgments

Stuart F. J. Le Grice and Jason W. Rausch were supported by the Intramural Research Program of the National Cancer Institute, National Institutes of Health, Department of Health and Human Services, USA.

Author Contributions

Jason W. Rausch and Stuart F. J. Le Grice co-wrote the manuscript.

References

1. Bray, M.; Prasad, S.; Dubay, J.W.; Hunter, E.; Jeang, K.T.; Rekosh, D.; Hammarskjold, M.L. A small element from the mason-pfizer monkey virus genome makes human immunodeficiency virus type 1 expression and replication Rev-independent. *Proc. Natl. Acad. Sci. USA* **1994**, *91*, 1256–1260. [CrossRef] [PubMed]
2. Legiewicz, M.; Badorrek, C.S.; Turner, K.B.; Fabris, D.; Hamm, T.E.; Rekosh, D.; Hammarskjold, M.L.; le Grice, S.F. Resistance to RevM10 inhibition reflects a conformational switch in the HIV-1 Rev response element. *Proc. Natl. Acad. Sci. USA* **2008**, *105*, 14365–14370. [CrossRef] [PubMed]
3. Pilkington, G.R.; Purzycka, K.J.; Bear, J.; Le Grice, S.F.; Felber, B.K. Gammaretrovirus mRNA expression is mediated by a novel, bipartite post-transcriptional regulatory element. *Nucl. Acids Res.* **2014**, *42*, 11092–11106. [CrossRef] [PubMed]
4. Lindtner, S.; Felber, B.K.; Kjems, J. An element in the 3′ untranslated region of human LINE-1

retrotransposon mRNA binds NXF1(TAP) and can function as a nuclear export element. *RNA* **2002**, *8*, 345–356. [CrossRef] [PubMed]

5. Malim, M.H.; Hauber, J.; Le, S.Y.; Maizel, J.V.; Cullen, B.R. The HIV-1 Rev trans-activator acts through a structured target sequence to activate nuclear export of unspliced viral mRNA. *Nature* **1989**, *338*, 254–257. [CrossRef] [PubMed]

6. Cullen, B.R. HIV-1 auxiliary proteins: Making connections in a dying cell. *Cell* **1998**, *93*, 685–692. [CrossRef]

7. Hope, T.J. The ins and outs of HIV Rev. *Arch. Biochem. Biophys.* **1999**, *365*, 186–191. [CrossRef] [PubMed]

8. Pollard, V.W.; Malim, M.H. The HIV-1 Rev protein. *Annu. Rev. Microbiol.* **1998**, *52*, 491–532. [CrossRef] [PubMed]

9. Vercruysse, T.; Daelemans, D. HIV-1 Rev multimerization: Mechanism and insights. *Curr. HIV Res.* **2013**, *11*, 623–634. [CrossRef] [PubMed]

10. Fernandes, J.; Jayaraman, B.; Frankel, A. The HIV-1 Rev response element: An RNA scaffold that directs the cooperative assembly of a homo-oligomeric ribonucleoprotein complex. *RNA Biol.* **2012**, *9*, 6–11. [CrossRef] [PubMed]

11. Heaphy, S.; Finch, J.T.; Gait, M.J.; Karn, J.; Singh, M. Human immunodeficiency virus type 1 regulator of virion expression, Rev, forms nucleoprotein filaments after binding to a purine-rich "bubble" located within the Rev-responsive region of viral mrnas. *Proc. Natl. Acad. Sci. USA* **1991**, *88*, 7366–7370. [CrossRef] [PubMed]

12. Iwai, S.; Pritchard, C.; Mann, D.A.; Karn, J.; Gait, M.J. Recognition of the high affinity binding site in Rev-response element RNA by the human immunodeficiency virus type-1 Rev protein. *Nucl. Acids Res.* **1992**, *20*, 6465–6472. [CrossRef] [PubMed]

13. Malim, M.H.; Tiley, L.S.; McCarn, D.F.; Rusche, J.R.; Hauber, J.; Cullen, B.R. HIV-1 structural gene expression requires binding of the Rev trans-activator to its RNA target sequence. *Cell* **1990**, *60*, 675–683. [CrossRef]

14. Olsen, H.S.; Nelbock, P.; Cochrane, A.W.; Rosen, C.A. Secondary structure is the major determinant for interaction of HIV Rev protein with RNA. *Science* **1990**, *247*, 845–848. [CrossRef] [PubMed]

15. Tiley, L.S.; Malim, M.H.; Tewary, H.K.; Stockley, P.G.; Cullen, B.R. Identification of a high-affinity RNA-binding site for the human immunodeficiency virus type 1 Rev protein. *Proc. Natl. Acad. Sci. USA* **1992**, *89*, 758–762. [CrossRef] [PubMed]

16. Battiste, J.L.; Mao, H.; Rao, N.S.; Tan, R.; Muhandiram, D.R.; Kay, L.E.; Frankel, A.D.; Williamson, J.R. A helix-RNA major groove recognition in an HIV-1 Rev peptide-RRE RNA complex. *Science* **1996**, *273*, 1547–1551. [CrossRef] [PubMed]

17. Bartel, D.P.; Zapp, M.L.; Green, M.R.; Szostak, J.W. HIV-1 Rev regulation involves recognition of non-watson-crick base pairs in viral RNA. *Cell* **1991**, *67*, 529–536. [CrossRef]

18. Battiste, J.L.; Tan, R.; Frankel, A.D.; Williamson, J.R. Binding of an HIV Rev peptide to Rev responsive element RNA induces formation of purine-purine base pairs. *Biochemistry* **1994**, *33*, 2741–2747. [CrossRef] [PubMed]

19. Peterson, R.D.; Bartel, D.P.; Szostak, J.W.; Horvath, S.J.; Feigon, J. 1 h NMR studies of the high-affinity Rev binding site of the Rev responsive element of HIV-1 mRNA: Base pairing in the core binding element. *Biochemistry* **1994**, *33*, 5357–5366. [CrossRef] [PubMed]

20. Williamson, J.R.; Battiste, J.L.; Mao, H.; Frankel, A.D. Interaction of HIV Rev peptides with the Rev response element RNA. *Nucl. Acids Symp. Ser.* **1995**, 46–48.

21. Tan, R.; Chen, L.; Buettner, J.A.; Hudson, D.; Frankel, A.D. RNA recognition by an isolated α helix. *Cell* **1993**, *73*, 1031–1040. [CrossRef]

22. Tan, R.; Frankel, A.D. Costabilization of peptide and RNA structure in an HIV Rev peptide-Rre complex. *Biochemistry* **1994**, *33*, 14579–14585. [CrossRef] [PubMed]

23. Ye, X.; Gorin, A.; Ellington, A.D.; Patel, D.J. Deep penetration of an α-helix into a widened RNA major groove in the HIV-1 Rev peptide-rna aptamer complex. *Nat. Struct. Biol.* **1996**, *3*, 1026–1033. [CrossRef] [PubMed]

24. Hung, L.W.; Holbrook, E.L.; Holbrook, S.R. The crystal structure of the Rev binding element of HIV-1 reveals novel base pairing and conformational variability. *Proc. Natl. Acad. Sci. USA* **2000**, *97*, 5107–5112. [CrossRef] [PubMed]

25. Ippolito, J.A.; Steitz, T.A. The structure of the HIV-1 rre high affinity rev binding site at 1.6 a resolution. *J. Mol. Biol.* **2000**, *295*, 711–717. [CrossRef] [PubMed]

26. Peterson, R.D.; Feigon, J. Structural change in Rev responsive element RNA of HIV-1 on binding Rev peptide. *J. Mol. Biol.* **1996**, *264*, 863–877. [CrossRef] [PubMed]

27. Jayaraman, B.; Crosby, D.C.; Homer, C.; Ribeiro, I.; Mavor, D.; Frankel, A.D. RNA-directed remodeling of the HIV-1 protein rev orchestrates assembly of the Rev-Rev response element complex. *eLife* **2014**, *3*, e04120. [CrossRef] [PubMed]

28. Zemmel, R.W.; Kelley, A.C.; Karn, J.; Butler, P.J. Flexible regions of RNA structure facilitate co-operative Rev assembly on the Rev-response element. *J. Mol. Biol.* **1996**, *258*, 763–777. [CrossRef] [PubMed]

29. Daugherty, M.D.; D'Orso, I.; Frankel, A.D. A solution to limited genomic capacity: Using adaptable binding surfaces to assemble the functional HIV Rev oligomer on RNA. *Mol. Cell* **2008**, *31*, 824–834. [CrossRef] [PubMed]

30. Bai, Y.; Tambe, A.; Zhou, K.; Doudna, J.A. RNA-guided assembly of Rev-RRE nuclear export complexes. *eLife* **2014**, *3*, e03656. [CrossRef] [PubMed]

31. Malim, M.H.; Cullen, B.R. HIV-1 structural gene expression requires the binding of multiple Rev monomers to the viral RRE: Implications for HIV-1 latency. *Cell* **1991**, *65*, 241–248. [CrossRef]

32. Mann, D.A.; Mikaelian, I.; Zemmel, R.W.; Green, S.M.; Lowe, A.D.; Kimura, T.; Singh, M.; Butler, P.J.; Gait, M.J.; Karn, J. A molecular rheostat. Co-operative rev binding to stem I of the Rev-response element modulates human immunodeficiency virus type-1 late gene expression. *J. Mol. Biol.* **1994**, *241*, 193–207. [CrossRef] [PubMed]

33. Wingfield, P.T.; Stahl, S.J.; Payton, M.A.; Venkatesan, S.; Misra, M.; Steven, A.C. HIV-1 REV expressed in recombinant escherichia coli: Purification, polymerization, and conformational properties. *Biochemistry* **1991**, *30*, 7527–7534. [CrossRef] [PubMed]

34. Zapp, M.L.; Green, M.R. Sequence-specific rna binding by the HIV-1 Rev protein. *Nature* **1989**, *342*, 714–716. [CrossRef] [PubMed]

35. Havlin, R.H.; Blanco, F.J.; Tycko, R. Constraints on protein structure in HIV-1 Rev and Rev-RNA supramolecular assemblies from two-dimensional solid state nuclear magnetic resonance. *Biochemistry* **2007**, *46*, 3586–3593. [CrossRef] [PubMed]

36. Watts, N.R.; Misra, M.; Wingfield, P.T.; Stahl, S.J.; Cheng, N.; Trus, B.L.; Steven, A.C.; Williams, R.W. Three-dimensional structure of HIV-1 Rev protein filaments. *J. Struct. Biol.* **1998**, *121*, 41–52. [CrossRef] [PubMed]

37. Jain, C.; Belasco, J.G. Structural model for the cooperative assembly of HIV-1 Rev multimers on the RRE as deduced from analysis of assembly-defective mutants. *Mol. Cell* **2001**, *7*, 603–614. [CrossRef]

38. DiMattia, M.A.; Watts, N.R.; Stahl, S.J.; Rader, C.; Wingfield, P.T.; Stuart, D.I.; Steven, A.C.; Grimes, J.M. Implications of the HIV-1 Rev dimer structure at 3.2 a resolution for multimeric binding to the rev response element. *Proc. Natl. Acad. Sci. USA* **2010**, *107*, 5810–5814. [CrossRef] [PubMed]

39. Daugherty, M.D.; Liu, B.; Frankel, A.D. Structural basis for cooperative rna binding and export complex assembly by HIV Rev. *Nat. Struct. Mol. Biol.* **2010**, *17*, 1337–1342. [CrossRef] [PubMed]

40. Edgcomb, S.P.; Aschrafi, A.; Kompfner, E.; Williamson, J.R.; Gerace, L.; Hennig, M. Protein structure and oligomerization are important for the formation of export-competent HIV-1 Rev-RRE complexes. *Protein Sci.* **2008**, *17*, 420–430. [CrossRef] [PubMed]

41. Kjems, J.; Brown, M.; Chang, D.D.; Sharp, P.A. Structural analysis of the interaction between the human immunodeficiency virus Rev protein and the Rev response element. *Proc. Natl. Acad. Sci. USA* **1991**, *88*, 683–687. [CrossRef] [PubMed]

42. Huang, X.J.; Hope, T.J.; Bond, B.L.; McDonald, D.; Grahl, K.; Parslow, T.G. Minimal Rev-response element for type 1 human immunodeficiency virus. *J. Virol.* **1991**, *65*, 2131–2134. [PubMed]

43. Pollom, E.; Dang, K.K.; Potter, E.L.; Gorelick, R.J.; Burch, C.L.; Weeks, K.M.; Swanstrom, R. Comparison of SIV and HIV-1 genomic RNA structures reveals impact of sequence evolution on conserved and non-conserved structural motifs. *PLoS Pathogens* **2013**, *9*, e1003294. [CrossRef] [PubMed]

44. Watts, J.M.; Dang, K.K.; Gorelick, R.J.; Leonard, C.W.; Bess, J.W., Jr.; Swanstrom, R.; Burch, C.L.; Weeks, K.M. Architecture and secondary structure of an entire HIV-1 RNA genome. *Nat.* **2009**, *460*, 711–716. [CrossRef] [PubMed]

45. Kenyon, J.C.; Prestwood, L.J.; Le Grice, S.F.; Lever, A.M. In-gel probing of individual RNA conformers within a mixed population reveals a dimerization structural switch in the HIV-1 leader. *Nucl. Acids Res.* **2013**, *41*, e174. [CrossRef] [PubMed]

46. Sherpa, C.; Rausch, J.W.; SF, J.L.G.; Hammarskjold, M.L.; Rekosh, D. The HIV-1 Rev response element (RRE) adopts alternative conformations that promote different rates of virus replication. *Nucl. Acids Res.* **2015**, *43*, 4676–4686. [CrossRef] [PubMed]

47. Lusvarghi, S.; Sztuba-Solinska, J.; Purzycka, K.J.; Pauly, G.T.; Rausch, J.W.; Grice, S.F. The HIV-2 Rev-response element: Determining secondary structure and defining folding intermediates. *Nucl. Acids Res.* **2013**, *41*, 6637–6649. [CrossRef] [PubMed]

48. Cunyat, F.; Beerens, N.; Garcia, E.; Clotet, B.; Kjems, J.; Cabrera, C. Functional analyses reveal extensive RRE plasticity in primary HIV-1 sequences selected under selective pressure. *PLoS ONE* **2014**, *9*, e106299. [CrossRef] [PubMed]

49. Sloan, E.A.; Kearney, M.F.; Gray, L.R.; Anastos, K.; Daar, E.S.; Margolick, J.; Maldarelli, F.; Hammarskjold, M.L.; Rekosh, D. Limited nucleotide changes in the rev response element (RRE) during HIV-1 infection alter overall Rev-RRE activity and Rev multimerization. *J. Virol.* **2013**, *87*, 11173–11186. [CrossRef] [PubMed]

50. Popenda, M.; Szachniuk, M.; Antczak, M.; Purzycka, K.J.; Lukasiak, P.; Bartol, N.; Blazewicz, J.; Adamiak, R.W. Automated 3D structure composition for large RNAs. *Nucl. Acids Res.* **2012**, *40*, e112. [CrossRef] [PubMed]

51. Fang, X.; Wang, J.; OÊijCarroll, I.P.; Mitchell, M.; Zuo, X.; Wang, Y.; Yu, P.; Liu, Y.; Rausch, J.W.; Dyba, M.A.; *et al.* An unusual topological structure of the HIV-1 Rev response element. *Cell* **2013**, *155*, 594–605. [CrossRef] [PubMed]

52. Lucks, J.B.; Mortimer, S.A.; Trapnell, C.; Luo, S.; Aviran, S.; Schroth, G.P.; Pachter, L.; Doudna, J.A.; Arkin, A.P. Multiplexed rna structure characterization with selective 2′-hydroxyl acylation analyzed by primer extension sequencing (shape-seq). *Proc. Natl. Acad. Sci. USA* **2011**, *108*, 11063–11068. [CrossRef] [PubMed]

53. Merino, E.J.; Wilkinson, K.A.; Coughlan, J.L.; Weeks, K.M. Rna structure analysis at single nucleotide resolution by selective 2′-hydroxyl acylation and primer extension (shape). *J. Am. Chem. Soc.* **2005**, *127*, 4223–4231. [CrossRef] [PubMed]

54. Holland, S.M.; Chavez, M.; Gerstberger, S.; Venkatesan, S. A specific sequence with a bulged guanosine residue(s) in a stem-bulge-stem structure of Rev-responsive element RNA is required for trans activation by human immunodeficiency virus type 1 Rev. *J. Virol.* **1992**, *66*, 3699–3706. [PubMed]

55. Kjems, J.; Sharp, P.A. The basic domain of rev from human immunodeficiency virus type 1 specifically blocks the entry of u4/u6.U5 small nuclear ribonucleoprotein in spliceosome assembly. *J. Virol.* **1993**, *67*, 4769–4776. [PubMed]

56. Brice, P.C.; Kelley, A.C.; Butler, P.J. Sensitive *in vitro* analysis of HIV-1 Rev multimerization. *Nucl. Acids Res.* **1999**, *27*, 2080–2085. [CrossRef] [PubMed]

57. Cook, K.S.; Fisk, G.J.; Hauber, J.; Usman, N.; Daly, T.J.; Rusche, J.R. Characterization of HIV-1 Rev protein: Binding stoichiometry and minimal rna substrate. *Nucl. Acids Res.* **1991**, *19*, 1577–1583. [CrossRef] [PubMed]

58. Daly, T.J.; Cook, K.S.; Gray, G.S.; Maione, T.E.; Rusche, J.R. Specific binding of HIV-1 recombinant Rev protein to the rev-responsive element *in vitro*. *Nature* **1989**, *342*, 816–819. [CrossRef] [PubMed]

59. Daly, T.J.; Doten, R.C.; Rennert, P.; Auer, M.; Jaksche, H.; Donner, A.; Fisk, G.; Rusche, J.R. Biochemical characterization of binding of multiple HIV-1 Rev monomeric proteins to the Rev responsive element. *Biochemistry* **1993**, *32*, 10497–10505. [CrossRef] [PubMed]

60. Holland, S.M.; Ahmad, N.; Maitra, R.K.; Wingfield, P.; Venkatesan, S. Human immunodeficiency virus Rev protein recognizes a target sequence in Rev-responsive element RNA within the context of rna secondary structure. *J. Virol.* **1990**, *64*, 5966–5975. [PubMed]

61. Jeong, K.S.; Nam, Y.S.; Venkatesan, S. Deletions near the *N*-terminus of HIV-1 Rev reduce RNA binding affinity and dominantly interfere with Rev function irrespective of the RNA target. *Arch. Virol.* **2000**, *145*, 2443–2467. [CrossRef] [PubMed]

62. Pallesen, J.; Dong, M.; Besenbacher, F.; Kjems, J. Structure of the HIV-1 Rev response element alone and in complex with regulator of virion (Rev) studied by atomic force microscopy. *FEBS J.* **2009**, *276*, 4223–4232. [CrossRef] [PubMed]

63. Van Ryk, D.I.; Venkatesan, S. Real-time kinetics of HIV-1 Rev-Rev response element interactions. Definition of minimal binding sites on RNA and protein and stoichiometric analysis. *J. Biol. Chem.* **1999**, *274*, 17452–17463. [CrossRef] [PubMed]

64. Pond, S.J.; Ridgeway, W.K.; Robertson, R.; Wang, J.; Millar, D.P. HIV-1 Rev protein assembles on viral RNA one molecule at a time. *Proc. Natl. Acad. Sci. USA* **2009**, *106*, 1404–1408. [CrossRef] [PubMed]

65. Booth, D.S.; Cheng, Y.; Frankel, A.D. The export receptor crm1 forms a dimer to promote nuclear export of HIV RNA. *eLife* **2014**, *3*, e04121. [CrossRef] [PubMed]

66. Malim, M.H.; Bohnlein, S.; Hauber, J.; Cullen, B.R. Functional dissection of the HIV-1 Rev trans-activator—Derivation of a trans-dominant repressor of rev function. *Cell* **1989**, *58*, 205–214. [CrossRef]

67. Bevec, D.; Dobrovnik, M.; Hauber, J.; Bohnlein, E. Inhibition of human immunodeficiency virus type 1 replication in human T cells by retroviral-mediated gene transfer of a dominant-negative Rev trans-activator. *Proc. Natl. Acad. Sci. USA* **1992**, *89*, 9870–9874. [CrossRef] [PubMed]

68. Chittapragada, M.; Roberts, S.; Ham, Y.W. Aminoglycosides: Molecular insights on the recognition of RNA and aminoglycoside mimics. *Perspect. Med. Chem.* **2009**, *3*, 21–37.

69. Tok, J.B.; Dunn, L.J.; Des Jean, R.C. Binding of dimeric aminoglycosides to the HIV-1 Rev responsive element (RRE) rna construct. *Bioorganic Med. Chem. Lett.* **2001**, *11*, 1127–1131. [CrossRef]

70. Ahn, D.G.; Shim, S.B.; Moon, J.E.; Kim, J.H.; Kim, S.J.; Oh, J.W. Interference of hepatitis C virus replication in cell culture by antisense peptide nucleic acids targeting the X-RNA. *J. Viral Hepat.* **2011**, *18*, e298–306. [CrossRef] [PubMed]

71. Mills, N.L.; Daugherty, M.D.; Frankel, A.D.; Guy, R.K. An α-helical peptidomimetic inhibitor of the HIV-1 Rev-RRE interaction. *J. Am. Chem. Soc.* **2006**, *128*, 3496–3497. [CrossRef] [PubMed]

72. Jin, Y.; Cowan, J.A. Cellular activity of Rev response element rna targeting metallopeptides. *J. Biol. Inorganic Chem.* **2007**, *12*, 637–644. [CrossRef] [PubMed]

73. Jin, Y.; Lewis, M.A.; Gokhale, N.H.; Long, E.C.; Cowan, J.A. Influence of stereochemistry and redox potentials on the single- and double-strand DNA cleavage efficiency of Cu(ii) and Ni(ii) Lys-Gly-His-derived atcun metallopeptides. *J. Am. Chem. Soc.* **2007**, *129*, 8353–8361. [CrossRef] [PubMed]

74. Jin, Y.; Cowan, J.A. Targeted cleavage of HIV Rev response element rna by metallopeptide complexes. *J. Am. Chem. Soc.* **2006**, *128*, 410–411. [CrossRef] [PubMed]

75. Joyner, J.C.; Cowan, J.A. Target-directed catalytic metallodrugs. *Braz. J. Med. Biol. Res.* **2013**, *46*, 465–485. [CrossRef] [PubMed]

76. Tam, J.P. Synthetic peptide vaccine design: Synthesis and properties of a high-density multiple antigenic peptide system. *Proc. Natl. Acad. Sci. USA* **1988**, *85*, 5409–5413. [CrossRef] [PubMed]

77. Bryson, D.I.; Zhang, W.; McLendon, P.M.; Reineke, T.M.; Santos, W.L. Toward targeting RNA structure: Branched peptides as cell-permeable ligands to tar RNA. *ACS Chem. Biol.* **2012**, *7*, 210–217. [CrossRef] [PubMed]

78. Zhang, W.; Bryson, D.I.; Crumpton, J.B.; Wynn, J.; Santos, W.L. Targeting folded RNA: A branched peptide boronic acid that binds to a large surface area of HIV-1 Rre RNA. *Organic Biomol. Chem.* **2013**, *11*, 6263–6271. [CrossRef] [PubMed]

79. Stelzer, A.C.; Frank, A.T.; Kratz, J.D.; Swanson, M.D.; Gonzalez-Hernandez, M.J.; Lee, J.; Andricioaei, I.; Markovitz, D.M.; Al-Hashimi, H.M. Discovery of selective bioactive small molecules by targeting an rna dynamic ensemble. *Nat. Chem. Biol.* **2011**, *7*, 553–559. [CrossRef] [PubMed]

80. Sztuba-Solinska, J.; Shenoy, S.R.; Gareiss, P.; Krumpe, L.R.; Le Grice, S.F.; OÊijKeefe, B.R.; Schneekloth, J.S., Jr. Identification of biologically active, HIV tar RNA-binding small molecules using small molecule microarrays. *J. Am. Chem. Soc.* **2014**, *136*, 8402–8410. [CrossRef] [PubMed]

81. Velagapudi, S.P.; Gallo, S.M.; Disney, M.D. Sequence-based design of bioactive small molecules that target precursor micrornas. *Nat. Chem. Biol.* **2014**, *10*, 291–297. [CrossRef] [PubMed]

82. Campos, N.; Myburgh, R.; Garcel, A.; Vautrin, A.; Lapasset, L.; Nadal, E.S.; Mahuteau-Betzer, F.; Najman, R.; Fornarelli, P.; Tantale, K.; *et al.* Long lasting control of viral rebound with a new drug ABX464 targeting Rev—Mediated viral RNA biogenesis. *Retrovirology* **2015**, *12*. [CrossRef] [PubMed]

7

RNA-Dependent RNA Polymerases of Picornaviruses: From the Structure to Regulatory Mechanisms

Cristina Ferrer-Orta, Diego Ferrero and Núria Verdaguer *

Molecular Biology Institute of Barcelona (CSIC), Barcelona Science Park (PCB), Baldiri i Reixac 10, Barcelona E-08028, Spain; E-Mails: cfocri@ibmb.csic.es (C.F.-O.); dfecri@ibmb.csic.es (D.F.)

* Author to whom correspondence should be addressed; E-Mail: nvmcri@ibmb.csic.es

Academic Editor: David Boehr

Abstract: RNA viruses typically encode their own RNA-dependent RNA polymerase (RdRP) to ensure genome replication within the infected cells. RdRP function is critical not only for the virus life cycle but also for its adaptive potential. The combination of low fidelity of replication and the absence of proofreading and excision activities within the RdRPs result in high mutation frequencies that allow these viruses a rapid adaptation to changing environments. In this review, we summarize the current knowledge about structural and functional aspects on RdRP catalytic complexes, focused mainly in the *Picornaviridae* family. The structural data currently available from these viruses provided high-resolution snapshots for a range of conformational states associated to RNA template-primer binding, rNTP recognition, catalysis and chain translocation. As these enzymes are major targets for the development of antiviral compounds, such structural information is essential for the design of new therapies.

Keywords: viral replication; RNA-dependent RNA polymerase; positive-strand RNA viruses; picornaviruses; replication fidelity

1. Introduction

RNA dependent RNA polymerases (RdRPs) are the catalytic components of the RNA replication and transcription machineries and the central players in the life cycle of RNA viruses. RdRPs belong to the superfamily of template-directed nucleic acid polymerases, including DNA-dependent DNA polymerases (DdDP), DNA-dependent RNA polymerases and Reverse Transcriptases (RT). All these enzymes share a cupped right hand structure, including fingers, palms and thumb domains, and catalyze phosphodiester bond formation through a conserved two-metal ion mechanism [1]. A structural feature unique to RdRPs is the "closed-hand" conformation, in opposition to the "open-hand" found in other polynucleotide polymerases. This "close-hand" conformation is accomplished by interconnecting the finger and thumb domains through the N-terminal portion of the protein and several loops protruding from fingers, named the fingertips that completely encircle the active site of the enzyme [2,3]. In the prototypic RdRPs the closed "right hand" architecture encircles seven motifs (A to G) conserved in sequence and structure (Figure 1), playing critical roles in substrate recognition and catalysis. Three well-defined channels have been identified in the RdRP structures, serving as: the entry path for template (template channel) and for nucleoside triphosphates (NTP channel) and the exit path for the dsRNA product (central channel) (Figure 1B).

The *Picornaviridae* family is one of the largest virus families known, including many important human and animal pathogens. Picornaviruses are non-enveloped RNA viruses possessing a single-stranded RNA genome (7–8 kb) of positive polarity, with a small peptide (VPg; from 19 to 26 amino acids long) linked to its 5'-end. Their genomes have a long highly structured 5' nontranslated region (NTR), a single large open reading frame (ORF) and a short 3' NTR, terminated with a poly(A) tail. The ORF is translated in the cytoplasm of the host cell into a polyprotein, which is proteolytically processed by viral proteases to release the structural proteins (VP1-4), needed to assemble virus capsids and the nonstructural proteins (2A-2B-2C-3A-3B-3C^{pro}-3D^{pol} and in some genera L) as well as some stable precursors necessary for virus replication in host cells [4]. The picornavirus genome is replicated via a negative-sense RNA intermediate by the viral RdRP, named 3D^{pol}. This enzyme uses VPg (the product of 3B) as a primer to initiate the replication process. The structure and function of 3D^{pol} has been studied extensively in the past decades and, to date, the 3D^{pol} crystal structures have been reported for six different members of the enterovirus genus [poliovirus (PV), coxsackievirus B3 (CVB3), enterovirus 71 (EV71) and the human rhinoviruses HRV1B, HRV14, and HRV16], for the aphthovirus FMDV and for the cardiovirus EMCV, either isolated or bound to different substrates [5–15]. These structures provided insights into both initiation of RNA synthesis and the replication elongation processes. Furthermore, mutational analyses in PV and FMDV also have demonstrated that some substitutions in residues located far from the active site, in particular at the polymerase N terminus, have significant effects on catalysis and fidelity. All of these observations suggest that nucleotide binding and incorporation are modulated by a long-distance network of interactions [5,16–22].

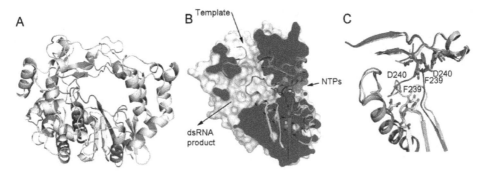

Figure 1. *Cont.*

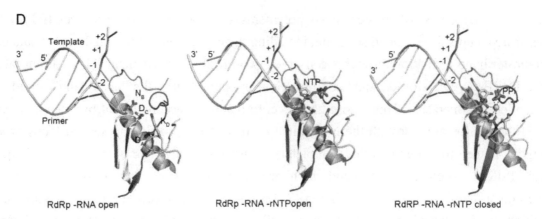

Figure 1. Overall structure of a viral RdRP. (**A**) Ribbon representation of a typical picornaviral RdRP (model from the cardiovirus EMCV 3D^{pol}, PDB id. 4NZ0). The seven conserved motifs are indicated in different colours: motif A, red; motif B, green; motif C, yellow; motif D, sand; motif E, cyan; motif F, blue; motif G, pink; (**B**) Lateral view of a surface representation of the enzyme (grey) that has been cut to expose the three channels that are the entry and exit sites of the different substrates and reaction products. The structural elements that support motifs A–G are also shown as ribbons. This panel also shows the organization of the palm sub-domain with motif A shown in two alternative conformations: the standard conformation (PDB id. 4NZ0) found in the apo-form of most crystallized 3D^{pol} proteins and the altered conformation found int the tetragonal crystal form of the EMCV enzyme (PDB id. 4NYZ). The alterations affect mainly Asp240, the amino acid in charge of incoming ribonucleotide triphosphate (rNTP) selection, and the neighboring Phe239 that move ~10 Å away from its position in the enzyme catalytic cavity directed towards the entrance of the nucleotide channel, approaching to motif F; (**C**) Close up of the structural superimposition of the two alternative conformations of the EMCV motif A; (**D**) The PV replication-elongation complexes. Sequential structures illustrating the movement of the different palm residues from a binary PV 3D^{pol}-RNA open complex (left) to an open 3D^{pol}-RNA-rNTP ternary complex (middle) where the incoming rNTP is positioned in the active site for catalysis and, a closed ternary complex (right) after nucleotide incorporation and pyrophosphate (PPi) release. The residues D$_A$ (involved in rNTP selection through an interaction with the 2′ hydroxyl group), D$_C$ (the catalytic aspartate of motif C), K$_D$ (the general acid residue of motif D that can coordinate the export of the PPi group) and N$_B$ (a conserved Asn of motif B, interacting with D$_A$) have been highlighted as sticks. The different structures correspond to the 3D^{pol}-RNA (PDB id. 3OL6), 3D^{pol}-RNA-CTP open complex (PDB id. 3OLB) and 3D^{pol}-RNA-CTP closed complex (PDB id. 3OL7) structures of PV elongation complexes, respectively [7].

2. VPg Binding to 3D^{pol} and Initiation of RNA Synthesis

Correct initiation of RNA synthesis is essential for the integrity of the viral genome. There are two main mechanisms by which viral replication can be initiated: primer-independent or *de novo*, and primer-dependent initiation, reviewed in [23]. Briefly, in the *de novo* synthesis, one initiation nucleotide provides the 3'-hydroxyl for the addition of the next nucleotide whereas the primer dependent initiation requires the use of either an oligonucleotide or a protein primer as provider of the hydroxyl nucleophile. It is remarkable that the RdRPs of viruses that initiate replication using *de novo* mechanisms (*i.e.*, members of the *Flaviviridae* family) share a number of unique features which ensure efficient and accurate initiation, including a larger thumb subdomain containing structural elements that fill most of the active site cavity, providing a support platform for the primer nucleotides (reviewed in [24,25]). These protrusions also serve as a physical barrier preventing chain elongation. Therefore, it is necessary that the initiation platform can move away from the active site after stabilizing the initiation complex, allowing the transition from initiation to elongation [26–28]. By contrast, the members of the *Picornaviridae* and *Caliciviridae* families use exclusively the protein-primed mechanism of initiation. The RNA polymerases of these viruses use VPg as primer for both minus and plus strand RNA synthesis. These enzymes display a more accessible active site cavity, enabling them to accommodate the primer protein for RNA synthesis [13,29].

The very first step in protein-primed initiation in picornavirus is the uridylylation of a strictly conserved tyrosine residue of VPg [30]. In this process, the viral polymerase 3D catalyzes the binding of two uridine monophosphate (UMP) molecules to the hydroxyl group of this tyrosine using as template a cis-replicating element (cre) that is located at different positions of the RNA genome, in the different picornaviridae genera (see [31] for an extensive review). The nucleotidylylation reaction can, however, also occur in a template-independent manner in other viruses, for example in caliciviruses [32].

The picornaviral proteins VPg, 3D^{pol} and 3C^{pro}, alone or in the 3CD precursor form, together with the viral RNA *cre* elements comprise the so-called "VPg uridylylation complex" responsible for VPg uridylylation *in vivo*. Despite extensive structural and biochemical studies, there are several different models for the interactions established between VPg and 3D^{pol} or 3CD in the uridylylation complex and the precise mechanism of uridylylation remains uncertain [31].

Biochemical and structural studies performed for different members of the family: PV [33], HRV16 [12], FMDV [13], CVB3 [8] and EV71 [34] revealed three distinct VPg binding sites on 3D^{pol} (Figure 2). Strikingly, whereas most picornaviruses express only a single VPg protein, FMDV possesses three similar but not identical copies of VPg: VPg1, VPg2 and VPg3 [35], all of which are found linked to viral RNA [36]. Although not all the copies are needed to maintain infectivity [37,38], there are no reports of naturally occurring FMDV strains with fewer than three copies of 3B, suggesting that there is a strong selective pressure towards maintaining this redundancy [39,40].

The structure of two complexes between FMDV 3D^{pol} and VPg1: 3D^{pol}-VPg1 and 3D^{pol}-VPg1-UMP revealed a number of residues in the active site cleft of the polymerase involved in VPg binding and in the uridylylation reaction. Functional assays performed with 3D^{pol} and VPg mutants with substitutions in residues involved in interactions, according to the structural data, showed important effects in uridylylation [13]. The position of VPg in complex with the FMDV 3D^{pol} is remarkably similar to the position of the primer and RNA duplex product found in the complex with the same enzyme [14,15]. Most of the amino acids of 3D^{pol} seen in contact with the RNA primer and duplex product are also involved in interactions with VPg. In fact, the structure shows how the VPg protein accesses the active site cavity from the front of the molecule through the large RNA binding cleft mimicking, at least in part, the RNA molecule. The N-terminal position of VPg projects into the active site where the hydroxyl moiety of the residue Tyr3 is in good proximity to the catalytic aspartates of motifs A and C (Figure 2). In this position, Tyr3 essentially mimics the 3' OH of the primer strand during the RNA elongation. Conserved residues in the fingers, palm and thumb domains of the polymerase were identified as being responsible for stabilizing VPg in its binding cavity. In the $3D^{pol}$-VPg1-UMP complex, the hydroxyl group of Tyr3 side chain was found covalently attached to the α-phosphate moiety of the uridine-monophosphate (UMP) molecule [13]. The positively charged residues of motif F also participate in the uridylylation process, stabilizing Tyr3 and UMP in a proper conformation for the reaction [13] (Figure 2B). Two divalent cations, together with the catalytic aspartic acid residues of motifs A and C, participate in VPg uridylylation. All the observed structural features suggest a conservation of the catalytic mechanism described for all polymerases [1]. Mutational analyses at the conserved FMDV $3D^{pol}$ residues that strongly interact with VPg in the crystal structures show a drastic defect in VPg uridylylation [13]. This "front-loading" model for VPg binding, compatible with a *cis* mechanism of VPg uridylylation was further supported by the crystal structures of HRV16 $3D^{pol}$ [12] and of the PV 3CD precursor [20]. In the latter structure, the extensive crystal packing contacts found between symmetry-related 3CD molecules and the proximity of the N-terminal domain of 3C to the VPg binding site, in the way that VPg was positioned in the FMDV 3D-VPg complex, suggests a possible role of the contacting interfaces in forming and regulating the VPg uridylylation complex during the initiation of viral replication [20].

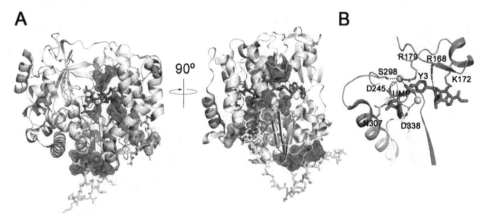

Figure 2. (**A**) Comparison of identified VPg binding sites in picornavirus $3D^{pols}$. Because all reported structures of picornavirus $3D^{pols}$ share high structural similarities, we used the structure of the FMDV $3D^{pol}$ (PDB id. 2F8E, [2]) as a representative model in this figure and colored it with a light-blue cartoon. The bound VPgs with FMDV [13], CVB3 (PDB id. 3CDW, [8]) and EV71 (PDB id. IKA4, [34]) are shown as red, yellow and cyan sticks, respectively. The residues for VPg (or 3AB) binding in PV (F377, R379, E382 and V391) [33], FMDV (E166, R179, D338, D387 and R388) [13] and EV71 (T313, F314, I317, L319, D320, Y335 and P337) [34] $3D^{pol}$ are represented as a surface and colored as sand, magenta and blue, respectively, in the cartoon representation; (**B**) Details of the interactions described in the active site of the FMDV $3D^{pol}$ during the uridylylation reaction. The VPg residues and UMP covalently linked are shown in red sticks, the divalent cations are shown as light-blue spheres and the amino acids involved in uridylylation reaction are shown as sticks. The motifs A, B, C and F are colored in red, green, yellow and blue, respectively.

After nucleotidylylation of VPg, some structural rearrangements of the $3D^{pol}$ will follow, marking the transition from initiation to the elongation phase of RNA synthesis. There is experimental evidence supporting possible structural differences in $3D^{pol}$ when involved in the priming reaction *vs.* elongation of RNA. *i.e.*, the nucleoside analog 5-Fluorouridine triphosphate (FUTP) is a potent inhibitor of VPg uridylylation but not of RNA elongation [19]. Furthermore, for the poliovirus, it has been suggested, based on the structure of the FMDV $3D^{pol}$-VPg complex that a conserved Asn in the polymerase motif B (Asn297) (equivalent to Asn307 in FMDV) interacts with the 3'-OH of the incoming nucleotide in the uridylylation complex, but with the 2'-OH in the elongation complex [41]. The interaction of Asn307 with the incoming rNTP 2' OH during RNA elongation has been confirmed in a number of picornavirus elongation complexes [6,7,15].

A second binding site for VPg was found in the structure of the CVB3 polymerase [8]. The VPg fragment solved, corresponding to the C-terminal half of the peptide, was bound at the base of the thumb sub-domain in an orientation that did not allow its uridylylation by its own carrier $3D^{pol}$ (Figure 2A). This VPg binding, partially agreed with previous data reported for PV, showing that a number of amino acids located in motif E of poliovirus (PV) were required for VPg or their precursor, 3AB, binding and affecting VPg uridylylation [33]. In light of these results, the authors proposed that VPg bound at this position was either uridylylated by another 3D molecule or that it played a stabilizing role within the uridylylation complex [8].

Finally, a third VPg binding site was discovered in the structure of the EV71 $3D^{pol}$-VPg complex. In this complex, VPg is anchored at the bottom of the palm domain of the polymerase, showing a V-shape conformation that crosses from the front side of the catalytic site to the back side of the enzyme (Figure 2A). Similarly to that occurring in the previously studied viruses, the mutational analyses of the interacting residues evidenced a reduced binding of VPg to the EV71 $3D^{pol}$ affecting uridylylation [34,42]. Additional experiments performed by the same authors, by mixing the VPg-binding-defective mutants with catalytic defective mutant of the EV71 polymerase, demonstrated *trans* complementation of VPg uridylylation *in vitro*. However, the structure of the EV71 $3D^{pol}$-VPg complex showed that the VPg Tyr3 is buried at the base of the polymerase palm indicating that a conformational change should occur to expose the side chain of Tyr3 for uridylylation.

Taking into account the important sequence homology between the picornaviral VPg sequences [13,43] and the high similarities existing in the $3D^{pol}$ structures, it is tempting to speculate that the three different VPg binding sites observed in the different crystal structures might reflect distinct binding positions of VPg to both $3D^{pol}$ or its precursor 3CD at different stages of the virus replication initiation process. As discussed above, the FMDV genome codes for three VPg molecules, all of them are present in naturally occurring viruses [39,40]. A global picture of the assembly of the multicomponent complexes involved in replication initiation in Picornaviruses and its regulation would require the structural and functional analyses of higher order complexes formed by the polymerase $3D^{pol}$, involving different proteins or protein precursors (VPg, 3AB, $3C^{pro}$, 3CD) and RNA templates. Such structures would shed new light on the molecular events underlying the initiation of RNA genome replication in these viruses, and should provide crucial information for the design of new antiviral strategies.

3. Structural Elements Regulating Replication Elongation in RdRPs

3.1. Subtle Conformational Changes Associated with Nucleotide Selection and Active Site Closure for Catalysis

The replication elongation process can be roughly divided in three steps, including nucleotide selection, phosphodiester bond formation and translocation to the next nucleotide for the subsequent round of nucleotide addition. Structural biology has been crucial to elucidate the structural changes associated with each phase of catalysis for a wide number of polymerases [25,29,44,45]. Extensive biochemical and structural studies in the A- and B-families of open-hand nucleic acid polymerases indicate that the movement of an α-helix of the fingers sub-domain would control each step of the nucleotide-addition cycle and facilitates translocation along the template after catalysis [46,47]. In contrast in the "closed-hand" RdRPs, the presence of the fingertips encircling the catalytic site constrains the fingers' movement relative to the thumb, avoiding the swinging movement of the fingers that is associated with active site closure in open-hand polymerases.

The structures of a large number of RdRP-RNA-rNTP replication-elongation complexes determined, for different members of both, the *Picornaviridae* and the *Caliciviridae* families, have provided important insights into the structural changes associated to each catalytic step [3,6,7,25,29,48–51]. These structures indicate that RdRPs use subtle rearrangements within the palm domain to fully structure the active site for catalysis upon correct rNTP binding. Briefly, in a first state, the RdRP-RNA complex in the absence of an incoming rNTP, shows an open conformation of the polymerase active site characterized by a partially formed three-stranded β-sheet of the palm domain motifs A and C (Figure 1B,C). A second feature characterizing this state is the presence of a fully prepositioned templating nucleotide (t+1), sitting above the active site and stacked on the upstream duplex and ready for the binding of the incoming rNTP (Figure 1C). In a second state, an incoming rNTP reaches the active site and establishes base-pairing interactions with the template base (t+1), but catalysis has not taken place because the catalytic site is still in open conformation. The third state occurs after binding of the correct nucleotide to the active site. This binding induces the realignment of β-strands in the palm subdomain that includes the structural motifs A and C, resulting in the repositioning of the motif A aspartate to allow interactions with both metal ions required for catalysis [6,7,48,49]. It is important to remark that the active site closure in RdRPs is triggered by the correct nucleotide binding, suggesting that nucleotide selection in RdRPs is a simple process in which base-pairing interactions control the initial rNTP binding geometry and the resulting positioning of the ribose hydroxyls becomes the major checkpoint for proper incoming nucleotide selection. In particular, two residues within motif B (Ser and Asn) and, a second Asp residue at the C-terminus of motif A, strictly conserved among picorna- and caliciviruses form the ribose binding pocket (Figure 1C). The interactions between these amino acids and the ribose hydroxyl groups of the incoming rNTP would stabilize the subtle restructuring of the palm domain that results in the formation of a functional active site. An incorrect nucleotide can bind, but its ribose hydroxyls will not be correctly positioned for active site closure, in consequence, the incorporation efficiency will be reduced.

In addition, growing amounts of data indicate that the conformational changes in motif D determine both efficiency and fidelity of nucleotide addition [52,53]. Biochemical studies of nucleotidyl transfer reactions catalyzed by RdRPs, RTs and single subunit DNA polymerases made the unexpected observation that two protons, not just one, are transferred during the reaction and that the second proton derives from a basic amino acid of the polymerase (termed general acid) and is transferred to the PPi leaving group [52]. PPi protonation is not essential but contributes from 50-fold to 1000-fold the rate of nucleotide addition. Additional data from mutagenesis and kinetics of nucleotide incorporation showed that the general acid was a lysine located in the conserved motif D of RdRPs and RTs [53]. Solution NMR studies were used to analyze the changes that occurred during nucleotide addition. A methionine within motif D, located in the vicinity of the conserved lysine, was found to be a very informative probe for the positioning of the motif. Authors have found that the constitution of the catalytically competent elongation complex (RdRP-RNA-NTP) required the formation of a hydrogen bond between the β-phosphate of incoming rNTP and motif D lysine [54]. The protonation state of this lysine was also observed to be critical to achieve the closed conformation of the active site. Moreover, the ability of motif D to reach the catalytically competent conformation seems to be hindered by the binding of an incorrect nucleotide and this ability continues to be affected after nucleotide misincorporation. Indeed, the NMR data correlates the conformational dynamics of motif D to the efficiency and fidelity of nucleotide incorporation [54].

A number of structures of enterovirus elongation complexes have been trapped in a post-catalysis state in which the newly incorporated nucleotide and the pyrophosphate product are still in the active site, and the active site opens its conformation [6,7]. This state will be followed by the translocation of the polymerase by one base pair to position the next templating nucleotide in the active site for the next round of nucleotide addition. A number of crystal structures have also been solved in a post-translocation state [6,7,15]; however, no structures of translocation intermediates are currently available for RdRPs and the precise mechanism is not yet known. Recent data also suggest that a conserved lysine residue within motif D can coordinate the export of the PPi group from the active site once catalysis has taken place [16], thereby triggering the end of the reaction cycle and allowing enzyme translocation. Very recently, structural and functional data in enteroviruses indicate that steric clashes between the motif-B loop and the template RNA would also promote translocation [51] (see below).

Viral RdRPs are considered to be low fidelity enzymes, generating mutations that allow the rapid adaptation of these viruses to different tissue types and host cells. Based on X-ray data of CVB3 and PV catalytic complexes, the laboratories of Peersen and Vignuzzi engineered different point mutations in these viral polymerases and studied their effects on *in vitro* nucleotide discrimination as well as virus growth and genome replication fidelity. Data obtained revealed that the palm mutations produced the greatest effects on *in vitro* nucleotide discrimination and that these effects appeared strongly correlated with elongation rates and *in vivo* mutation frequencies, with faster polymerases having lower fidelity. These findings suggested that picornaviral polymerases have retained a unique palm domain-based active-site closure as a mechanism for the evolutionary fine-tuning replication fidelity and provide a pathway for developing live attenuated virus vaccines based on engineering the polymerase to reduce virus fitness [55,56].

Finally, recent structural data on calicivirus RdRPs have provided evidence of new conformational changes occurring during catalysis. Structural comparison of the human Norovirus (NV) RdRP determined in multiple crystal forms, in the presence and absence of divalent metal cations, nucleoside triphosphates, inhibitors and primer-template duplex RNAs, revealed that in addition to the active site closure, the NV RdRP exhibits two additional key changes: a rotation of the central helix in the thumb domain by 22°, resulting in the formation of a binding pocket for the primer RNA strand and the displacement of the C-terminal tail region away from the central active-site groove, which also allows the rotation of the thumb helix [57].

3.2. Conformational Plasticity of the Motif B Loop Regulates RdRP Activity

The central role of the motif B loop in template binding, incoming nucleotide recognition and correct positioning of the sugar in the ribose-binding pocket was evidenced in the first structures of the FMDV catalytic complexes [15,29]. This loop, connecting the base of the middle finger to the α-helix of motif B, is able to adopt different conformations when it binds to different template and incoming nucleotides, being one the most flexible elements of the active site of RdRPs in picornaviruses, as well as in other viral families (reviewed in [50]). In fact, structural comparisons evidenced large movements of the B-loop, ranging from a conformation in which the loop is packed against the fingers domain leaving the active site cavity fully accessible for template entry, to a configuration where the loop protrudes towards the catalytic cavity and clashes with the template RNA (Figure 3). The key residue of this flexible region is a strictly conserved glycine, which acts as a hinge for the movement. The critical role of the B-loop dynamics was previously anticipated by site-directed mutagenesis in the picornavirus EMCV. Substitutions of the hinge glycine in $3D^{pol}$ essentially abolished RNA synthesis *in vitro* [58]. Furthermore, additional interactions established between the B-loop and the RNA phosphodiester backbone of the upstream duplex, between the −1 and −2 nucleotides, prompted researchers to hypothesize a function of the loop in modulating polymerase activity through effects on translocation [6,7,15].

Extensive structural and functional work in PV, using several polymerase mutants, harboring substitutions within the B-loop sequence Ser288-Gly289-Cys290, evidenced a major role of these residues in the $3D^{pol}$ catalytic cycle [51]. The work concluded that the B-loop is able to adopt mainly three major conformations, termed *in/up*, *in/down* and *out/down* and that each alternative conformation is important for the correct NTP binding and for the post-catalysis translocation step. The terms *in/out* refer to the loop conformations, packed against the fingers (in), or protruding into the catalytic cavity (out) (Figure 3A). The designation is based on whether the residue Cys290 is buried "in" a hydrophobic pocket directly behind the loop or is "out" of the pocket and exposed to solvent [59]. Moreover, the Sholders and Peersen work highlighted the role of the PV Ser288. The side chain of this residue may also adopt two alternative conformations: pointing "up" toward the ring finger and away from the active site, or "down" pointing toward the active site. These authors propose a sequential model for the structural changes occurred during the PV RdRP catalytic cycle where initially, in the apo-form structure of PV $3D^{pol}$, the B-loop is in the *in/up* conformation, allowing rNTP entry. Equivalent conformations of the B-loop were observed in the apo-forms of $3D^{pol}$ in FMDV [14], Rhinovirus [11] and Coxsackievirus [8,9]. On nucleotide binding, the Ser288 flips down toward the active site (*in/down* conformation), establishing a hydrogen bond with the Aspartic acid residue of motif A, involved in the selection of the incoming ribonucleotide (Asp_A). It is also important to remark that in the previous step, Asp_A was hydrogen bonded to a conserved Asn from the motif B (Asn_B) (Figure 1C,D). Following these changes, new rearrangements occur, including the realignment of the palm motif A, required for catalysis. After the phosphodiester bond formation, the Sholders and Peersen model proposes a movement of the B-loop from "in/down" to "out/down" configuration, resulting in a steric clash between the B-loop and the backbone of the RNA template strand that would facilitate translocation along the RNA and prevent backtracking after translocation. In addition, the finding that the PV 3D^{pol} G289A mutant was able to catalyze single nucleotide addition but was defective for processive elongation provided more evidence of the crucial role of this glycine that confers the flexibility required for the loop movements.

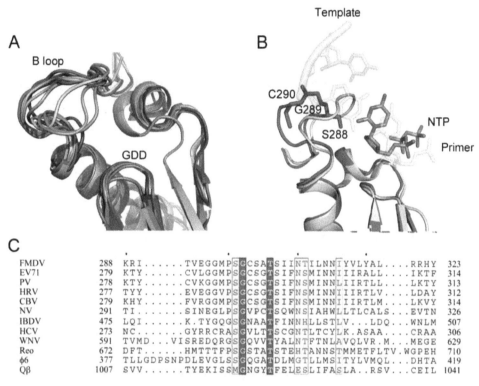

Figure 3. The conformational changes in the B-loop of RdRPs. (**A**) Superposition of the different conformations described for the B-loop. Motifs A, B and C are represented as ribbons and colored in gray tones. The B-loop is shown in different colors for each observed conformation, from red (up) to blue (down): NV NS7, Mg^{2+} bound (PDB id. 1SH3, chain A) chocolate; PV apo-form (PDB id, 1RA6) red; FMDV-RNA complex (PDB id, 1WNE) magenta; PV C290V mutant (PDB id. 4NLP) light-orange; IBDV VP1 + VP3 C-ter peptide (PDB id. 2R70) orange; NV NS7, Mg^{2+} bound (PDB id. 1SH3, chain B) yellow; PV C290F mutant (PDB id. 4NLQ) light-blue; IBDV VP1 apoform (PDB id. 2PUS) slate; FMDV K18E mutant (PDB id. 4WYL) blue; (**B**) Superimposition of the up conformation of PV apo-form (PDB id. 1RA6) red and the down conformation of PV C290F mutant (PDB id. 4NLQ) slate with the RNA template-primer and an incoming rNTP molecule are represented as sticks in semi-transparent representation; (**C**) Sequence alignment of the B-loop region of all the RdRPs from dsRNA and +ssRNA.

The high sequence and structural conservation of the B-loop among viral polymerases (Figure 3C) strongly suggest that its conformational dynamics would be a common feature of the RNA-dependent RNA polymerases from positive-strand RNA viruses.

3.3. Unusual Conformation of Motif A Captured in the Structure of the Cardiovirus EMCV 3D^{pol}

The crystal structure of the EMCV 3D^{pol} in its unbound state has been recently solved in two different crystal forms [60]. As expected, the overall architecture of the enzyme was similar to that of the known RdRPs of other members of the *Picornaviridae* family. However, structural comparisons revealed a large reorganization of the active-site cavity in one of the crystal forms. The rearrangement affects mainly the C-terminal loop of motif A, containing the aspartic acid residue involved in incoming rNTP selection (Asp240 in EMCV) (Figure 1B). The heart of this conformational change is that the Asp240 neighbor residue, Phe239, made a drastic movement whereby it is popped out of a hydrophobic pocket in the palm domain to participate in an intriguing set of cation-π interactions in the fingers domain at the edge of the NTP entry tunnel [60]. Another important feature of this altered active site conformation is that the active site of the enzyme was captured in a closed-like state, with the β-sheet supporting motif A totally formed and the catalytic Asp235 positioned in front of the motif C Asp333 (Figure 1B). This active-site conformation has never been observed before in the absence of RNA and a correctly base-paired rNTP. In addition, the N-terminal Gly1 residue was moved out of its binding site, anchored in the fingers domain, toward a totally exposed orientation in the polymerase surface. This is extremely intriguing because like most of the picornaviral RdRPs, the EMCV enzyme is only active when cleaved from the polyprotein to generate an N-terminus with a Gly1 residue [61]. Those observations prompt to hypothesize that this EMCV 3D^{pol} crystal form might represent the structure of the inactive form of the enzyme that would be present in the precursor protein, where Gly1 cannot be buried because is not a terminal residue. However, this hypothesis seems to be in conflict with the structural data currently available for the poliovirus precursor 3CD that also showed Gly1 exposed as part of the flexible linker joining 3C and 3D^{pol} but with the active site in the standard open conformation found in the pre-catalytic complexes [20]. The possible role of the altered conformation of the motif A loop, in particular, of the positioning of the rNTP binding residue Asp240 at the edge of the rNTP entry tunnel, is another mystery to decipher in order to gain insight on the regulation of this enzyme activity.

3.4. Conformational Flexibility in the Template Channel

The template channel also exhibits substantial flexibility as visualized by comparing the X-ray structures of different replication elongation complexes [6,7,14,15,19,29,62], as well as predicted by molecular dynamics simulations in a number of picornaviral polymerases [22]. This flexibility appears directly correlated with the role of this channel in driving the template nucleotides toward the catalytic cavity. Of particular importance is the flexible nature of a region included at the 3D^{pol} N-terminus (residues 16 to 20; FMDV numbering), lining the channel that appears to be interacting with the RNA near the single-strand/double-strand junction (Figure 4). The conformational changes occurring in this region would assist the movement of the template nucleotides at the +2 and +3 positions. Interestingly, structural comparisons of the wild type FMDV 3D^{pol} catalytic complexes showed that the basic side chain of Arg17 is involved in different interactions with the template nucleotide t+2 in all complexes analyzed [14,15]. In these complexes, the t+2 nucleotide points towards the active site cavity, stacked with the t+1 nucleotide that is located in the opening of the central cavity, in close contact with the motif B loop (Figure 4B). The equivalent residue of Arg17 in enteroviral polymerases is Pro20

(PV numbering). Structural comparisons of distinct enteroviruses elongation complexes show that Pro20 and its surrounding residues form a conserved pocket where the t+2 nucleotide binds [6,7] (Figure 4). This pocket found in the enterovirus 3D^{pol} seems to be a preformed structure that is also present in the unbound enzymes. In contrast, the FMDV wild type enzyme lacks a preformed pocket in the template channel and, as mentioned above, the t+2 nucleotide is oriented towards the active site cavity (Figure 4), constituting an important structural difference between the enterovirus and FMDV catalytic complexes. Surprisingly, the comparative structural analyses of FMDV 3D^{pol} mutants presenting alterations in RNA binding affinity and incoming nucleotide incorporation, including a remarkable increase or decrease in the incorporation of the nucleoside analog ribavirin, showed important movements in this polymerase region that result in the formation of distinct pockets where the t+2 nucleotide binds [19,62] (Figure 4).

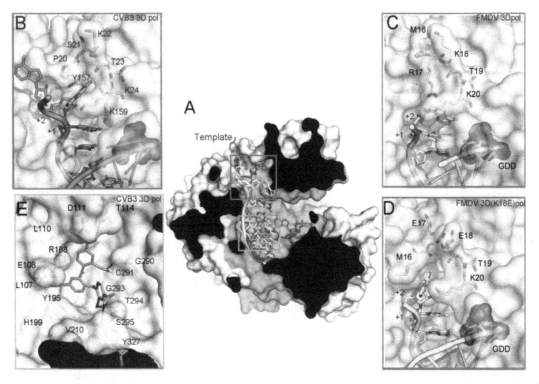

Figure 4. Structure and interactions in the template channel of a picornavirus 3D^{pol}. (**A**) The structure of the CVB3 3D^{pol} (PDB id. 4K4Y) has been used as a model, the molecular surface of the polymerase is shown in grey with the acidic residues of the active site in red and the RNA depicted as a cartoon in orange and the FMDV RNA is superimposed in yellow. The non-nucleoside analogue inhibitor is also superimposed in green; (**B**) Structure and interactions in the template channel at the entrance of the active site of CVB3 3D^{pol} (PDB id. 4K4Y), the N-terminal residues 20–24 depicted as sticks in cyan and the RNA in orange and others residues involved in the binding RNA are represented as grey sticks; (**C**) The wild type FMDV 3D^{pol}-RNA complex (PDB id. 1WNE); and (**D**) the FMDV 3D^{pol} (K18E)-RNA complex (PDB id. 4WZM); (**E**) Interaction network between GPC-N114 and its binding pocket of CVB3 3D^{pol} represented by surfaces (PDB id. 4Y2A). The polymerase residues in direct contact with the inhibitor are shown with sticks in atom type color with carbon in slate and explicitly labeled. Hydrogen bonds are depicted as dashed lines.

Besides facilitating specific contacts with the RNA template, the dynamic nature of residues lining the template channel should permit the access of the t+1 nucleotide into the 3D catalytic site. The base of the template channel is built mainly by residues of the motif B loop. These B-loop residues are involved in interactions with the t+1 nucleotide in the active site, as well as with the incoming rNTP [15]. Putting together all data is tentative to speculate that the rearrangements in the template channel and the B-loop occur in a concerted manner and that these concerted changes serve to regulate both RNA replication processivity and fidelity.

4. Polymerase Oligomerization

Proteins can oligomerize through reversible associations mediated by electrostatic and hydrophobic interactions, hydrogen bonds or by covalent stabilization by disulfide bonds. RdRPs, the enzymes that exclusively belong to the RNA virus world are not the exception. In recent years, the X-ray and Cryo-electron microscopy (cryo-EM) analyses revealed the quaternary structures of a large number of RdRP oligomers, defining the critical residues that lead these associations. Otherwise, complementary biochemical analyses allowed deciphering of the functional roles in most of these arrangements.

RdRP-RdRP interactions to form dimers or higher order oligomers have been predominantly reported for (+) ssRNA viruses, including several Picornavirus [63–68], Flavivirus [68,69] and Calicivirus [67,70] enzymes, as well as, in RdRPs of plants [71] and Insect viruses [72]. The homo-interaction of RdRPs was also described in replicases of (−) ssRNA viruses such as influenza A virus [73], and in more distant dsRNA viruses like infectious pancreatic necrosis virus [74]. RdRP oligomerization has been predominantly observed *in vitro* during crystallization, probably produced by the high protein concentration, as well as by other environmental changes like pH or ionic strength. However, intracellular accumulation of oligomeric polymerases was also observed during viral infection of different RNA viruses including PV [64], Sendai virus [75], Rift Valley Fever virus [76] and norovirus [70], among others. These observations suggest that RdRP oligomerization can also be a natural event as a sort of post translational modification. The specificity for the dimerization/multimerization involves distinct surfaces depending on the enzyme, *i.e.*, HCV RdRP dimerization has been proposed to be mediated by the thumb domain [69], whereas, in PV, the polymerase fingers appear to be crucially involved [66]. Contacts between these domains during oligomer formation may cause small conformational changes that are transferred to the active site as an allosteric regulation or could even modify the accessibility of the substrate channels.

The first oligomerization state of an RdRP was described for PV and the nature of the molecular contacts at two different polymerase interfaces, termed I and II, were postulated from the first crystal structure of PV 3D^{pol} [63,64]. The Interface I derived from interactions between the front of the thumb subdomain of one molecule and the back of the palm subdomain of the neighbour molecule in the crystal (Figure 5) whereas interface II involved two N-terminal regions of the polymerase that appeared disordered in this structure [63]. Later on, Lyle *et al.*, using cryo-EM demonstrated that the purified PV 3D^{pol} was able to organize two-dimensional lattices and tubular arrangements formed by polymerase fibres [65] and, recently, the structure of these assemblies has been characterized at the pseudo-atomic level [77,78]. The planar lattices, forming a ribbon-like structure, consist of linear arrays of dimeric RdRPs supported by strong interactions through the interface-I as defined in the PV 3D^{pol} crystal structure [77]. The tubular structure is also formed via interface-I but is also assisted by a second set of interactions placed in interface-II, involving interactions between the fingertips of one molecule and the palm of its contacting neighbour. The fitting of the 3D^{pol} coordinates into the cryo-EM reconstructions showed that interface I connects adjacent dimers by head-to-tail contacts [78] (Figure 5). The relevance of a number of interface II residues in lattice formation was further confirmed by mutagenic analysis. In particular, mutations at residues Tyr32 and Ser438 involved interface II contacts both in planar and tubular array results in a disruption of PV 3D^{pol} lattice formation [77]. Furthermore, several lines of evidence suggested that the PV polymerase can change the conformation upon forming oligomers and, in the tubular assemblies, the porous nature of the polymerase lattice is likely to allow the participation of other viral and cellular proteins [78].

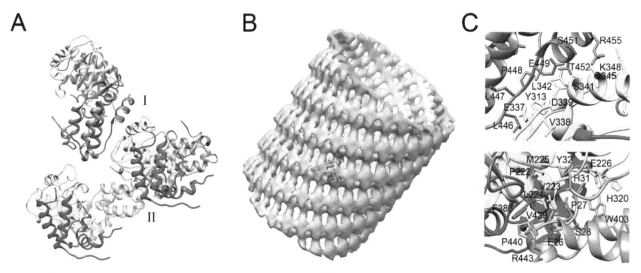

Figure 5. Oligomerization of the PV 3D^{pol}. (**A**) Polymerase-polymerase interactions mediated by interfaces I and II, explicitly marked (PDB id. 1RDR) in yellow palm subdomain, in red and blue fingers subdomain and in light colors thumb subdomain; (**B**) Volume map of the reconstructed 3D^{pol} tubes with the crystallographic model positioned inside. The volume map was reproduced from [78] (EM code emd2270); (**C**) Close up of the interface I (upper panel) and interfase II (bottom panel) in the oligomeric tubular array of PV 3D^{pol} according to [78].

All examples of high-order RdRP assemblies point out that oligomerization may be an advantageous feature providing a functional control, such as allosteric regulation in addition to increasing the stability against degradation and denaturation. The functioning of RdRPs in oligomeric arrays also has an additional advantage, to concentrate the reaction substrates in a physical place and dispose the active sites to use them iteratively.

5. Implications for Antiviral Drug Discovery

RdRP synthesize RNA using an RNA template. This biochemical activity, almost exclusive of RNA virures offers the opportunity to identify very selective inhibitors of this viral enzyme. Antiviral drugs targeting the RdRPs may either directly inhibit polymerase activity or essential interactions with the RNA template, or the RdRP-RdRP contacts promoting oligomerization, or interactions with other regulatory proteins. The detailed structural and mechanistic understanding of the conformational changes occurring during catalysis is essential not only for understanding of viral replication at the molecular level but also for the design of novel inhibitors capable of trapping the enzyme in specific conformational states. The Flaviviruses, Hepatitis C virus, Dengue virus and West Nile virus, as well as the calicivirus NV are clear examples of how much effort has been directed towards developing drugs that inhibit viral replication [79–82]. In general Direct-Acting Antivirals (DAAs), inhibiting RNA replication can be classified into two groups on the basis of their chemical structure and mechanisms of action: Nucleoside Analog (NA) inhibitors and Non-Nucleoside Inhibitors (NNIs). NAs target the active site of the polymerase and need to be converted by the host cell machinery to the corresponding nucleotides, which can either induce premature termination of RNA synthesis [83,84] or be incorporated by the viral polymerase into the nascent RNA, causing accumulation of mutations and contributing to virus extinction through lethal mutagenesis [85].

Conversely, NNIs bind mainly to allosteric pockets of the target polymerase causing alterations in the enzyme dynamics. They might either stabilize an inactive conformation or trap the enzyme in a functional conformation but impeding either the transition between initiation and elongation or the processivity of polymerase elongation [86,87].

Allosteric inhibitors directed against protein-RNA or protein-protein interactions involving viral polymerases are less explored as antiviral drugs. However, effective antiviral molecules that seem to inhibit interactions of the viral polymerase within the replicative complex have been for pestiviruses [88,89] and, a new compound against Dengue virus have been identified that appears to block the RdRP activity through binding to the RNA template channel [20].

In a very recent study, we have identified a novel non-nucleoside inhibitor of 3D^{pol}, the compound GPC-N114 (2,2′-[(4-chloro-1,2-phenylene)bis(oxy)]bis(5-nitro-benzonitrile), with broad-spectrum antiviral activity against both enteroviruses and cardioviruses [90]. The X-ray analysis of CVB3 3D^{pol}-GPC-N144 co-crystals revealed that the binding site of the compound was located at the junction of the palm and the fingers domains, partially overlapping with the binding site of the templating nucleotide (Figure 4E). The polymerase-inhibitor interactions involved different residues of the conserved motifs G, F, B and A, most found in direct contact with the RNA templates in all picornaviral 3D^{pol}-RNA complexes determined so far. Structural comparisons between unbound and GPC-N114 bound CVB3 3D^{pol} revealed that the polymerase did not undergo any major conformational change upon binding of the compound.

Surprisingly, GPC-N114-resistant enterovirus variants could not be obtained, but two EMCV resistance mutations (Met300Val and Ile303Val, in the motif B-loop) were readily selected in the presence of suboptimal concentration of GPC-N114 [90]. The reason for the inability of enteroviruses to develop resistance against GPC-N114 remains to be established. A possible explanation is that mutations that would confer resistance to GPC-N114 also impair binding of the template-primer, thereby preventing replication. Although the exact reason remains to be determined, the structural data suggest that, in contrast to most allosteric binding sites, the GPC-N114-binding cavity in enterovirus 3D^{pol} lacks the conformational plasticity required to develop resistance. In contrast, EMCV 3D^{pol} appears to be sufficiently plastic to allow for compound-resistance substitutions. As expected, structural comparisons showed high similarity between the CVB3 and the EMCV enzymes, but a major difference existed in the main interactions established between these polymerases with the inhibitor that could possibly underlie the differences observed in the emergence of resistant mutations. A key interaction of the CVB3 $3D^{pol}$-GPC-N114 was mediated by Tyr195 (Figure 4E). In contrast, the equivalent residue in EMCV $3D^{pol}$ is Ala, resulting in a weaker interaction with the compound. This weaker interaction together with an increment of the flexibility in the compound-binding area, induced by the B loop resistance mutations, might result in a decrease of GPC-N114 binding to the EMCV enzyme.

In summary, the identification of this novel drug-binding pocket in the picornaviral $3D^{pol}$ might serve as a starting point for the design of new antiviral compounds targeting the template-binding channel.

6. Conclusions

RNA-dependent RNA polymerases (RdRPs) play central roles in both transcription and viral genome replication. In picornaviruses, these functions are catalyzed by the virally encoded RdRP, termed $3D^{pol}$. $3D^{pol}$ also catalyzes the covalent binding of two UMP molecules to a tyrosine on the small protein VPg. Uridylylated VPg then serves as a protein primer for the initiation of RNA synthesis.

The ever growing availability of structures of picornaviral catalytic complexes provided an increasingly accurate picture of the functional steps and regulation events underliying viral RNA genome replication. Data currently available provides high-resolution pictures for a range of conformational states associated to template and primer recognition, VPg uridylylation, rNTP recognition and binding, catalysis and chain translocation. Such structural information is providing new insights into the fidelity of RNA replication, and for the design of antiviral compounds.

Protein primed mechanism of replication initiation mediated by VPg appears to be a process that involves more than one VPg binding site in $3D^{pol}$ possibly at different stages of the virus replication initiation process. Although the number of $3D^{pol}$-VPg structures available show individual snapshots of the process, to obtain a global picture of the assembly, regulation and dynamics of complete replication initiation complexes, requires further analyses of high order assemblies formed not only by the polymerase and VPg but also involving different viral and host proteins, protein precursors and RNA templates. Such structures should provide a more detailed view of the molecular events underliying the initiation of picornavirus genome replication.

The structures of a large number of picornavirus replication-elongation complexes captured subtle conformational changes associated with nucleotide selection and active site closure. Among these movements, the motif B-loop assists in the positioning of the template nucleotide in the active site. Binding of the correct nucleotide then induces the β-strands realignment in the palm subdomain and repositioning of the motif A aspartate for catalysis. Steric clashes between the motif B-loop and the template RNA would finally promote translocation.

Acknowledgements

Núria Verdaguer acknowledges funding from the Spanish Ministry of Economy and Competitiveness (BIO2011-24333).

References

1. Steitz, T.A. A mechanism for all polymerases. *Nature* **1998**, *391*, 231–232. [CrossRef] [PubMed]

2. Ferrer-Orta, C.; Arias, A.; Escarmis, C.; Verdaguer, N. A comparison of viral RNA-dependent RNA polymerases. *Curr. Opin. Struct. Biol.* **2006**, *16*, 27–34. [CrossRef] [PubMed]

3. Ng, K.K.; Arnold, J.J.; Cameron, C.E. Structure-function relationships among RNA-dependent RNA polymerases. *Curr. Top. Microbiol. Immunol.* **2008**, *320*, 137–156. [PubMed]

4. Wimmer, E.; Paul, A.V. The making of a picornavirus genome. In *The Picornavirus*; Ehrenfeld, E., Domingo, E., Ross, R.P., Eds.; ASM Press: Washington, DC, USA, 2010; pp. 33–55.

5. Thompson, A.A.; Peersen, O.B. Structural basis for proteolysis-dependent activation of the poliovirus RNA-dependent RNA polymerase. *EMBO J.* **2004**, *23*, 3462–3471. [CrossRef] [PubMed]

6. Gong, P.; Kortus, M.G.; Nix, J.C.; Davis, R.E.; Peersen, O.B. Structures of coxsackievirus, rhinovirus, and poliovirus polymerase elongation complexes solved by engineering RNA mediated crystal contacts. *PLoS ONE* **2013**, *8*, e60272. [CrossRef] [PubMed]

7. Gong, P.; Peersen, O.B. Structural basis for active site closure by the poliovirus RNA-dependent RNA polymerase. *Proc. Natl. Acad. Sci. USA* **2010**, *107*, 22505–22510. [CrossRef] [PubMed]

8. Gruez, A.; Selisko, B.; Roberts, M.; Bricogne, G.; Bussetta, C.; Jabafi, I.; Coutard, B.; de Palma, A.M.; Neyts, J.; Canard, B. The crystal structure of coxsackievirus B3 RNA-dependent RNA polymerase in complex with its protein primer VPg confirms the existence of a second VPg binding site on *Picornaviridae* polymerases. *J. Virol.* **2008**, *82*, 9577–9590. [CrossRef] [PubMed]

9. Campagnola, G.; Weygandt, M.; Scoggin, K.; Peersen, O. Crystal structure of coxsackievirus B3 $3D^{pol}$ highlights the functional importance of residue 5 in picornavirus polymerases. *J. Virol.* **2008**, *82*, 9458–9464. [CrossRef] [PubMed]

10. Wu, Y.; Lou, Z.; Miao, Y.; Yu, Y.; Dong, H.; Peng, W.; Bartlam, M.; Li, X.; Rao, Z. Structures of EV71 RNA-dependent RNA polymerase in complex with substrate and analogue provide a drug target against the hand-foot-and-mouth disease pandemic in China. *Protein Cell* **2010**, *1*, 491–500. [CrossRef] [PubMed]

11. Love, R.A.; Maegley, K.A.; Yu, X.; Ferre, R.A.; Lingardo, L.K.; Diehl, W.; Parge, H.E.; Dragovich, P.S.; Fuhrman, S.A. The crystal structure of the RNA-dependent RNA polymerase from human rhinovirus: A dual function target for common cold antiviral therapy. *Structure* **2004**, *12*, 1533–1544. [CrossRef] [PubMed]

12. Appleby, T.C.; Luecke, H.; Shim, J.H.; Wu, J.Z.; Cheney, I.W.; Zhong, W.; Vogeley, L.; Hong, Z.; Yao, N. Crystal structure of complete rhinovirus RNA polymerase suggests front loading of protein primer. *J. Virol.* **2005**, *79*, 277–288. [CrossRef] [PubMed]

13. Ferrer-Orta, C.; Arias, A.; Agudo, R.; Perez-Luque, R.; Escarmis, C.; Domingo, E.; Verdaguer, N. The structure of a protein primer-polymerase complex in the initiation of genome replication. *EMBO J.* **2006**, *25*, 880–888. [CrossRef] [PubMed]

14. Ferrer-Orta, C.; Arias, A.; Perez-Luque, R.; Escarmis, C.; Domingo, E.; Verdaguer, N. Structure of foot-and-mouth disease virus RNA-dependent RNA polymerase and its complex with a template-primer RNA. *J. Biol. Chem.* **2004**, *279*, 47212–47221. [CrossRef] [PubMed]

15. Ferrer-Orta, C.; Arias, A.; Perez-Luque, R.; Escarmis, C.; Domingo, E.; Verdaguer, N. Sequential structures provide insights into the fidelity of RNA replication. *Proc. Natl. Acad. Sci. USA* **2007**, *104*, 9463–9468. [CrossRef] [PubMed]

16. Shen, H.; Sun, H.; Li, G. What is the role of motif D in the nucleotide incorporation catalyzed by the RNA-dependent RNA polymerase from poliovirus? *PLoS Comput. Biol.* **2012**, *8*, e1002851. [CrossRef] [PubMed]

17. Pfeiffer, J.K.; Kirkegaard, K. A single mutation in poliovirus RNA-dependent RNA polymerase confers resistance to mutagenic nucleotide analogs via increased fidelity. *Proc. Natl. Acad. Sci. USA* **2003**, *100*, 7289–7294. [CrossRef] [PubMed]

18. Arnold, J.J.; Vignuzzi, M.; Stone, J.K.; Andino, R.; Cameron, C.E. Remote site control of an active site fidelity checkpoint in a viral RNA-dependent RNA polymerase. *J. Biol. Chem.* **2005**, *280*, 25706–25716. [CrossRef] [PubMed]

19. Agudo, R.; Ferrer-Orta, C.; Arias, A.; de la Higuera, I.; Perales, C.; Perez-Luque, R.; Verdaguer, N.; Domingo, E. A multi-step process of viral adaptation to a mutagenic nucleoside analogue by modulation of transition types leads to extinction-escape. *PLoS Pathog.* **2010**, *6*, e1001072. [CrossRef] [PubMed]

20. Marcotte, L.L.; Wass, A.B.; Gohara, D.W.; Pathak, H.B.; Arnold, J.J.; Filman, D.J.; Cameron, C.E.; Hogle, J.M. Crystal structure of poliovirus 3CD protein: Virally encoded protease and precursor to the RNA-dependent RNA polymerase. *J. Virol.* **2007**, *81*, 3583–3596. [CrossRef] [PubMed]

21. Ferrer-Orta, C.; Sierra, M.; Agudo, R.; de la Higuera, I.; Arias, A.; Perez-Luque, R.; Escarmis, C.; Domingo, E.; Verdaguer, N. Structure of foot-and-mouth disease virus mutant polymerases with reduced sensitivity to ribavirin. *J. Virol.* **2010**, *84*, 6188–6199. [CrossRef] [PubMed]

22. Moustafa, I.M.; Shen, H.; Morton, B.; Colina, C.M.; Cameron, C.E. Molecular dynamics simulations of viral RNA polymerases link conserved and correlated motions of functional elements to fidelity. *J. Mol. Biol.* **2011**, *410*, 159–181. [CrossRef] [PubMed]

23. Van Dijk, A.A.; Makeyev, E.V.; Bamford, D.H. Initiation of viral RNA-dependent RNA polymerization. *J. Gen. Virol.* **2004**, *85*, 1077–1093. [CrossRef] [PubMed]

24. Choi, K.H.; Rossmann, M.G. RNA-dependent RNA polymerases from *Flaviviridae*. *Curr. Opin. Struct. Biol.* **2009**, *19*, 746–751. [CrossRef] [PubMed]

25. Lescar, J.; Canard, B. RNA-dependent RNA polymerases from flaviviruses and *Picornaviridae*. *Curr. Opin. Struct. Biol.* **2009**, *19*, 759–767. [CrossRef] [PubMed]

26. Butcher, S.J.; Grimes, J.M.; Makeyev, E.V.; Bamford, D.H.; Stuart, D.I. A mechanism for initiating RNA-dependent RNA polymerization. *Nature* **2001**, *410*, 235–240. [CrossRef] [PubMed]

27. Mosley, R.T.; Edwards, T.E.; Murakami, E.; Lam, A.M.; Grice, R.L.; Du, J.; Sofia, M.J.; Furman, P.A.; Otto, M.J. Structure of hepatitis C virus polymerase in complex with primer-template RNA. *J. Virol.* **2012**, *86*, 6503–6511. [CrossRef] [PubMed]

28. Appleby, T.C.; Perry, J.K.; Murakami, E.; Barauskas, O.; Feng, J.; Cho, A.; Fox, D., 3rd; Wetmore, D.R.; McGrath, M.E.; Ray, A.S.; *et al.* Viral replication. Structural basis for RNA replication by the hepatitis C virus polymerase. *Science* **2015**, *347*, 771–775. [CrossRef] [PubMed]

29. Ferrer-Orta, C.; Agudo, R.; Domingo, E.; Verdaguer, N. Structural insights into replication

initiation and elongation processes by the FMDV RNA-dependent RNA polymerase. *Curr. Opin. Struct. Biol.* **2009**, *19*, 752–758. [CrossRef] [PubMed]

30. Paul, A.V.; van Boom, J.H.; Filippov, D.; Wimmer, E. Protein-primed RNA synthesis by purified poliovirus RNA polymerase. *Nature* **1998**, *393*, 280–284. [CrossRef] [PubMed]

31. Paul, A.V.; Wimmer, E. Initiation of protein-primed picornavirus RNA synthesis. *Virus Res.* **2015**, *206*, 12–26. [CrossRef] [PubMed]

32. Goodfellow, I. The genome-linked protein VPg of vertebrate viruses—A multifaceted protein. *Curr. Opin. Virol.* **2011**, *1*, 355–362. [CrossRef] [PubMed]

33. Lyle, J.M.; Clewell, A.; Richmond, K.; Richards, O.C.; Hope, D.A.; Schultz, S.C.; Kirkegaard, K. Similar structural basis for membrane localization and protein priming by an RNA-dependent RNA polymerase. *J. Biol. Chem.* **2002**, *277*, 16324–16331. [CrossRef] [PubMed]

34. Chen, C.; Wang, Y.; Shan, C.; Sun, Y.; Xu, P.; Zhou, H.; Yang, C.; Shi, P.Y.; Rao, Z.; Zhang, B.; *et al.* Crystal structure of enterovirus 71 RNA-dependent RNA polymerase complexed with its protein primer VPg: Implication for a trans mechanism of VPg uridylylation. *J. Virol.* **2013**, *87*, 5755–5768. [CrossRef] [PubMed]

35. Forss, S.; Schaller, H. A tandem repeat gene in a picornavirus. *Nucleic Acids Res.* **1982**, *10*, 6441–6450. [CrossRef] [PubMed]

36. King, A.M.; Sangar, D.V.; Harris, T.J.; Brown, F. Heterogeneity of the genome-linked protein of foot-and-mouth disease virus. *J. Virol.* **1980**, *34*, 627–634. [PubMed]

37. Falk, M.M.; Sobrino, F.; Beck, E. VPg gene amplification correlates with infective particle formation in foot-and-mouth disease virus. *J. Virol.* **1992**, *66*, 2251–2260. [PubMed]

38. Pacheco, J.M.; Henry, T.M.; O'Donnell, V.K.; Gregory, J.B.; Mason, P.W. Role of nonstructural proteins 3A and 3B in host range and pathogenicity of foot-and-mouth disease virus. *J. Virol.* **2003**, *77*, 13017–13027. [CrossRef] [PubMed]

39. Carrillo, C.; Lu, Z.; Borca, M.V.; Vagnozzi, A.; Kutish, G.F.; Rock, D.L. Genetic and phenotypic variation of foot-and-mouth disease virus during serial passages in a natural host. *J. Virol.* **2007**, *81*, 11341–11351. [CrossRef] [PubMed]

40. MacKenzie, J.S.; Slade, W.R.; Lake, J.; Priston, R.A.; Bisby, J.; Laing, S.; Newman, J. Temperature-sensitive mutants of foot-and-mouth disease virus: The isolation of mutants and observations on their properties and genetic recombination. *J. Gen. Virol.* **1975**, *27*, 61–70. [CrossRef] [PubMed]

41. Korneeva, V.S.; Cameron, C.E. Structure-function relationships of the viral RNA-dependent RNA polymerase: Fidelity, replication speed, and initiation mechanism determined by a residue in the ribose-binding pocket. *J. Biol. Chem.* **2007**, *282*, 16135–16145. [CrossRef] [PubMed]

42. Sun, Y.; Wang, Y.; Shan, C.; Chen, C.; Xu, P.; Song, M.; Zhou, H.; Yang, C.; Xu, W.; Shi, P.Y.; *et al.* Enterovirus 71 VPg uridylation uses a two-molecular mechanism of 3D polymerase. *J. Virol.* **2012**, *86*, 13662–13671. [CrossRef] [PubMed]

43. Sun, Y.; Guo, Y.; Lou, Z. Formation and working mechanism of the picornavirus VPg uridylylation complex. *Curr. Opin. Virol.* **2014**, *9*, 24–30. [CrossRef] [PubMed]

44. Berdis, A.J. Mechanisms of DNA polymerases. *Chem. Rev.* **2009**, *109*, 2862–2879. [CrossRef] [PubMed]

45. Steitz, T.A. Visualizing polynucleotide polymerase machines at work. *EMBO J.* **2006**, *25*, 3458–3468. [CrossRef] [PubMed]

46. Kornberg, R.D. The molecular basis of eukaryotic transcription. *Proc. Natl. Acad. Sci. USA* **2007**, *104*, 12955–12961. [CrossRef] [PubMed]

47. Steitz, T.A. The structural changes of T7 RNA polymerase from transcription initiation to elongation. *Curr. Opin. Struct. Biol.* **2009**, *19*, 683–690. [CrossRef] [PubMed]

48. Zamyatkin, D.F.; Parra, F.; Alonso, J.M.; Harki, D.A.; Peterson, B.R.; Grochulski, P.; Ng, K.K. Structural insights into mechanisms of catalysis and inhibition in Norwalk virus polymerase. *J. Biol. Chem.* **2008**, *283*, 7705–7712. [CrossRef] [PubMed]

49. Zamyatkin, D.F.; Parra, F.; Machin, A.; Grochulski, P.; Ng, K.K. Binding of 2′-amino-2′-deoxycytidine-5′-triphosphate to norovirus polymerase induces rearrangement of the active site. *J. Mol. Biol.* **2009**, *390*, 10–16. [CrossRef] [PubMed]

50. Garriga, D.; Ferrer-Orta, C.; Querol-Audi, J.; Oliva, B.; Verdaguer, N. Role of motif B loop in allosteric regulation of RNA-dependent RNA polymerization activity. *J. Mol. Biol.* **2013**, *425*, 2279–2287. [CrossRef] [PubMed]

51. Sholders, A.J.; Peersen, O.B. Distinct conformations of a putative translocation element in poliovirus polymerase. *J. Mol. Biol.* **2014**, *426*, 1407–1419. [CrossRef] [PubMed]

52. Castro, C.; Smidansky, E.; Maksimchuk, K.R.; Arnold, J.J.; Korneeva, V.S.; Gotte, M.; Konigsberg, W.; Cameron, C.E. Two proton transfers in the transition state for nucleotidyl transfer catalyzed by RNA- and DNA-dependent RNA and DNA polymerases. *Proc. Natl. Acad. Sci. USA* **2007**, *104*, 4267–4272. [CrossRef] [PubMed]

53. Castro, C.; Smidansky, E.D.; Arnold, J.J.; Maksimchuk, K.R.; Moustafa, I.; Uchida, A.; Gotte, M.; Konigsberg, W.; Cameron, C.E. Nucleic acid polymerases use a general acid for nucleotidyl transfer. *Nat. Struct. Mol. Biol.* **2009**, *16*, 212–218. [CrossRef] [PubMed]

54. Yang, X.; Smidansky, E.D.; Maksimchuk, K.R.; Lum, D.; Welch, J.L.; Arnold, J.J.; Cameron, C.E.; Boehr, D.D. Motif D of viral RNA-dependent RNA polymerases determines efficiency and fidelity of nucleotide addition. *Structure* **2012**, *20*, 1519–1527. [CrossRef] [PubMed]

55. Gnädig, N.F.; Beaucourt, S.; Campagnola, G.; Bordería, A.V.; Sanz-Ramos, M.; Gong, P.; Blanc, H.; Peersen, O.B.; Vignuzzi, M. Coxsackievirus B3 mutator strains are attenuated *in vivo*. *Proc. Natl. Acad. Sci. USA* **2012**, *109*, E2294–E2303. [CrossRef] [PubMed]

56. Campagnola, G.; McDonald, S.; Beaucourt, S.; Vignuzzi, M.; Peersen, O.B. Structure-function relationships underlying the replication fidelity of viral RNA-dependent RNA polymerases. *J. Virol.* **2015**, *89*, 275–286. [CrossRef] [PubMed]

57. Zamyatkin, D.; Rao, C.; Hoffarth, E.; Jurca, G.; Rho, H.; Parra, F.; Grochulski, P.; Ng, K.K. Structure of a backtracked state reveals conformational changes similar to the state following nucleotide incorporation in human norovirus polymerase. *Acta Crystallogr. D Biol. Crystallogr.* **2014**, *70*, 3099–3109. [CrossRef] [PubMed]

58. Sankar, S.; Porter, A.G. Point mutations which drastically affect the polymerization activity of encephalomyocarditis virus RNA-dependent RNA polymerase correspond to the active site of *Escherichia coli* DNA polymerase I. *J. Biol. Chem.* **1992**, *267*, 10168–10176. [PubMed]

59. Boehr, D.D. The ins and outs of viral RNA polymerase translocation. *J. Mol. Biol.* **2014**, *426*,

1373–1376. [CrossRef] [PubMed]

60. Vives-Adrian, L.; Lujan, C.; Oliva, B.; van der Linden, L.; Selisko, B.; Coutard, B.; Canard, B.; van Kuppeveld, F.J.; Ferrer-Orta, C.; Verdaguer, N. The crystal structure of a cardiovirus RNA-dependent RNA polymerase reveals an unusual conformation of the polymerase active site. *J. Virol.* **2014**, *88*, 5595–5607. [CrossRef] [PubMed]

61. Hall, D.J.; Palmenberg, A.C. Cleavage site mutations in the encephalomyocarditis virus P3 region lethally abrogate the normal processing cascade. *J. Virol.* **1996**, *70*, 5954–5961. [PubMed]

62. Ferrer-Orta, C.; de la Higuera, I.; Caridi, F.; Sanchez-Aparicio, M.T.; Moreno, E.; Perales, C.; Singh, K.; Sarafianos, S.G.; Sobrino, F.; Domingo, E.; *et al.* Multifunctionality of a picornavirus polymerase domain: Nuclear localization signal and nucleotide recognition. *J. Virol.* **2015**, *89*, 6848–6859. [CrossRef] [PubMed]

63. Hansen, J.L.; Long, A.M.; Schultz, S.C. Structure of the RNA-dependent RNA polymerase of poliovirus. *Structure* **1997**, *5*, 1109–1122. [CrossRef]

64. Hobson, S.D.; Rosenblum, E.S.; Richards, O.C.; Richmond, K.; Kirkegaard, K.; Schultz, S.C. Oligomeric structures of poliovirus polymerase are important for function. *EMBO J.* **2001**, *20*, 1153–1163. [CrossRef] [PubMed]

65. Lyle, J.M.; Bullitt, E.; Bienz, K.; Kirkegaard, K. Visualization and functional analysis of RNA-dependent RNA polymerase lattices. *Science* **2002**, *296*, 2218–2222. [CrossRef] [PubMed]

66. Spagnolo, J.F.; Rossignol, E.; Bullitt, E.; Kirkegaard, K. Enzymatic and nonenzymatic functions of viral RNA-dependent RNA polymerases within oligomeric arrays. *RNA* **2010**, *16*, 382–393. [CrossRef] [PubMed]

67. Kaiser, W.J.; Chaudhry, Y.; Sosnovtsev, S.V.; Goodfellow, I.G. Analysis of protein-protein interactions in the feline calicivirus replication complex. *J. Gen. Virol.* **2006**, *87*, 363–368. [CrossRef] [PubMed]

68. Luo, G.; Hamatake, R.K.; Mathis, D.M.; Racela, J.; Rigat, K.L.; Lemm, J.; Colonno, R.J. *De novo* initiation of RNA synthesis by the RNA-dependent RNA polymerase (NS5B) of hepatitis C virus. *J. Virol.* **2000**, *74*, 851–863. [CrossRef] [PubMed]

69. Chinnaswamy, S.; Murali, A.; Li, P.; Fujisaki, K.; Kao, C.C. Regulation of *de novo*-initiated RNA synthesis in hepatitis C virus RNA-dependent RNA polymerase by intermolecular interactions. *J. Virol.* **2010**, *84*, 5923–5935. [CrossRef] [PubMed]

70. Hogbom, M.; Jager, K.; Robel, I.; Unge, T.; Rohayem, J. The active form of the norovirus RNA-dependent RNA polymerase is a homodimer with cooperative activity. *J. Gen. Virol.* **2009**, *90*, 281–291. [CrossRef] [PubMed]

71. Cevik, B. The RNA-dependent RNA polymerase of *Citrus tristeza* virus forms oligomers. *Virology* **2013**, *447*, 121–130. [CrossRef] [PubMed]

72. Ferrero, D.; Buxaderas, M.; Rodriguez, J.F. The structure of the RNA-dependent RNA polymerase of a Permutotetravirus suggests a link between primer-dependent and primer-independent polymerases. *PLoS Pathog.* **2015**. Submitted for publication.

73. Chang, S.; Sun, D.; Liang, H.; Wang, J.; Li, J.; Guo, L.; Wang, X.; Guan, C.; Boruah, B.M.; Yuan, L.; *et al.* Cryo-EM structure of influenza virus RNA polymerase complex at 4.3 Å resolution. *Mol. Cell* **2015**, *57*, 925–935. [CrossRef] [PubMed]

74. Graham, S.C.; Sarin, L.P.; Bahar, M.W.; Myers, R.A.; Stuart, D.I.; Bamford, D.H.; Grimes, J.M. The N-terminus of the RNA polymerase from infectious pancreatic necrosis virus is the determinant of genome attachment. *PLoS Pathog.* **2011**, *7*, e1002085. [CrossRef] [PubMed]

75. Smallwood, S.; Hovel, T.; Neubert, W.J.; Moyer, S.A. Different substitutions at conserved amino acids in domains II and III in the Sendai L RNA polymerase protein inactivate viral RNA synthesis. *Virology* **2002**, *304*, 135–145. [CrossRef] [PubMed]

76. Zamoto-Niikura, A.; Terasaki, K.; Ikegami, T.; Peters, C.J.; Makino, S. Rift valley fever virus L protein forms a biologically active oligomer. *J. Virol.* **2009**, *83*, 12779–12789. [CrossRef] [PubMed]

77. Tellez, A.B.; Wang, J.; Tanner, E.J.; Spagnolo, J.F.; Kirkegaard, K.; Bullitt, E. Interstitial contacts in an RNA-dependent RNA polymerase lattice. *J. Mol. Biol.* **2011**, *412*, 737–750. [CrossRef] [PubMed]

78. Wang, J.; Lyle, J.M.; Bullitt, E. Surface for catalysis by poliovirus RNA-dependent RNA polymerase. *J. Mol. Biol.* **2013**, *425*, 2529–2540. [CrossRef] [PubMed]

79. Powdrill, M.H.; Bernatchez, J.A.; Gotte, M. Inhibitors of the hepatitis C virus RNA-dependent RNA polymerase NS5B. *Viruses* **2010**, *2*, 2169–2195. [CrossRef] [PubMed]

80. Gentile, I.; Buonomo, A.R.; Zappulo, E.; Coppola, N.; Borgia, G. GS-9669: A novel non-nucleoside inhibitor of viral polymerase for the treatment of hepatitis C virus infection. *Expert Rev. Anti-Infect. Ther.* **2014**, *12*, 1179–1186. [CrossRef] [PubMed]

81. Caillet-Saguy, C.; Lim, S.P.; Shi, P.Y.; Lescar, J.; Bressanelli, S. Polymerases of hepatitis C viruses and flaviviruses: Structural and mechanistic insights and drug development. *Antivir. Res.* **2014**, *105*, 8–16. [CrossRef] [PubMed]

82. Eltahla, A.A.; Lim, K.L.; Eden, J.S.; Kelly, A.G.; Mackenzie, J.M.; White, P.A. Nonnucleoside inhibitors of norovirus RNA polymerase: Scaffolds for rational drug design. *Antimicrob. Agents Chemother.* **2014**, *58*, 3115–3123. [CrossRef] [PubMed]

83. Carfi, M.; Gennari, A.; Malerba, I.; Corsini, E.; Pallardy, M.; Pieters, R.; van Loveren, H.; Vohr, H.W.; Hartung, T.; Gribaldo, L. *In vitro* tests to evaluate immunotoxicity: A preliminary study. *Toxicology* **2007**, *229*, 11–22. [CrossRef] [PubMed]

84. De Clercq, E.; Neyts, J. Antiviral agents acting as DNA or RNA chain terminators. *Handb. Exp. Pharmacol.* **2009**, *189*, 53–84. [PubMed]

85. Domingo, E. Virus entry into error catastrophe as a new antiviral strategy. *Virus Res.* **2005**, *107*, 115–228. [CrossRef]

86. De Francesco, R.; Tomei, L.; Altamura, S.; Summa, V.; Migliaccio, G. Approaching a new era for hepatitis C virus therapy: Inhibitors of the NS3–4A serine protease and the NS5B RNA-dependent RNA polymerase. *Antivir. Res.* **2003**, *58*, 1–16. [CrossRef]

87. Biswal, B.K.; Wang, M.; Cherney, M.M.; Chan, L.; Yannopoulos, C.G.; Bilimoria, D.; Bedard, J.; James, M.N. Non-nucleoside inhibitors binding to hepatitis C virus NS5B polymerase reveal a novel mechanism of inhibition. *J. Mol. Biol.* **2006**, *361*, 33–45. [CrossRef] [PubMed]

88. Paeshuyse, J.; Leyssen, P.; Mabery, E.; Boddeker, N.; Vrancken, R.; Froeyen, M.; Ansari, I.H.; Dutartre, H.; Rozenski, J.; Gil, L.H.; *et al.* A novel, highly selective inhibitor of pestivirus replication that targets the viral RNA-dependent RNA polymerase. *J. Virol.* **2006**, *80*, 149–160.

[CrossRef] [PubMed]

89. Paeshuyse, J.; Chezal, J.M.; Froeyen, M.; Leyssen, P.; Dutartre, H.; Vrancken, R.; Canard, B.; Letellier, C.; Li, T.; Mittendorfer, H.; *et al.* The imidazopyrrolopyridine analogue AG110 is a novel, highly selective inhibitor of pestiviruses that targets the viral RNA-dependent RNA polymerase at a hot spot for inhibition of viral replication. *J. Virol.* **2007**, *81*, 11046–11053. [CrossRef] [PubMed]

90. Van der Linden, L.; Vives-Adrián, L.; Selisko, B.; Ferrer-Orta, C.; Liu, X.; Lanke, K.; Ulferts, R.; de Palma, A.M.; Tanchis, F.; Goris, N.; *et al.* The RNA template channel of the RNA-dependent RNA polymerase as a target for development of antiviral therapy of multiple genera within a virus family. *PLoS Pathog.* **2015**, *23*, e1004733. [CrossRef] [PubMed]

8

Amino Terminal Region of Dengue Virus NS4A Cytosolic Domain Binds to Highly Curved Liposomes

Yu-Fu Hung [1,2], **Melanie Schwarten** [1], **Silke Hoffmann** [1], **Dieter Willbold** [1,2], **Ella H. Sklan** [3] **and Bernd W. Koenig** [1,2,*]

[1] Institute of Complex Systems, Structural Biochemistry (ICS-6), Forschungszentrum Jülich, 52425 Jülich, Germany; E-Mails: y.hung@fz-juelich.de (Y.-F.H.); m.schwarten@fz-juelich.de (M.S.); si.hoffmann@fz-juelich.de (S.H.); d.willbold@fz-juelich.de (D.W.)

[2] Institut für Physikalische Biologie, Heinrich-Heine-Universität Düsseldorf, Universitätsstraße 1, 40255 Düsseldorf, Germany

[3] Department Clinical Microbiology and Immunology, Sackler School of Medicine, Tel Aviv University, Tel Aviv 69978, Israel; E-Mail: sklan@post.tau.ac.il

* Author to whom correspondence should be addressed; E-Mail: b.koenig@fz-juelich.de

Academic Editor: David Boehr

Abstract: Dengue virus (DENV) is an important human pathogen causing millions of disease cases and thousands of deaths worldwide. Non-structural protein 4A (NS4A) is a vital component of the viral replication complex (RC) and plays a major role in the formation of host cell membrane-derived structures that provide a scaffold for replication. The N-terminal cytoplasmic region of NS4A(1–48) is known to preferentially interact with highly curved membranes. Here, we provide experimental evidence for the stable binding of NS4A(1–48) to small liposomes using a liposome floatation assay and identify the lipid binding sequence by NMR spectroscopy. Mutations L6E;M10E were previously shown to inhibit DENV replication and to interfere with the binding of NS4A(1–48) to small liposomes. Our results provide new details on the interaction of the N-terminal region of NS4A with membranes and will prompt studies of the functional relevance of the curvature sensitive membrane anchor at the N-terminus of NS4A.

Keywords: Dengue virus (DENV); non-structural protein 4A (NS4A); amphipathic helix; curvature sensing; peptide membrane interaction

---

1. Introduction

Dengue virus (DENV), the causative agent of dengue fever, is a positive strand RNA, enveloped virus belonging to the *Flaviviridae* family. The viral RNA is translated into a single polyprotein which is processed by cellular and viral proteases into three structural proteins (capsid, premembrane, and envelope) and the seven non-structural (NS) proteins NS1, NS2A, NS2B, NS3, NS4A, NS4B, and NS5 [1]. The NS proteins are not found in the mature virion but are crucial for viral replication. Synthesis of viral RNA takes place in replication complexes (RCs) that contain essential NS proteins, viral RNA and host cell factors [2]. Upon DENV infection, a complex and continuous network of ER membrane-derived vesicular structures and convoluted membranes is formed. These structures contain the viral replication complexes and the sites of virion assembly [3–5].

NS4A is small integral membrane protein containing four predicted transmembrane segments (pTMSs) [1]. Although pTMS4, often referred to as the 2k fragment, is not part of the mature NS4A, it serves as a signal peptide for the ER localization of NS4B and is cleaved from the mature NS4A [6]. Experimental data verify that pTMS1 and pTMS3 span the membrane while pTMS2 is embedded in the luminal leaflet of the ER membrane and does not span it [1].

NS4A is crucial for the formation of the virus-induced membrane structures. Expression of NS4A lacking the 2k fragment alone is sufficient to induce membrane alterations that resemble the DENV-induced highly curved membranes that harbor the RCs [1]. Clearly, to induce these structures NS4A will have to closely interact with host membranes. However, the mechanism by which NS4A induces the curved morphology of these newly formed membranes is still unknown. Insertion of amphipathic helices into one leaflet of a membrane bilayer, as well as oligomerization of membrane proteins are among the mechanisms known to participate in the induction of membrane curvature [7]. Molecular dynamics (MD) simulations suggest that pTMS2 of NS4A could support membrane undulations upon stable association with the membrane [8]. Curved vesicular structures might also be induced via homooligomerization of NS4A [9,10]. While we have previously shown that the NS4A N-terminal cytoplasmic region is implicated in its oligomerization [10], a recent study demonstrated that pTMS1 is the major determinant in this process [9]. Introduction of two point mutations at the N-terminal of NS4A (L6E and M10E) reduced both the amphipathic character of this region and NS4A homooligomerization and abolished viral replication [10].

NS4A is an essential component of the viral RC [1]. Direct interaction of NS4A with the cytoskeletal protein vimentin was reported to be necessary for correct localization of the RC at the perinuclear site [11]. The vimentin binding site was found to be located at the N-terminal 50 residues of NS4A [11]. NS4A was also reported to bind NS4B, another component of the RC, via pTMS1 [9]. Mutational analysis suggests a functional relevance of this interaction for viral replication [9]. It was speculated that the interaction between NS4A and NS4B in concert with NS4A oligomerization and NS4B dimerization may play a role in the spatial and temporal regulation of distinct molecular complexes involved in the viral infection cycle [9].

Our previous circular dichroism (CD) data demonstrated that NS4A(1–48) interacts with highly curved small unilamellar liposomes under α-helix formation, while mutated NS4A(1–48, L6E;M10E) does not [12]. Surface plasmon resonance data indicated a seven fold-greater association of wild type NS4A(1–48) with immobilized liposomes compared to the mutant [12]. The structure of NS4A(1–48) in presence of membrane mimicking SDS micelles was characterized by NMR [12]. To further extend these results we used liposome flotation for direct proof of NS4A(1–48) binding to free liposomes. The exact location of lipid binding sites in the amino acid sequence of NS4A(1–48) was addressed by NMR spectroscopy. Our findings provide a basis for specific structure-function studies that will enhance our understanding of the role of NS4A and might provide future targets for anti-viral intervention.

2. Materials and Methods

2.1. Peptide Production

The peptide NS4A(1–48) corresponds to amino acid residues 1–48 from the N-terminal of NS4A of dengue virus serotype 2 (NCBI Protein database accession number: NP739588). A mutant peptide containing the mutations L6E and M10E was designated NS4A(1–48, L6E;M10E). The two NS4A peptides were recombinantly produced in E. coli BL21 cells and enzymatically cleaved from the affinity tag as described earlier [13]. Uniform isotope labeling with ^{15}N or ^{13}C, ^{15}N was achieved by expression in M9 medium containing ^{15}N ammonium chloride and ^{13}C glucose (Eurisotop, Saarbrücken, Germany) as the sole source of nitrogen and carbon, respectively. Unlabeled peptides were expressed in LB media.

2.2. Fluorescence Labeling

Alexa Fluor 488 succinimidyl ester (NHS ester) was purchased from Life Technologies, Darmstadt, Germany. The dye was dissolved in anhydrous DMSO at a concentration of 3 mM immediately prior to the labeling reaction. For the reaction 300 µL from a 100 µM NS4A(1–48) or NS4A(1–48, L6E;M10E) stock in sample buffer (50 mM sodium phosphate, pH 6.8, 150 mM NaCl) were combined with 100 µL of 0.4 M NaHCO$_3$ and the pH was adjusted to 8.3. This 400 µL peptide solution was supplemented with 100 µL of the Alexa Fluor 488 NHS ester in DMSO resulting in an approximately ten-fold excess of dye-over-peptide. The labeling reaction was wrapped in aluminum foil and incubated on a rocking platform shaker at 4 °C for 16 h. Labeled protein and free dye were separated on a Superdex 75 10/300 GL column (GE Healthcare, Freiburg, Germany) operated on an ÄKTApurifier system (GE Healthcare).

2.3. Liposome Preparation

The lipid 1-palmitoyl-2-oleoyl-sn-glycero-3-phosphocholine (POPC) in chloroform solution was purchased from Avanti Polar Lipids (Alabaster, AL, USA). Small unilamellar lipid vesicles (SUVs) were prepared from chloroform-free POPC dispersions (20 mg·mL^{-1}) in sample buffer as described earlier [12]. SUVs were obtained by sequential extrusion through 50 nm (15 times) and 30 nm (15 times) Nuclepore polycarbonate membranes (GE Healthcare) with nominal pore diameter of either 50 or 30 nm, followed by sonication with a 3 mm microtip of a Branson 250 sonifier (15 cycles of sonication, 20 s each, interrupted by cooling for 2 min after each cycle). Sonicated SUVs were centrifuged for 10 min at $16,100\times g$ and 10 °C in a refrigerated Eppendorf 5415 R tabletop centrifuge to remove any titanium abrasion of the microtip from the sample. The hydrodynamic radius of each liposome preparation was determined by dynamic light scattering (DLS) using a Dyna Pro instrument (Protein Solutions, Lakewood, NJ, USA) equipped with a 3 mm path length 45 μL quartz cell. Liposome solutions (20 mg of POPC per mL) were diluted 100-fold with buffer directly after extrusion or sonication and measured immediately. Data were analyzed with Dynamics V6 software distributed with the instrument. Experimental data were fitted to the model of Rayleigh spheres.

2.4. Liposome Floatation Assay

Equal volumes of 80 μM suspensions of Alexa Fluor 488-labeled peptides or free dye and sonicated POPC SUV (20 mg· mL^{-1}) in sample buffer were combined and mixed at room temperature for 5 min. 100 μL of each of the three resulting samples were thoroughly mixed with 100 μL of a 70% (*w/v*) sucrose solution to obtain homogeneous solutions containing 20 μM of either Alexa Fluor 488-labeled peptide or free Alexa Fluor 488 dye, POPC liposomes (5 mg· mL^{-1}) and 35% (*w/v*) sucrose in sample buffer. Sucrose solution (400 μL of a 70% (*w/v*) solution in sample buffer) was transferred to the bottom of a Polyallomer centrifuge tube (11 mm × 34 mm; Beckman Coulter) followed by a second layer formed by the 200 μL of 35% (*w/v*) sucrose solution containing one of the labeled NS4A peptides or the free dye and POPC liposomes.

Finally, each sample was carefully overlaid with two cushions of decreasing sucrose concentration, *i.e.*, 1.2 mL of 20% (*w/v*) followed by 200 μL of 10% (*w/v*) sucrose in sample buffer (cf. scheme in Figure 1, left). Samples were centrifuged for 14 h at 259,000× *g* and 4 °C in a Beckman Coulter Optima Max-XP ultracentrifuge using a TLS 55 swinging bucket rotor. Fluorescence images were taken in front of a Mini Transilluminator (Bio-Rad, Munich, Germany) prior to and immediately after centrifugation.

2.5. Nuclear Magnetic Resonance (NMR) Spectroscopy

NMR experiments were conducted at 30 °C on Bruker Avance III HD NMR and Varian VNMRS instruments, equipped with cryogenic Z-axis pulse-field-gradient (PFG) triple resonance probes operating at proton frequencies of 700 and 900 MHz, respectively. Samples for resonance assignment contained 300 μM [U-^{15}N, ^{13}C]-labeled NS4A(1–48) in sample buffer (50 mM sodium phosphate, pH 6.8, 150 mM NaCl) as used for the liposome flotation experiments but supplemented with 10% (*v/v*) deuterium oxide and 0.03% (*w/v*) NaN$_3$ (referred to as NMR buffer). Assignment of protein backbone resonances was accomplished using a combined set of heteronuclear multidimensional NMR experiments: 2D (^{1}H–^{15}N)-HSQC [14,15], 2D (^{1}H–^{13}C)-HSQC [16], 3D HNCA [17], 3D BT-HNCO [18], and 3D HNcaCO [19]. ^{1}H and ^{13}C chemical shifts were referenced directly to internal 4,4-dimethyl-4-silapentane-1-sulfonic acid (DSS) at 0 ppm and ^{15}N chemical shifts were referenced indirectly to DSS using the absolute ratio of the ^{15}N and ^{1}H zero point frequencies [20]. NMR data were processed using NMRPipe, v.8.1 [21] and evaluated with CcpNmr v.2.4 [22].

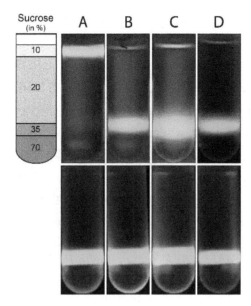

Figure 1. Liposome floatation assay of wild type and mutant NS4A(1–48). Alexa-488-labeled NS4A(1–48) (**A**); Alexa-488-labeled NS4A(1–48, L6E;M10E) (**B**); or free Alexa Fluor 488 dye (**C**) were mixed with sonicated POPC liposomes and loaded with the 35% (*w/v*) sucrose layer of a sucrose step gradient schematically shown on the left; Alexa-488-labeled NS4A(1–48) without liposomes was loaded with the 35% (*w/v*) sucrose layer in lane (**D**). Note: The narrow green line at the top of tubes (**B**), (**C**) and (**D**) results from reflection of fluorescence light at the air buffer interface rather than the presence of dye. Fluorescence images of the four tubes recorded prior to ultracentrifugation are shown in the lower row.

The interaction of NS4A(1–48) and NS4A(1–48, L6E;M10E) with sonicated POPC liposomes was studied based on a series of 2D (^{1}H–^{15}N) Heteronuclear Single Quantum Coherence (HSQC) spectra recorded with 40 μM peptide in NMR buffer but with different amounts of liposomes (0; 2.5; 5, or 10 mg of POPC per mL). Spectra were acquired with 150 complex data points in the ^{15}N time domain, up to 128 scans per *t*1 increment and a recycle delay of 1.5 s. Data were processed and analyzed using NMRPipe.

3. Results

3.1. Wild Type NS4A(1–48) Binds to Highly Curved POPC SUVs

Binding of fluorescently labeled proteins to liposomes can be visualized using a simple floatation assay [23]. POPC bilayers have a mass density very close to 1 g·cm^{-3} [24] at room temperature. Thus, POPC liposomes will migrate to the top layer containing the lowest sucrose concentration in a step gradient of decreasing sucrose concentrations upon centrifugation. In contrast, a small protein of ~5 kDa molecular weight is expected to have a mass density well above 1.41 g·cm^{-3} [25] and thus will accumulate in a sucrose rich layer of high mass density. Alexa-488-labeled NS4A(1–48) apparently binds to small sonicated POPC liposomes and migrates with the liposomes to the top of the gradient (Figure 1A). In contrast, the mutated peptide NS4A(1–48, L6E;M10E) remains in the high density layer with 35% (*w/v*) sucrose even after 14 h of centrifugation (Figure 1B), indicating that this peptide does not bind to liposomes. As a negative control, the Alexa-488 free dye was loaded to the 35% (w/v) sucrose layer with POPC liposomes. The free dye remained in the high-density region after centrifugation (Figure 1C) indicating a lack of association with the POPC SUVs. The fluorescent band of the free dye has a larger vertical extension than that of the labeled NS4A(1–48, L6E;M10E) after 14 h of centrifugation (Figure 1B,C). This probably reflects the larger diffusion coefficient of the low molecular weight dye molecule (643.4 g· mol^{-1}). In a second control experiment Alexa-488-labeled NS4A(1–48) was loaded to the 35% (w/v) sucrose layer without adding POPC liposomes. As expected, the peptide did not significantly migrate during the 14 h of centrifugation and remained almost completely in the 35% (w/v) sucrose layer (Figure 1D). The floatation results confirm the binding of the wild type peptide to small liposomes.

3.2. NMR Identifies Regions of NS4A(1–48) Associated with POPC Liposomes

NMR spectroscopy was used to identify the regions of NS4A(1–48) associated with POPC liposomes. Almost complete assignment of the expected ^{1}HN, ^{15}N, ^{13}Cα and ^{13}C' backbone resonances was accomplished for both NS4A(1–48) and NS4A(1–48, L6E;M10E) in NMR buffer. No assignments were obtained for S1 and N42. In addition to P14 there are four residues which only lack amide (^{1}HN, ^{15}N) assignments. Assignments have been deposited at the Biological Magnetic Resonance Data Bank (BMRB) under accession number 25586 for NS4A(1–48) and accession number 25676 for NS4A(1–48, L6E;M10E).

Amide ^{1}HN, ^{15}N cross peaks in HSQC spectra of NS4A(1–48) were used to monitor peptide interaction with sonicated POPC liposomes in an amino acid residue resolved manner. The hydrodynamic radius of the POPC liposomes was ~26 nm based on DLS measurements. Backbone cross peaks for most of the 48 amino acid residues of NS4A(1–48) were identified in buffer without liposomes except for S1, L2, P14, M17, H32, N42 and H43. The observed cross peaks characterize the free peptide conformation. Addition of increasing amounts of sonicated POPC liposomes at constant peptide concentration caused gradual peak intensity reductions in a peptide region specific manner (Figure 2A). Interestingly, peak positions did not significantly change, except for R12, which showed a small shift. Such a behavior is typical for slow or intermediate exchange of the peptide between the free and liposome-bound state. Strongest peak intensity reduction is observed in the N-terminal region extending up to K20 (Figure 2A). Some peaks in this region completely disappear already at 2.5 mg· mL^{-1} POPC (N5, I7, E9, G11, K20) while the others are reduced to less than 40%. At the highest liposome concentration studied (10 mg· mL^{-1}) only two cross peaks from this peptide region remain visible and show low intensity. It is likely, that a number of amino acid residues of the N-terminal region bind directly to the liposome. Binding will change the chemical environment and strongly reduce the rotational correlation time of the amino acid residues in direct contact with lipids. NMR signals of bound residues are likely broadened beyond detection. Exchange dynamics may differ somewhat among the amino acid residues in this region, explaining the variable intensity reduction of the free state cross peaks.

Cross peaks of the central region from A21 through L31 of NS4A(1–48) show rather uniform intensity reduction upon gradual liposome addition (Figure 2A). All free state peaks remain visible and retain about 10% of their original intensity even in the presence of 10 mg· mL^{-1} POPC. Different scenarios might contribute to the reduction of the free state peak intensities. Peptides that are anchored with their N-terminal region in the liposome might retain their free state conformation in the central domain, albeit with a reduced overall rotational correlation time and thus lower peak intensities. In addition, amino acid residues of the central region of some NS4A peptides might bind directly to the liposome leading to the disappearance of the corresponding NMR signals. The uniform peak intensity reduction pattern in the central region may suggest a concerted binding of this amino acid stretch, e.g., as one secondary structure element, to the liposome.

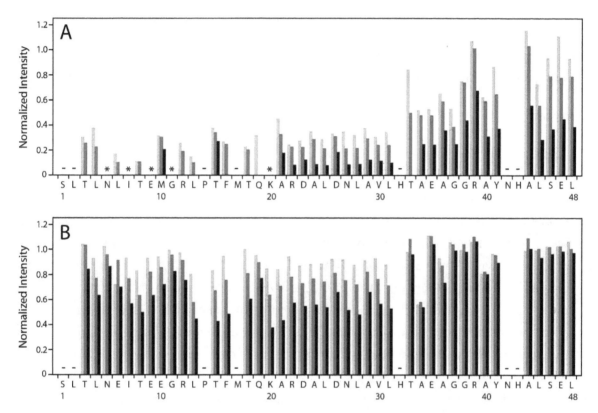

Figure 2. Intensity of backbone amide cross peaks in HSQC spectra of NS4A(1–48) (**A**) and NS4A(1–48, L6E;M10E) (**B**) recorded at various lipid concentrations. Peak intensities measured at 2.5 (green), 5 (red) and 10 mg·mL^{-1} POPC (black) in the sample were normalized to the intensity of the same signal observed in lipid-free buffer and are shown as a function of the amino acid sequence of the studied peptide. Cross peaks not observed in the lipid-free sample are indicated by minus signs. Cross peaks that are present in buffer but completely disappear after addition of 2.5 mg·mL^{-1} POPC are indicated by asterisk.

Finally, amino acid residues in the C-terminal region of NS4A(1–48) from T33 to L48 show the smallest reduction in the free state peak intensities upon liposome addition (Figure 2A). Perhaps, these small intensity reductions might be entirely caused by anchoring of the peptides via amino acid residues in the N-terminal and perhaps the central regions.

3.3. NMR Data Suggest Diminished Interaction of NS4A(1–48, L6E;M10E) with POPC Liposomes

Inspection of HSQC spectra of mutant NS4A(1–48, L6E;M10E) in buffer and with increasing amounts of sonicated POPC liposomes revealed no changes in cross peak positions and only a minor influence of lipid addition on cross peak intensities of backbone amide signals. All cross peaks observed in buffer remain visible in presence of liposomes and retain at least 40% of their original intensity upon addition of 10 mg· mL^{-1} of POPC liposomes (Figure 2B). Almost no peak reduction is observed for the C-terminal region of NS4A(1–48, L6E;M10E). Peak intensity reductions in the N-terminal and central regions are quite moderate, in particular at low liposome concentration (2.5 and 5 mg· mL^{-1} POPC). Even at 10 mg· mL^{-1} POPC liposomes cross peak intensities remain between 40% and 80% (N-terminal region) or between 40% and 60% (central region) with respect to intensities measured in buffer. Apparently some interaction of the mutant peptide with liposomes is retained at the highest lipid concentration studied. Amino acid residues in the central and perhaps in the N-terminal part of the mutant peptide are likely to make transient contact with the liposome. However, the observed peak intensity reductions are much weaker in case of NS4A(1–48, L6E;M10E) than for wild type.

4. Discussion

The L6E and M10E mutations that disrupt the amphipathic character of the NS4A *N*-terminal abolish viral replication, indicating that the 48 N-terminal residues of NS4A play a crucial role in replication. Furthermore, these mutations had a similar effect when inserted as single mutations [10]. We have previously shown that NS4A(1–48) interacts preferentially with highly curved liposomes. CD spectroscopy demonstrated that these two point mutations severely compromise this interaction [12]. The interaction of wild type NS4A(1–48) with highly curved membranes has been demonstrated for three different lipid compositions, *i.e.*, pure POPC, a POPC/DOPS mixture at a molar ratio of 4:1, and a blend of synthetic lipids resembling the composition of membranes in the ER (ER lipid mix), but no dependence on lipid composition was detected [12]. Here, we confirm these results using a liposome floatation assay and NMR. Initial floatation experiments were conducted with pure POPC and with the ER lipid mix. Again, no influence of lipid composition on NS4A(1–48) binding was observed. The detailed analysis presented in the current manuscript was conducted with single component POPC SUVs. Liposome floatation experiments (Figure 1) clearly show that wild type NS4A(1–48) binds to highly curved POPC liposomes, this interaction was not observed with the mutant peptide. The interaction between NS4A(1–48) and liposomes was further characterized using NMR. These experiments indicate that the main lipid binding sites of the peptide are located at the N-terminal 20 amino acid residues of NS4A(1–48). Residues A21 through L31 may also be involved in liposome binding while the remaining C-terminal residues are only weakly affected by liposome binding and do not seem to play a direct role in this process. The backbone resonance assignment of NS4A(1–48) and NS4A(1–48, L6E;M10E) reported in this manuscript contains information on the secondary structure of the two peptides in lipid-free buffer. Analysis with the TALOS-N software [26] clearly shows the lack of secondary structure for both peptides in buffer. The NMR data on NS4A(1–48) recorded in presence of liposomes do not allow any straightforward conclusion on the structure of the liposome-bound peptide. However, interaction of NS4A(1–48) with sodium dodecyl sulfate (SDS) micelles induces the formation of two amphipathic helices (AH) encompassing residues N5 to E9 (AH1) and T15 to L31 (AH2) of NS4A(1–48) [12]. CD spectra of NS4A(1–48) recorded in presence of either SDS micelles or small POPC liposomes are very similar [12]. Therefore, it is conceivable that the two amphipathic helices AH1 and AH2 are also formed in the liposome-bound peptide. The *N*-terminal region of NS4A(1–48) that forms AH1, an interhelical linker and the N-terminal half of AH2 in SDS micelles seem to be crucial for peptide binding to liposomes. Interestingly, this region also contains the two mutations L6E and M10E, which abolished liposome binding in the floatation experiment (Figure 1). Moreover, mutagenesis of other residues in this region including P14A [10], R12A and K20A [9] was also shown to reduce or abolish DENV replication.

The central part of NS4A(1–48) encompassing residues A21 to L31 shows less pronounced NMR signal intensity reductions upon titration with POPC liposomes than its N-terminal 20 amino acid residues (Figure 2A). A comparatively weak NMR signal intensity reduction is observed for both the N-terminal and central regions of NS4A(1–48, L6E;M10E) (Figure 2B) indicating some residual interaction with liposomes. However, the liposome floatation assay clearly shows that this interaction of the mutant peptide with POPC liposomes is too weak for stable anchoring of the peptide at the membrane. The amino acid sequence of A21 through L31 is identical in both peptides. We conclude that this amino acid stretch A21 to L31 is not sufficient for stable membrane anchoring of NS4A(1–48).

DENV NS4A apparently contains two separate membrane anchors. The membrane spanning helices pTMS1 and pTMS3 stably integrate the protein into the membrane. The N-terminal region of the cytosolic domain specifically binds to the convex surface of highly curved membranes [12] and may serve as a second membrane anchor. Therefore, one hypothesis might be that NS4A can bridge two adjacent membranes or connect separate patches of the same membrane that come into close proximity due to membrane convolution. We speculate that membrane bridging by NS4A might play a crucial role in stabilizing the complex morphology of DENV-induced ER-derived membrane structures, which include stacks of convoluted membranes (CM), double membrane vesicles and tubes [5]. The vesicles were described as invaginations of the ER, which are connected to the cytosol via pore-like openings [5]. NS4A may play different roles in the reorganization of these ER-derived membranes. Asymmetric insertion of pTMS2 into the luminal leaflet of the inner membrane of the vesicles as well as oligomerization of NS4A may induce concave membrane curvature required for vesicle formation [7]. Binding of the N-terminal region of NS4A to the saddle-shaped neck region connecting the vesicle and the pore may further stabilize the vesicular structures.

All positive-strand RNA viruses form their replication complexes on modified host membranes. However, the source of the membranes and the nature of the modifications vary (for review see [2]). In general, the role of these modifications is twofold; to provide a scaffold concentrating and correctly positioning the viral and host factors for efficient viral replication and to protect the replicating virus from detection by the host immune system. The mechanisms driving the formation of these structures are still incompletely understood. Convoluted membranes are a form of membrane modification induced by several positive-strand RNA viruses including, in addition to DENV, severe acute respiratory syndrome coronavirus (SARS-CoV) and Kunjin virus (KUNV) [5,27,28], for example. In KUNV convoluted membranes are thought to be the site of polyprotein processing [27]. While in DENV the role of these structures is still unclear, they are thought to be a depot for factors required for replication [5].

In summary, the liposome floatation data provide direct proof for specific binding of NS4A(1–48) to highly curved free liposomes. The main lipid binding sites in NS4A(1–48) are located within the N-terminal 20 amino acid residues. The exact role of this specific interaction in the viral life cycle is still under investigation. Nevertheless, this important structural information may assist in further understanding of the role of NS4A and the mechanism by which it induces the membrane alterations underlying the viral RC formation.

Acknowledgments

We would like to thank Matthias Stoldt for excellent NMR support. D.W. acknowledges funding from the Sonderforschungsbereich SFB974.

Author Contributions

Bernd W. Koenig, Silke Hoffmann, Yu-Fu Hung and Melanie Schwarten conceived and designed the study; Yu-Fu Hung and Melanie Schwarten performed the experiments and analyzed the data, Ella H. Sklan, Silke Hoffmann and Dieter Willbold discussed the results and revised the manuscript; Bernd W. Koenig wrote the manuscript.

References

1. Miller, S.; Kastner, S.; Krijnse-Locker, J.; Buhler, S.; Bartenschlager, R. The non-structural protein 4A of dengue virus is an integral membrane protein inducing membrane alterations in a 2K-regulated manner. *J. Biol. Chem.* **2007**, *282*, 8873–8882. [CrossRef] [PubMed]

2. Miller, S.; Krijnse-Locker, J. Modification of intracellular membrane structures for virus replication. *Nat. Rev. Microbiol.* **2008**, *6*, 363–374. [CrossRef] [PubMed]

3. Salonen, A.; Ahola, T.; Kaariainen, L. Viral RNA replication in association with cellular membranes. *Curr. Top. Mirobiol. Immunol.* **2005**, *285*, 139–173.

4. Mackenzie, J. Wrapping things up about virus RNA replication. *Traffic* **2005**, *6*, 967–977. [CrossRef] [PubMed]

5. Welsch, S.; Miller, S.; Romero-Brey, I.; Merz, A.; Bleck, C.K.E.; Walther, P.; Fuller, S.D.; Antony, C.; Krijnse-Locker, J.; Bartenschlager, R. Composition and three-dimensional architecture of the dengue virus replication and assembly sites. *Cell Host Microbe* **2009**, *5*, 365–375. [CrossRef] [PubMed]

6. Lin, C.; Amberg, S.M.; Chambers, T.J.; Rice, C.M. Cleavage at a novel site in the NS4A region by the yellow fever virus NS2B-3 proteinase is a prerequisite for processing at the downstream 4A/4B signalase site. *J. Virol.* **1993**, *67*, 2327–2335. [PubMed]

7. McMahon, H.T.; Gallop, J.L. Membrane curvature and mechanisms of dynamic cell membrane remodelling. *Nature* **2005**, *438*, 590–596. [CrossRef] [PubMed]

8. Lin, M.H.; Hsu, H.J.; Bartenschlager, R.; Fischer, W.B. Membrane undulation induced by NS4A of Dengue virus: A molecular dynamics simulation study. *J. Biomol. Struct. Dyn.* **2013**, *32*, 1552–1562. [CrossRef] [PubMed]

9. Lee, C.M.; Xie, X.; Zou, J.; Li, S.H.; Lee, M.Y.; Dong, H.; Qin, C.F.; Kang, C.; Shi, P.Y. Determinants of dengue virus NS4A protein oligomerization. *J. Virol.* **2015**, *89*, 6171–6183. [CrossRef] [PubMed]

10. Stern, O.; Hung, Y.F.; Valdau, O.; Yaffe, Y.; Harris, E.; Hoffmann, S.; Willbold, D.; Sklan, E.H. An

N-Terminal amphipathic helix in dengue virus nonstructural protein 4A mediates oligomerization and is essential for replication. *J. Virol.* **2013**, *87*, 4080–4085. [CrossRef] [PubMed]

11. Teo, C.S.; Chu, J.J. Cellular vimentin regulates construction of dengue virus replication complexes through interaction with NS4A protein. *J. Virol.* **2014**, *88*, 1897–1913. [CrossRef] [PubMed]

12. Hung, Y.F.; Schwarten, M.; Schunke, S.; Thiagarajan-Rosenkranz, P.; Hoffmann, S.; Sklan, E.H.; Willbold, D.; Koenig, B.W. Dengue virus NS4A cytoplasmic domain binding to liposomes is sensitive to membrane curvature. *Biochim. Biophys. Acta* **2015**, *1848*, 1119–1126. [CrossRef] [PubMed]

13. Hung, Y.F.; Valdau, O.; Schunke, S.; Stern, O.; Koenig, B.W.; Willbold, D.; Hoffmann, S. Recombinant production of the amino terminal cytoplasmic region of dengue virus non-structural protein 4A for structural studies. *PLoS ONE* **2014**, *9*, e86482. [CrossRef] [PubMed]

14. Bodenhausen, G.; Ruben, D.J. Natural abundance nitrogen-15 NMR by enhanced heteronuclear spectroscopy. *Chem. Phys. Lett.* **1980**, *69*, 185–189. [CrossRef]

15. Grzesiek, S.; Bax, A. Amino acid type determination in the sequential assignment procedure of uniformly 13C/15N-enriched proteins. *J. Biomol. NMR* **1993**, *3*, 185–204. [CrossRef] [PubMed]

16. Kay, L.E.; Keifer, P.; Saarinen, T. Pure absorption gradient enhanced heteronuclear single quantum correlation spectroscopy with improved sensitivity. *J. Am. Chem. Soc.* **1992**, *114*, 10663–10665. [CrossRef]

17. Ikura, M.; Kay, L.E.; Bax, A. A novel approach for sequential assignment of 1H, 13C, and 15N spectra of proteins: Heteronuclear triple-resonance three-dimensional NMR spectroscopy. Application to calmodulin. *Biochemistry* **1990**, *29*, 4659–4667. [CrossRef] [PubMed]

18. Solyom, Z.; Schwarten, M.; Geist, L.; Konrat, R.; Willbold, D.; Brutscher, B. BEST-TROSY experiments for time-efficient sequential resonance assignment of large disordered proteins. *J. Biomol. NMR* **2013**, *55*, 311–321. [CrossRef] [PubMed]

19. Yamazaki, T.; Lee, W.; Arrowsmith, C.H.; Muhandiram, D.R.; Kay, L.E. A suite of triple resonance NMR experiments for the backbone assignment of ^{15}N, ^{13}C, ^{2}H labeled proteins with high sensitivity. *J. Am. Chem. Soc.* **1994**, *116*, 11655–11666. [CrossRef]

20. Wishart, D.S.; Bigam, C.G.; Holm, A.; Hodges, R.S.; Sykes, B.D. ^{1}H, ^{13}C and ^{15}N random coil NMR chemical shifts of the common amino acids. I. Investigations of nearest-neighbor effects. *J. Biomol. NMR* **1995**, *5*, 67–81. [CrossRef] [PubMed]

21. Delaglio, F.; Grzesiek, S.; Vuister, G.W.; Zhu, G.; Pfeifer, J.; Bax, A. NMRPipe: A multidimensional spectral processing system based on UNIX pipes. *J. Biomol. NMR* **1995**, *6*, 277–293. [CrossRef] [PubMed]

22. Vranken, W.F.; Boucher, W.; Stevens, T.J.; Fogh, R.H.; Pajon, A.; Llinas, M.; Ulrich, E.L.; Markley, J.L.; Ionides, J.; Laue, E.D. The CCPN data model for NMR spectroscopy: Development of a software pipeline. *Proteins* **2005**, *59*, 687–696. [CrossRef] [PubMed]

23. Bigay, J.; Casella, J.F.; Drin, G.; Mesmin, B.; Antonny, B. ArfGAP1 responds to membrane curvature through the folding of a lipid packing sensor motif. *EMBO J.* **2005**, *24*, 2244–2253. [CrossRef] [PubMed]

24. Koenig, B.W.; Gawrisch, K. Specific volumes of unsaturated phosphatidylcholines in the liquid

crystalline lamellar phase. *Biochim. Biophys. Acta* **2005**, *1715*, 65–70. [CrossRef] [PubMed]

25. Fischer, H.; Polikarpov, I.; Craievich, A.F. Average protein density is a molecular-weight-dependent function. *Protein Sci.* **2004**, *13*, 2825–2828. [CrossRef] [PubMed]

26. Shen, Y.; Bax, A. Protein backbone and sidechain torsion angles predicted from NMR chemical shifts using artificial neural networks. *J. Biomol. NMR* **2013**, *56*, 227–241. [CrossRef] [PubMed]

27. Westaway, E.G.; Khromykh, A.A.; Kenney, M.T.; Mackenzie, J.M.; Jones, M.K. Proteins C and NS4B of the flavivirus Kunjin translocate independently into the nucleus. *Virology* **1997**, *234*, 31–41. [CrossRef] [PubMed]

28. Knoops, K.; Kikkert, M.; Worm, S.H.; Zevenhoven-Dobbe, J.C.; van der Meer, Y.; Koster, A.J.; Mommaas, A.M.; Snijder, E.J. SARS-coronavirus replication is supported by a reticulovesicular network of modified endoplasmic reticulum. *PLoS Biol.* **2008**, *6*, e226. [CrossRef] [PubMed]

9

Flaviviral Replication Complex: Coordination between RNA Synthesis and 5'-RNA Capping

Valerie J. Klema [1], Radhakrishnan Padmanabhan [2] and Kyung H. Choi [1],*

[1] Department of Biochemistry and Molecular Biology, Sealy Center for Structural Biology and Molecular Biophysics, University of Texas Medical Branch at Galveston, Galveston, TX 77555-0647, USA; E-Mail: vaklema@utmb.edu

[2] Department of Microbiology and Immunology, Georgetown University School of Medicine, Washington, DC 20057, USA; E-Mail: rp55@georgetown.edu

* Author to whom correspondence should be addressed; E-Mail: kychoi@utmb.edu

Academic Editor: David D. Boehr

Abstract: Genome replication in flavivirus requires (−) strand RNA synthesis, (+) strand RNA synthesis, and 5′-RNA capping and methylation. To carry out viral genome replication, flavivirus assembles a replication complex, consisting of both viral and host proteins, on the cytoplasmic side of the endoplasmic reticulum (ER) membrane. Two major components of the replication complex are the viral non-structural (NS) proteins NS3 and NS5. Together they possess all the enzymatic activities required for genome replication, yet how these activities are coordinated during genome replication is not clear. We provide an overview of the flaviviral genome replication process, the membrane-bound replication complex, and recent crystal structures of full-length NS5. We propose a model of how NS3 and NS5 coordinate their activities in the individual steps of (−) RNA synthesis, (+) RNA synthesis, and 5′-RNA capping and methylation.

Keywords: viral replication complex; flavivirus; RNA-dependent RNA polymerase; RNA synthesis; 5′-RNA capping; NS5; NS3

1. Flavivirus Genome and Viral Non-Structural (NS) Proteins

Flaviviruses are positive (+) sense RNA viruses belonging to the family *Flaviviridae*, which also includes *Hepacivirus* and *Pestivirus*. The *Flavivirus* genus includes over 70 viruses, many of which cause arboviral diseases in humans, such as dengue (DENV), West Nile (WNV), tick-borne encephalitis (TBEV), and yellow fever virus. The 11 kb flaviviral RNA genome consists of a 5′-cap, a 5′-untranslated region (5′-UTR), a single open reading frame (ORF), and a 3′-UTR. The 5′- and 3′-UTRs contain conserved RNA secondary structures that are important for viral replication, including sequences that mediate long range 5′- and 3′-RNA interactions [1–8]. The viral ORF is translated into a polyprotein, C-prM-E-NS1-NS2A-NS2B-NS3-NS4A-NS4B-NS5, that is subsequently cleaved into individual proteins by viral and host proteases. Three structural proteins (C, prM, and E) form capsids and seven non-structural (NS) proteins (NS1, NS2A, NS2B, NS3, NS4A, NS4B, NS5) are involved in the assembly of the viral replication complex [9]. Among NS proteins, the functions of NS3 and NS5 in viral replication are well established. NS3 consists of an N-terminal serine protease and a C-terminal helicase. NS3 protease activity requires NS2B as a cofactor, and cleaves the viral polyprotein at several positions between NS proteins [10–14]. The NS3 helicase domain has helicase, RNA-stimulated nucleoside triphosphate hydrolase and 5′-RNA triphosphatase activities [15–19]. Helicase activity would be required for unwinding the double-stranded (ds) RNA intermediate formed during genome synthesis, and 5′-RNA triphosphatase activity is required for 5′-RNA cap formation [20,21]. NS5, the largest viral protein (103 kDa), consists of an N-terminal methyltransferase (MTase) and a C-terminal RNA-dependent RNA polymerase (RdRp). The RdRp is involved in viral genome replication and carries out both (−) and (+) strand RNA synthesis [22]. The NS5 MTase has RNA guanylyltransferase and methyltransferase activities necessary for 5′-RNA capping and cap methylations [23,24]. Little is known about functions of the membrane proteins NS2A, NS4A, and NS4B, but they are likely involved in membrane alterations and assembly of the viral replication complex on the cellular membrane [25–29]. NS1 may be involved in multiple steps in the viral life cycle, including viral replication [30–33]. NS1 exists as two forms, either a membrane-bound dimer in the viral replication complex or a secreted hexameric form with as-yet-unknown function [34,35]. More extensive reviews of NS protein functions can be found elsewhere [36,37].

2. Flavivirus Genome Replication

The flavivirus replication complex carries out RNA synthesis, RNA capping and RNA methylation steps to produce the genome with a type 1 cap structure (^{m7}GpppN^m-RNA) at its 5′ end. RNA synthesis in flavivirus is semi-conservative and asymmetric. The (+) sense RNA, which is the same polarity as the viral genome, is predominantly formed over the (−) sense RNA; single-stranded (ss) RNA found in flavivirus-infected cells is (+) sense RNA, and (−) strand RNA is only detected in the dsRNA form [38]. When DENV or Kunjin virus-infected cells were incubated with a radiolabeled NTP (^{3}H-UTP or ^{32}P-labeled guanosine-5'-triphosphate (GTP)), three radiolabeled RNA products were identified: a double-stranded replicative form (RF) RNA, single-stranded genome length RNA, and the slowest migrating, replicative intermediate (RI) [22,39,40]. RNase treatment of the RI suggested that the RI likely contains growing nascent RNAs on the RF template, which displace a pre-existing strand of the same polarity (Figure 1A). One to ten nascent RNA strands are synthesized on one RF template at a time [29,39]. Based on these observations, a general scheme of RNA synthesis and the associated required enzymatic activities are outlined in Figure 1A. The genomic (+) sense RNA is first used as a template by NS5 RdRp to synthesize a complementary (−) sense RNA. The (−) strand RNA remains base-paired with the (+) strand RNA, resulting in a dsRNA intermediate. The (−) strand within the dsRNA intermediate then serves as the template to generate (+) sense RNA. NS3 helicase activity may be required to unwind the dsRNA. The nascent (+) strand synthesized on the (−) RNA template displaces a pre-existing (+) strand and is released as a dsRNA product. The newly generated dsRNA is then recycled as a template to generate additional copies of (+) sense RNA.

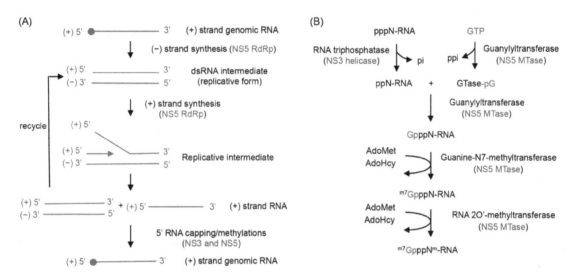

Figure 1. RNA synthesis by the flavivirus replication complex. (**A**) RNA replication by flaviviral NS3 and NS5 proteins. A (+) strand genomic RNA serves as a template to produce (−) strand RNA. The (−) strand RNA exists as a dsRNA intermediate (replicative form). The (−) strand within the dsRNA intermediate is then used as a template for (+) strand RNA synthesis. The dsRNA product is released and recycled for additional (+) strand synthesis. Flavivirus replication is asymmetric, and multiple copies of (+) strand RNA are synthesized from a (−) strand template. The (+) strand RNA is then capped and methylated by NS3 and NS5 to form (+) strand genomic RNA. The identities of the enzymes involved in each step are shown in purple. (**B**) 5'-RNA cap synthesis by flaviviral NS3 and NS5 proteins. The type 1 cap is formed on (+) strand RNA via the four sequential enzyme activities of RNA triphosphatase, guanylyltransferase, guanine-N7-MTase, and RNA 2'O-MTase. First, triphosphatase activity of the NS3 helicase releases the terminal phosphate from the 5'-triphosphate end of (+) strand RNA. A guanosine monophosphate (GMP) moiety from GTP is transferred to the 5' end of the now-diphosphorylated RNA through guanylyltransferase activity of the NS5 MTase. The capped RNA is then methylated first at the N7 position of the guanine cap and subsequently at the ribose 2'-O position of the first RNA nucleotide. The MTase domain of NS5 carries out both methylations using S-adenosyl-L-methionine (AdoMet) as a methyl donor. AdoMet is converted to S-adenosyl-L-homosysteine (AdoHcy) during this process.

The (+) strand progeny RNA is subsequently capped at its 5′ end and methylated to form a type 1 cap (Figure 1B). The cap is shown to be present only on genomic RNA, and not on the dsRNA intermediate (RF form) in WNV-infected cells [38]. The RNA-capping process is likely to occur as the (+) strand RNA is synthesized during the initial stages of RNA synthesis (Figure 1), but little is known about how flavivirus coordinates RNA synthesis and 5′ end RNA capping. The RNA capping and methylation processes require three enzymatic activities. First, the 5′-triphosphate end of (+) RNA is converted into a 5′-diphosphate by the RNA triphosphatase activity of the NS3 helicase. Second, a GMP moiety from GTP is transferred to the 5′-diphosphate RNA by a guanylyltransferase (GTase). The NS5 MTase domain has weak guanylyltransferase activity that transfers a GMP cap from GTP to the 5′ end of (+) sense RNA [23]. Finally, the capped RNA is first methylated at the N7 position of guanine and then at the ribose 2′-OH position of the first nucleotide of the RNA. The MTase domain of NS5 functions as both the N7-MTase and the 2′O-MTase, and transfers a methyl group from the cofactor AdoMet to the substrate capped RNA in both reactions [24,41].

3. Flaviviral Replication Complex

Genome replication in flavivirus is carried out by a membrane-bound viral replication complex consisting of viral NS proteins, viral RNA and unidentified host proteins [9]. Membrane fractions prepared from the lysates of flavivirus-infected cells contained virus-specific RNA and proteins, and retained all of the RdRp activity [29]. Flaviviruses alter host cellular membrane structures, presumably to protect the viral RNA and replication proteins from triggering the host immune response and being degraded by cytoplasmic enzymes. Using electron microscopy and tomography, flavivirus (DENV, WNV, and TBEV)-infected cells have been shown to form spherical single-membrane vesicles of 80–100 nm in diameter via invagination of the ER membrane into the ER lumen [25,42,43] (Figure 2). This contrasts to the related hepatitis C virus, which shows the double membrane vesicle structures with varied sizes between 150–1000 nm [44,45]. Viral NS proteins NS1, NS2B, NS3, NS4A, NS4B and NS5 along with dsRNA are located in the membrane vesicles, suggesting that these vesicles are the sites of RNA replication [28,45]. The vesicles have a pore connecting the interior of the vesicle to the cytoplasm, which is thought to allow exchange of nucleotides and RNA product with the cytoplasm [25,42]. In addition, virus particles have been shown to bud out on ER membranes next to the replication vesicles, suggesting that the viral replication and encapsidation of the viral genome/virion assembly are likely coordinated.

Although the exact composition of the replication complex is not known, all flaviviral NS proteins have been shown to be a component of the replication complex. Using immunoprecipitation, yeast two-hybrid, and fluorescence resonance energy transfer (FRET) assays, interactions among NS proteins have been identified [27,46–48] (Figure 2). Membrane proteins NS2B, NS4A, and NS4B are likely involved in membrane alterations and/or anchoring the viral replication complex to the membrane [26]. NS2B interacts with the three other membrane proteins, NS2A, NS4A, and NS4B [47]. NS4A and NS4B have been proposed to interact with NS1 by genetic studies [49,50]. Oligomerization of NS4A, and dimerizations of NS4B and NS1 have also been reported [51,52]. NS3 itself does not have any membrane-association or transmembrane region. However, the active protease function of NS3 (in the N-terminal domain) requires the cofactor NS2B, thus NS3 localizes to the membrane as an NS3-NS2B complex [15,47]. NS3 has also been shown to interact with NS4B through its C-terminal helicase domain [53]. NS5 also does not have any membrane-associated region and interacts only with NS3. Thus, NS5 likely accumulates to the membrane via NS3-NS5 interactions. Unidentified host proteins are also involved in viral replication. It has been shown that DENV NS5 protein alone was not able to use a viral dsRNA intermediate (RF form) as a template for RNA synthesis. However, the addition of uninfected cell lysate to the NS5 reaction could restore polymerase activity, suggesting that host proteins are also involved in viral replication [40]. Consequently, an *in vitro* flaviviral replication system, which can use the viral genome to synthesize the methylated, capped RNA product, is currently unavailable.

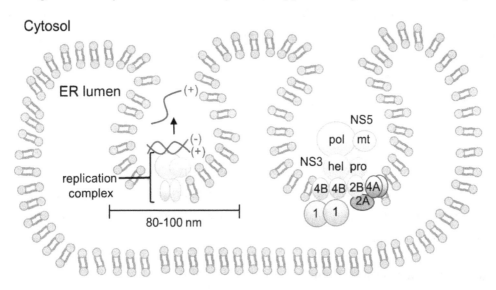

Figure 2. Flaviviral replication complex assembled on the cytoplasmic side of the ER membrane. The viral replication complex is associated with the virus vesicle (single membrane vesicle) formed by invagination of the ER membrane. Interactions between flaviviral NS proteins are indicated schematically. Viral membrane proteins NS2A, NS2B, NS4A, and NS4B form a scaffold for the assembly of NS3 and NS5 proteins. Oligomerization of NS4A and dimerization of NS4B are depicted [51,52]. NS3 interacts with NS2B through its protease domain and with NS4B through its helicase domain. NS5 interacts only with NS3. The NS1 dimer is located in the ER lumen, and associates with NS4A and NS4B.

4. Inter- and Intramolecular Coordination between NS5 and NS3 during Genome Replication

At the heart of the flaviviral replication complex are NS3 and NS5, which together account for all enzymatic activities required to amplify the RNA genome and to attach a type 1 cap to its 5' end (Figure 1). Other viral and host proteins in the viral replication complex would provide additional efficiency and specificity for viral replication. NS5 is involved in RNA synthesis, 5'-RNA cap transfer, and cap methylations. The NS3 helicase domain is involved in dsRNA unwinding and removal of the γ-phosphate at the 5'-RNA prior to RNA capping. However, how NS3 and NS5 activities are coordinated during viral replication is not known. It is possible that individual activities of the two proteins are modulated by mutual interaction with each other as well as with other proteins and viral RNA [23,54–56]. Interactions between NS3 and NS5 proteins from several flaviviral species have been shown using pull-down assays from infected cell extracts, *in vivo* fluorescent measurements, and yeast two-hybrid studies [46,47,57]. Their respective binding sites map to the C-terminal domain of NS3 (residues 303–618) and the RdRp domain of NS5 (residues 320–368, which also includes a nuclear localization sequence) [58–60].

The physical linkage between the MTase and RdRp domains within NS5 suggests that viral genome replication and 5′-RNA capping may be coordinated between the two domains. Several genetic studies show that mutations in the MTase domain impact RNA replication by the RdRp domain, suggesting that the RdRp and MTase interact during viral genome replication [61–64]. In addition, a full-length DENV type 2 RNA containing an NS5 chimera, in which the MTase from DENV type 4 is fused to the RdRp from DENV type 2, cannot carry out replication although the MTase from DENV type 2 and 4 NS5 share high sequence identity (~70%) [65]. This suggests that sequence-specific interactions between the MTase and RdRp domains are necessary for viral replication. Consequently, there has been great interest in carrying out structural and biochemical studies to define whether and/or how coordination between these activities may occur, and how coordination may be exploited to design novel antiviral agents effective against flavivirus infection [66,67]. The individual MTase and RdRp domains from several flaviviral sources have been structurally characterized, but until recently there were no available structures of full-length flavivirus NS5. The crystal structures of full-length NS5 from Japanese encephalitis virus (JEV) and DENV were recently reported, and provided a first glimpse into how the MTase and RdRp domains interact.

5. Full-Length NS5 Structure and Function

The recently reported structures of full-length NS5 from JEV and DENV offer details of specific interactions between the MTase and RdRp domains and suggest a structural mechanism by which their activities may be coordinated within the flavivirus replication complex [68,69]. We have recently determined the crystal structure of DENV NS5 as a dimer (in preparation), and will provide an overview of the three structures (Figure 3). Briefly, the MTase domain contains a central core structure characteristic of AdoMet-dependent MTases consisting of a 7-stranded β-sheet surrounded by 4 α-helices. The RdRp domain adopts the shape of a closed right hand and consists of palm, fingers, and thumb subdomains. The inner surfaces of the fingers and thumb subdomains form the entrance to a template-binding channel that leads to the active site in the palm subdomain. The locations of respective active sites in the MTase and the RdRp are indicated in Figure 3. Surprisingly, in both JEV and DENV full-length NS5 the active sites of the MTase and RdRp domains are located on opposite faces and do not interact with each other. The MTase domain sits "behind" the RdRp domain when viewed in its right hand orientation, opposite the dsRNA exit channel and close to the template-binding channel (Figure 3). In all three structures, the individual MTase and RdRp domains are nearly superimposable with one another. The major difference between the JEV and DENV NS5 structures is the relative orientation of the MTase and RdRp domains, primarily due to different conformations of the inter-domain linker (residues 263-272, DENV type 3 numbering). The MTase domains in JEV and DENV NS5 are rotated relative to one another by ~100°. A strictly conserved ^{260}Gly-Thr-Arg^{262} (GTR) motif, N-terminal to the inter-domain linker, is proposed to act as a hinge that allows movement of the MTase and RdRp domains relative to one another [68–70]. Accordingly, the relative orientation of the MTase and RdRp active sites are quite different in JEV and DENV NS5, and the domain interfaces involve different sets of MTase residues to interact with the same area of the RdRp. In the JEV NS5, the domain interface centers around a hydrophobic core consisting of residues P113, L115, and W121 from the MTase domain, and F467, F351, and P585 from the RdRp. In contrast, the domain interface in DENV NS5 is stabilized mainly by polar and electrostatic interactions involving MTase residues Q63, E252, and D256 to interact with relatively the same area of RdRp (Figure 3). Interestingly, the MTase residues involved in the inter-domain interactions in the JEV NS5 monomer (P113, P115, and W121 in DENV numbering) are present at the dimer interface in our DENV NS5 dimer structure. Thus, the same MTase region mediates intra- and inter-molecular interactions with the RdRp domain in JEV and DENV NS5, respectively, suggesting that NS5 may also function as a dimer (Figure 3). Taken together, comparison of the full-length NS5 structures from JEV and DENV shows that flavivirus NS5 can alter relative MTase and RdRp domain orientations and its monomer-dimer state by modulating its linker region. This may endow NS5 with the flexibility it needs to carry out multiple functions in the replication complex and interact with other proteins at different steps of viral replication (see below).

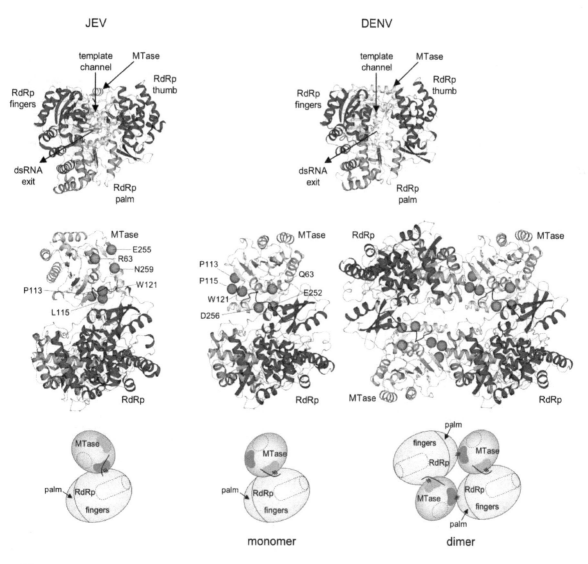

Figure 3. Comparison of full-length NS5 structures from JEV and DENV. Overall fold of JEV and DENV full-length NS5 (Protein Data Bank entries 4K6M and 4VOQ) are shown from

the canonical right hand view of RdRp (top) and side view (bottom). Side views of both monomer and dimer forms of DENV NS5 are shown. The RdRp domain consists of fingers (green), palm (blue), and thumb (red) subdomains. The MTase domain (cyan) sits opposite the dsRNA exit channel of the RdRp and near the template-binding channel. The methyl donor used during MTase reaction, AdoMet, is shown in yellow. In each structure, the same region of the RdRp contacts the MTase. The MTase residues at the domain interface from JEV and DENV are indicated by pink and orange spheres, respectively. For comparison, corresponding residues are also mapped on the structures. Domain interactions are also indicated in schematics by a series of three lines below the side view. The MTase active site is indicated by dashed ovals, and the RdRp template channel is shown as a dashed tube.

6. Model of Flaviviral Replication

The multifunctional flavivirus NS5 carries out several reactions. These include synthesis of (−) strand RNA, synthesis of (+) strand RNA, addition of a guanine cap to the 5′ end of (+) strand RNA, and two methylation reactions to form a type 1 cap structure at the 5′ end of the nascent RNA. Consequently, NS5 interacts with other viral proteins and several different forms of viral RNA, including both single- and double-stranded, (+) and (−) strand, capped and uncapped, and methylated and unmethylated forms of capped RNA. Thus, NS5 likely needs to adopt multiple conformations during different stages of genome replication and processing, as suggested by the conformational variation observed for the structures of full-length NS5 from JEV and DENV. Below, we summarize viral replication steps in terms of coordination of NS3 and NS5 functions.

1. (−) strand RNA synthesis: (−) sense RNA synthesis is carried out by NS5 polymerase using the viral (+) RNA genome as a template. The function of NS3 does not seem to be required for (−) strand RNA synthesis, since NS5 (either the full-length protein or the polymerase domain) alone is capable of synthesizing RNA using viral subgenomic RNA. The (+) sense RNA genome contains cyclization sequences at the 5′- and 3′-UTR, and the cyclized genome serves as the template for (−) strand RNA synthesis [1–8,71]. NS5 polymerase recognizes a conserved RNA structure called stem loop A (SLA) within the 5′-UTR as a promoter, and initiates (−) strand RNA synthesis at the 3′ end of the genome. The nascent (−) RNA product exists as dsRNA, base-paired with the (+) strand RNA template [29].

2. (+) strand RNA synthesis: The viral replication complex uses the (−) strand RNA in the dsRNA intermediate as the template to synthesize multiple copies of (+) sense RNA. Both NS5 and NS3 activities are required for (+) RNA synthesis (Figure 4). NS3 helicase first unwinds the dsRNA intermediate into (+) and (−) sense RNAs. Following strand separation, the 3′ end of (−) sense RNA will bind the template-binding channel of NS5 RdRp and serve as a template for (+) strand RNA synthesis. Because (−) strand RNA only exists in the dsRNA form, specific RNA interactions, such as those between the cyclization sequences or stem loop structures necessary for (−) strand synthesis, are not required for (+) sense RNA synthesis [29,72]. Upon completion of nascent (+) sense RNA synthesis, a dsRNA product (consisting of the nascent (+) strand and the

(−) strand template) will be released from the RdRp and recycled for another round of (+) strand RNA synthesis [29,36].

3. 5'-RNA capping and methylation of (+) sense RNA: The (+) sense RNA will be capped and methylated to form a type 1 cap at the 5' end (Figure 1B). NS3 helicase hydrolyzes the 5'-terminal phosphate of (+) sense RNA and converts it to a diphosphate using its RNA triphosphatase activity. The 5'-diphosphorylated (+) sense RNA then binds the NS5 MTase for capping and methylations. For 5'-RNA capping, NS5 MTase first reacts with GTP to form a covalently linked GMP-enzyme intermediate, and then transfers the GMP moiety to the 5'-diphosphate of the (+) strand RNA [23]. Next, the MTase methylates the N7 position of the guanine cap and the ribose 2'-OH position of the first nucleotide. Each methylation reaction requires distinct RNA sequences and lengths, and N7 cap methylation requires the presence of the stem loop A in the 5'-UTR of the viral genome [73]. Since the NS5 MTase carries out three separate reactions using one active site, the 5' end of (+) RNA needs to dissociate and then re-associate with the MTase at each step [74].

4. Coordination of (+) RNA synthesis and 5'-RNA capping: The 5'-RNA capping and methylation steps are likely coupled with (+) sense RNA synthesis, but during which step of RNA synthesis 5'-RNA capping and methylation occur is not clear. Capping of 5'-RNA could occur on the nascent (+) strand RNA while it is being synthesized by the NS5 RdRp domain (co-transcriptional model in Figure 4A). After a short stretch of (+) sense RNA is synthesized, the 5' end of the nascent (+) RNA could be dephosphorylated by NS3 helicase, and capped and methylated by NS5 MTase. The 5'-capped (+) sense RNA will then be continuously synthesized until the entire (−) strand is copied. Alternatively, the 5'-RNA capping could occur on the preexisting full-length (+) RNA that is separated from the (−) strand template RNA by NS3 helicase (post-transcriptional model in Figure 4A). Upon unwinding of dsRNA, NS3 helicase hydrolyzes the 5'-terminal phosphate of (+) sense RNA and NS5 MTase subsequently attaches the type 1 cap to the 5' end. The consequential difference between the two mechanisms would be whether the (+) strand RNA in the dsRNA form is capped. In the co-transcriptional model, all (+) strand RNA, either ssRNA or dsRNA, would be capped and methylated because the nascent RNA is co-transcriptionally capped and methylated. In the post-transcriptional model, only displaced (+) strand RNA, and not the (+) sense RNA in the dsRNA form, would be capped and methylated. Since the cap is shown to be present only on genomic (+) sense RNA, and not on the dsRNA form in WNV-infected cell [29,38], it seems likely that 5'-capping occurs on the fully synthesized (+) sense RNA. In this case, upon completion of a cycle of nascent (+) sense RNA synthesis, a dsRNA product and a capped (+) strand RNA (identical to the viral genome) will be released from the RdRp and MTase domains of NS5 (Figure 4B).

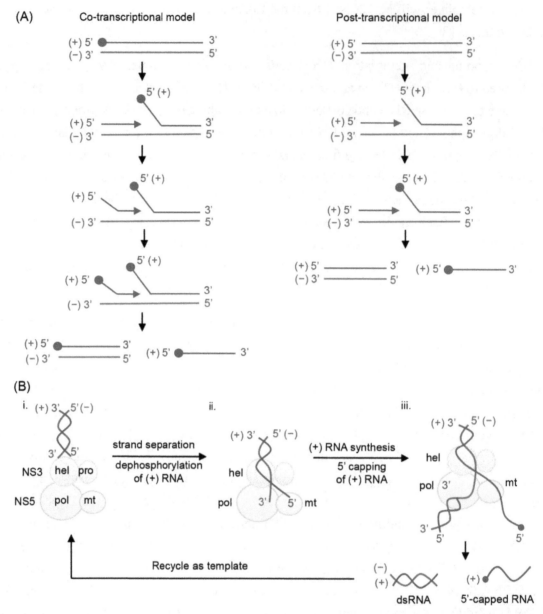

Figure 4. Possible mechanisms of coordination between RNA synthesis and 5′-RNA capping. (**A**) Two possible mechanisms are proposed for 5′-RNA capping for (+) sense RNA. In a co-transcriptional model, 5′-RNA capping occurs co-transcriptionally while nascent (+) sense RNA is being synthesized from dsRNA replicative form. After a short stretch of (+) sense RNA being synthesized, the 5′ end of the nascent (+) RNA would be dephosphorylated by NS3 helicase, and capped and methylated by NS5 MTase. The cap structure is depicted with a closed circle. The 5′-capped (+) sense RNA would then be continuously synthesized until the entire (−) strand is copied. Upon completion of a cycle, a dsRNA containing a capped (+) strand and a capped (+) RNA are synthesized. In a post-transcriptional model, 5′-RNA capping occurs on the fully synthesized (+) strand RNA. Following strand separation of dsRNA, the (+) sense RNA could be dephoshphorylated by NS3 helicase, and capped and methylated by NS5 MTase. Upon completion of a cycle, a dsRNA and a capped (+) RNA are synthesized. (**B**) Coordination of RNA synthesis and 5′-RNA capping by flavivirus NS3 and NS5 in the post-transcriptional model. (**i**) NS3 helicase unwinds the dsRNA intermediate into

(+) and (−) sense RNA, and hydrolyzes the 5' end of (+) sense RNA by its 5'-RNA triphosphatase activity. (**ii**) The 3' end of the (−) strand RNA will enter the NS5 RdRp template-binding channel, and serve as a template for (+) strand RNA synthesis. The 5'-dephosphorylated (+) RNA enters the NS5 MTase active site for 5'-RNA capping and methylations. (**iii**) Upon completion of nascent (+) sense RNA synthesis, a dsRNA and a 5'-capped (+) strand RNA are released from the NS5 RdRp and MTase domains. The dsRNA product, consisting of the (−) strand template and nascent (+) strand, is then recycled for another round of (+) strand RNA synthesis.

7. Conclusions and Future Perspectives

The identification and visualization of subcellular membrane structures used by flavivirus as replication factories have provided insight into how flavivirus may orchestrate viral replication and virion assembly in infected cells. Additionally, crystal structures of the full-length NS5 have provided details of intra- and inter-molecular interactions that stabilize different forms of NS5. These forms may interact with RNA and protein components of the replication complex differently and thus serve as a platform to promote different steps during genome replication. Future studies will be geared toward understanding how these membrane vesicles are related to the assembly and function of the viral replication complex at the molecular level. In particular, how flaviviral and host proteins establish such elaborate membrane vesicles for replication, how flaviviral NS proteins are assembled in the membrane-bound viral replication complex, and how the viral replication complex carries out individual steps of RNA replication are of great interest.

Acknowledgments

This work was supported by NIH Research Grants AI087856 (to K.H.C. and R.P) and AI105985 (to K.H.C.).

Author Contributions

V.J.K., R.P. and K.H.C. wrote the paper.

References

1. Lodeiro, M.F.; Filomatori, C.V.; Gamarnik, A.V. Structural and functional studies of the promoter element for dengue virus RNA replication. *J. Virol.* **2009**, *83*, 993–1008. [CrossRef] [PubMed]
2. Filomatori, C.V.; Lodeiro, M.F.; Alvarez, D.E.; Samsa, M.M.; Pietrasanta, L.; Gamarnik, A.V. A 5' RNA element promotes dengue virus RNA synthesis on a circular genome. *Genes Dev.* **2006**, *20*, 2238–2249. [CrossRef] [PubMed]
3. Ackermann, M.; Padmanabhan, R. De novo synthesis of RNA by the dengue virus RNA-dependent RNA polymerase exhibits temperature dependence at the initiation but not elongation phase. *J. Biol. Chem.* **2001**, *276*, 39926–39937. [CrossRef] [PubMed]

4. You, S.; Falgout, B.; Markoff, L.; Padmanabhan, R. *In vitro* RNA synthesis from exogenous dengue viral rna templates requires long range interactions between 5′- and 3′-terminal regions that influence RNA structure. *J. Biol. Chem.* **2001**, *276*, 15581–15591. [CrossRef] [PubMed]

5. Hahn, C.S.; Hahn, Y.S.; Rice, C.M.; Lee, E.; Dalgarno, L.; Strauss, E.G.; Strauss, J.H. Conserved elements in the 3′ untranslated region of flavivirus RNAs and potential cyclization sequences. *J. Mol. Biol.* **1987**, *198*, 33–41. [CrossRef]

6. Khromykh, A.A.; Meka, H.; Guyatt, K.J.; Westaway, E.G. Essential role of cyclization sequences in flavivirus RNA replication. *J. Virol.* **2001**, *75*, 6719–6728. [CrossRef] [PubMed]

7. Corver, J.; Lenches, E.; Smith, K.; Robison, R.A.; Sando, T.; Strauss, E.G.; Strauss, J.H. Fine mapping of a cis-acting sequence element in yellow fever virus RNA that is required for RNA replication and cyclization. *J. Virol.* **2003**, *77*, 2265–2270. [CrossRef] [PubMed]

8. Lo, M.K.; Tilgner, M.; Bernard, K.A.; Shi, P.Y. Functional analysis of mosquito-borne flavivirus conserved sequence elements within 3′ untranslated region of West Nile virus by use of a reporting replicon that differentiates between viral translation and RNA replication. *J. Virol.* **2003**, *77*, 10004–10014. [CrossRef] [PubMed]

9. Lindenbach, B.D.; Thiel, H.J.; Rice, C.M. *Flaviviridae: The Viruses and Their Replication*; Lippincott-Raven Publishers: Philadelphia, PA, USA, 2007; Volume 1, pp. 1101–1152.

10. Falgout, B.; Pethel, M.; Zhang, Y.M.; Lai, C.J. Both nonstructural proteins NS2B and NS3 are required for the proteolytic processing of dengue virus nonstructural proteins. *J. Virol.* **1991**, *65*, 2467–2475. [PubMed]

11. Yusof, R.; Clum, S.; Wetzel, M.; Murthy, H.M.; Padmanabhan, R. Purified NS2B/NS3 serine protease of dengue virus type 2 exhibits cofactor NS2B dependence for cleavage of substrates with dibasic amino acids *in vitro*. *J. Biol. Chem.* **2000**, *275*, 9963–9969. [CrossRef] [PubMed]

12. Chambers, T.J.; Weir, R.C.; Grakoui, A.; McCourt, D.W.; Bazan, J.F.; Fletterick, R.J.; Rice, C.M. Evidence that the N-terminal domain of nonstructural protein NS3 from yellow fever virus is a serine protease responsible for site-specific cleavages in the viral polyprotein. *Proc. Natl. Acad. Sci. USA* **1990**, *87*, 8898–8902. [CrossRef] [PubMed]

13. Jan, L.R.; Yang, C.S.; Trent, D.W.; Falgout, B.; Lai, C.J. Processing of japanese encephalitis virus non-structural proteins: NS2B-NS3 complex and heterologous proteases. *J. Gen. Virol.* **1995**, *76* (Pt 3), 573–580. [CrossRef] [PubMed]

14. Clum, S.; Ebner, K.E.; Padmanabhan, R. Cotranslational membrane insertion of the serine proteinase precursor NS2B-NS3(pro) of dengue virus type 2 is required for efficient *in vitro* processing and is mediated through the hydrophobic regions of NS2B. *J. Biol. Chem.* **1997**, *272*, 30715–30723. [CrossRef] [PubMed]

15. Li, H.; Clum, S.; You, S.; Ebner, K.E.; Padmanabhan, R. The serine protease and RNA-stimulated nucleoside triphosphatase and RNA helicase functional domains of dengue virus type 2 NS3 converge within a region of 20 amino acids. *J. Virol.* **1999**, *73*, 3108–3116. [PubMed]

16. Matusan, A.E.; Pryor, M.J.; Davidson, A.D.; Wright, P.J. Mutagenesis of the dengue virus type 2 NS3 protein within and outside helicase motifs: Effects on enzyme activity and virus replication. *J. Virol.* **2001**, *75*, 9633–9643. [CrossRef] [PubMed]

17. Bartelma, G.; Padmanabhan, R. Expression, purification, and characterization of the RNA

5'-triphosphatase activity of dengue virus type 2 nonstructural protein 3. *Virology* **2002**, *299*, 122–132. [CrossRef] [PubMed]

18. Benarroch, D.; Selisko, B.; Locatelli, G.A.; Maga, G.; Romette, J.L.; Canard, B. The RNA helicase, nucleotide 5'-triphosphatase, and RNA 5'-triphosphatase activities of dengue virus protein NS3 are Mg^{2+}-dependent and require a functional walker B motif in the helicase catalytic core. *Virology* **2004**, *328*, 208–218. [CrossRef] [PubMed]

19. Wengler, G.; Wengler, G. The NS 3 nonstructural protein of flaviviruses contains an RNA triphosphatase activity. *Virology* **1993**, *197*, 265–273. [CrossRef] [PubMed]

20. Xu, T.; Sampath, A.; Chao, A.; Wen, D.; Nanao, M.; Chene, P.; Vasudevan, S.G.; Lescar, J. Structure of the dengue virus helicase/nucleoside triphosphatase catalytic domain at a resolution of 2.4 a. *J. Virol.* **2005**, *79*, 10278–10288. [CrossRef] [PubMed]

21. Wang, C.C.; Huang, Z.S.; Chiang, P.L.; Chen, C.T.; Wu, H.N. Analysis of the nucleoside triphosphatase, RNA triphosphatase, and unwinding activities of the helicase domain of dengue virus NS3 protein. *FEBS Lett.* **2009**, *583*, 691–696. [CrossRef] [PubMed]

22. Chu, P.W.; Westaway, E.G. Replication strategy of kunjin virus: Evidence for recycling role of replicative form RNA as template in semiconservative and asymmetric replication. *Virology* **1985**, *140*, 68–79. [CrossRef]

23. Issur, M.; Geiss, B.J.; Bougie, I.; Picard-Jean, F.; Despins, S.; Mayette, J.; Hobdey, S.E.; Bisaillon, M. The flavivirus NS5 protein is a true RNA guanylyltransferase that catalyzes a two-step reaction to form the RNA cap structure. *RNA* **2009**, *15*, 2340–2350. [CrossRef] [PubMed]

24. Egloff, M.P.; Benarroch, D.; Selisko, B.; Romette, J.L.; Canard, B. An RNA cap (nucleoside-2'-o-)-methyltransferase in the flavivirus RNA polymerase NS5: Crystal structure and functional characterization. *EMBO J.* **2002**, *21*, 2757–2768. [CrossRef] [PubMed]

25. Welsch, S.; Miller, S.; Romero-Brey, I.; Merz, A.; Bleck, C.K.; Walther, P.; Fuller, S.D.; Antony, C.; Krijnse-Locker, J.; Bartenschlager, R. Composition and three-dimensional architecture of the dengue virus replication and assembly sites. *Cell. Host Microbe* **2009**, *5*, 365–375. [CrossRef] [PubMed]

26. Miller, S.; Kastner, S.; Krijnse-Locker, J.; Buhler, S.; Bartenschlager, R. The non-structural protein 4A of dengue virus is an integral membrane protein inducing membrane alterations in a 2k-regulated manner. *J. Biol. Chem.* **2007**, *282*, 8873–8882. [CrossRef] [PubMed]

27. Westaway, E.G.; Mackenzie, J.M.; Kenney, M.T.; Jones, M.K.; Khromykh, A.A. Ultrastructure of kunjin virus-infected cells: Colocalization of NS1 and NS3 with double-stranded RNA, and of NS2b with NS3, in virus-induced membrane structures. *J. Virol.* **1997**, *71*, 6650–6661. [PubMed]

28. Mackenzie, J.M.; Khromykh, A.A.; Jones, M.K.; Westaway, E.G. Subcellular localization and some biochemical properties of the flavivirus kunjin nonstructural proteins NS2A and NS4A. *Virology* **1998**, *245*, 203–215. [CrossRef] [PubMed]

29. Westaway, E.G.; Mackenzie, J.M.; Khromykh, A.A. Kunjin RNA replication and applications of kunjin replicons. *Adv. Virus Res.* **2003**, *59*, 99–140. [PubMed]

30. Youn, S.; Ambrose, R.L.; Mackenzie, J.M.; Diamond, M.S. Non-structural protein-1 is required for West Nile virus replication complex formation and viral RNA synthesis. *Virol. J.* **2013**, *10*, e339. [CrossRef] [PubMed]

31. Mackenzie, J.M.; Jones, M.K.; Young, P.R. Immunolocalization of the dengue virus nonstructural glycoprotein NS1 suggests a role in viral RNA replication. *Virology* **1996**, *220*, 232–240. [CrossRef] [PubMed]

32. Khromykh, A.A.; Sedlak, P.L.; Guyatt, K.J.; Hall, R.A.; Westaway, E.G. Efficient trans-complementation of the flavivirus kunjin NS5 protein but not of the NS1 protein requires its coexpression with other components of the viral replicase. *J. Virol.* **1999**, *73*, 10272–10280. [PubMed]

33. Lindenbach, B.D.; Rice, C.M. Trans-complementation of yellow fever virus NS1 reveals a role in early RNA replication. *J. Virol.* **1997**, *71*, 9608–9617. [PubMed]

34. Flamand, M.; Megret, F.; Mathieu, M.; Lepault, J.; Rey, F.A.; Deubel, V. Dengue virus type 1 nonstructural glycoprotein NS1 is secreted from mammalian cells as a soluble hexamer in a glycosylation-dependent fashion. *J. Virol.* **1999**, *73*, 6104–6110. [PubMed]

35. Winkler, G.; Maxwell, S.E.; Ruemmler, C.; Stollar, V. Newly synthesized dengue-2 virus nonstructural protein NS1 is a soluble protein but becomes partially hydrophobic and membrane-associated after dimerization. *Virology* **1989**, *171*, 302–305. [CrossRef]

36. Roby, J.A.; Funk, A.; Khromykh, A.A. Flavivirus replication and assembly. In *Molecular Virology and Control of Flaviviruses*; Shi, P.Y., Ed.; Caister Academic Press: Norfolk, UK, 2012.

37. Brinton, M.A. Replication cycle and molecular biology of the West Nile virus. *Viruses* **2014**, *6*, 13–53. [CrossRef] [PubMed]

38. Wengler, G.; Wengler, G.; Gross, H.J. Studies on virus-specific nucleic acids synthesized in vertebrate and mosquito cells infected with flaviviruses. *Virology* **1978**, *89*, 423–437. [CrossRef]

39. Cleaves, G.R.; Ryan, T.E.; Schlesinger, R.W. Identification and characterization of type 2 dengue virus replicative intermediate and replicative form RNAs. *Virology* **1981**, *111*, 73–83. [CrossRef]

40. Raviprakash, K.; Sinha, M.; Hayes, C.G.; Porter, K.R. Conversion of dengue virus replicative form RNA (RF) to replicative intermediate (RI) by nonstructural proteins NS-5 and NS-3. *Am. J. Trop. Med. Hyg.* **1998**, *58*, 90–95. [PubMed]

41. Ray, D.; Shah, A.; Tilgner, M.; Guo, Y.; Zhao, Y.; Dong, H.; Deas, T.S.; Zhou, Y.; Li, H.; Shi, P.Y. West Nile virus 5′-cap structure is formed by sequential guanine N-7 and ribose 2′-O methylations by nonstructural protein 5. *J. Virol.* **2006**, *80*, 8362–8370. [CrossRef] [PubMed]

42. Miorin, L.; Romero-Brey, I.; Maiuri, P.; Hoppe, S.; Krijnse-Locker, J.; Bartenschlager, R.; Marcello, A. Three-dimensional architecture of tick-borne encephalitis virus replication sites and trafficking of the replicated RNA. *J. Virol.* **2013**, *87*, 6469–6481. [CrossRef] [PubMed]

43. Gillespie, L.K.; Hoenen, A.; Morgan, G.; Mackenzie, J.M. The endoplasmic reticulum provides the membrane platform for biogenesis of the flavivirus replication complex. *J. Virol.* **2010**, *84*, 10438–10447. [CrossRef] [PubMed]

44. Romero-Brey, I.; Merz, A.; Chiramel, A.; Lee, J.Y.; Chlanda, P.; Haselman, U.; Santarella-Mellwig, R.; Habermann, A.; Hoppe, S.; Kallis, S.; *et al.* Three-dimensional architecture and biogenesis of membrane structures associated with hepatitis C virus replication. *PLoS Pathog.* **2012**, *8*, e1003056. [CrossRef] [PubMed]

45. Harak, C.; Lohmann, V. Ultrastructure of the replication sites of positive-strand RNA viruses. *Virology* **2015**, *479–480*. [CrossRef] [PubMed]

46. Kapoor, M.; Zhang, L.; Ramachandra, M.; Kusukawa, J.; Ebner, K.E.; Padmanabhan, R. Association between NS3 and NS5 proteins of dengue virus type 2 in the putative RNA replicase is linked to differential phosphorylation of NS5. *J. Biol. Chem.* **1995**, *270*, 19100–19106. [PubMed]

47. Yu, L.; Takeda, K.; Markoff, L. Protein-protein interactions among West Nile non-structural proteins and transmembrane complex formation in mammalian cells. *Virology* **2013**, *446*, 365–377. [CrossRef] [PubMed]

48. Chen, C.J.; Kuo, M.D.; Chien, L.J.; Hsu, S.L.; Wang, Y.M.; Lin, J.H. RNA-protein interactions: Involvement of NS3, NS5, and 3′ noncoding regions of japanese encephalitis virus genomic RNA. *J. Virol.* **1997**, *71*, 3466–3473. [PubMed]

49. Youn, S.; Li, T.; McCune, B.T.; Edeling, M.A.; Fremont, D.H.; Cristea, I.M.; Diamond, M.S. Evidence for a genetic and physical interaction between nonstructural proteins NS1 and NS4B that modulates replication of West Nile virus. *J. Virol.* **2012**, *86*, 7360–7371. [CrossRef] [PubMed]

50. Lindenbach, B.D.; Rice, C.M. Genetic interaction of flavivirus nonstructural proteins NS1 and NS4A as a determinant of replicase function. *J. Virol.* **1999**, *73*, 4611–4621. [PubMed]

51. Stern, O.; Hung, Y.F.; Valdau, O.; Yaffe, Y.; Harris, E.; Hoffmann, S.; Willbold, D.; Sklan, E.H. An N-terminal amphipathic helix in dengue virus nonstructural protein 4A mediates oligomerization and is essential for replication. *J. Virol.* **2013**, *87*, 4080–4085. [CrossRef] [PubMed]

52. Zou, J.; Xie, X.; Lee le, T.; Chandrasekaran, R.; Reynaud, A.; Yap, L.; Wang, Q.Y.; Dong, H.; Kang, C.; Yuan, Z.; *et al.* Dimerization of flavivirus NS4B protein. *J. Virol.* **2014**, *88*, 3379–3391. [CrossRef] [PubMed]

53. Umareddy, I.; Chao, A.; Sampath, A.; Gu, F.; Vasudevan, S.G. Dengue virus NS4B interacts with NS3 and dissociates it from single-stranded RNA. *J. Gen. Virol.* **2006**, *87*, 2605–2614. [CrossRef] [PubMed]

54. Cui, T.; Sugrue, R.J.; Xu, Q.; Lee, A.K.; Chan, Y.C.; Fu, J. Recombinant dengue virus type 1 NS3 protein exhibits specific viral RNA binding and NTPase activity regulated by the NS5 protein. *Virology* **1998**, *246*, 409–417. [CrossRef] [PubMed]

55. Yon, C.; Teramoto, T.; Mueller, N.; Phelan, J.; Ganesh, V.K.; Murthy, K.H.; Padmanabhan, R. Modulation of the nucleoside triphosphatase/RNA helicase and 5′-RNA triphosphatase activities of dengue virus type 2 nonstructural protein 3 (NS3) by interaction with NS5, the RNA-dependent RNA polymerase. *J. Biol. Chem.* **2005**, *280*, 27412–27419. [CrossRef] [PubMed]

56. Liu, W.J.; Sedlak, P.L.; Kondratieva, N.; Khromykh, A.A. Complementation analysis of the flavivirus kunjin NS3 and NS5 proteins defines the minimal regions essential for formation of a replication complex and shows a requirement of NS3 in cis for virus assembly. *J. Virol.* **2002**, *76*, 10766–10775. [CrossRef] [PubMed]

57. Johansson, M.; Brooks, A.J.; Jans, D.A.; Vasudevan, S.G. A small region of the dengue virus-encoded RNA-dependent RNA polymerase, NS5, confers interaction with both the nuclear transport receptor importin-beta and the viral helicase, NS3. *J. Gen. Virol.* **2001**, *82*, 735–745. [PubMed]

58. Tay, M.Y.; Saw, W.G.; Zhao, Y.; Chan, K.W.; Singh, D.; Chong, Y.; Forwood, J.K.; Ooi, E.E.;

Gruber, G.; Lescar, J.; *et al.* The C-terminal 50 amino acid residues of dengue NS3 protein are important for NS3-NS5 interaction and viral replication. *J. Biol. Chem.* **2015**, *290*, 2379–2394. [CrossRef] [PubMed]

59. Moreland, N.J.; Tay, M.Y.; Lim, E.; Rathore, A.P.; Lim, A.P.; Hanson, B.J.; Vasudevan, S.G. Monoclonal antibodies against dengue NS2B and NS3 proteins for the study of protein interactions in the flaviviral replication complex. *J. Virol. Methods* **2012**, *179*, 97–103. [CrossRef] [PubMed]

60. Zou, G.; Chen, Y.L.; Dong, H.; Lim, C.C.; Yap, L.J.; Yau, Y.H.; Shochat, S.G.; Lescar, J.; Shi, P.Y. Functional analysis of two cavities in flavivirus NS5 polymerase. *J. Biol. Chem.* **2011**, *286*, 14362–14372. [CrossRef] [PubMed]

61. Tan, C.S.; Hobson-Peters, J.M.; Stoermer, M.J.; Fairlie, D.P.; Khromykh, A.A.; Hall, R.A. An interaction between the methyltransferase and RNA dependent RNA polymerase domains of the West Nile virus NS5 protein. *J. Gen. Virol.* **2013**, *94*, 1961–1971. [CrossRef] [PubMed]

62. Zhang, B.; Dong, H.; Zhou, Y.; Shi, P.Y. Genetic interactions among the West Nile virus methyltransferase, the RNA-dependent RNA polymerase, and the 5′ stem-loop of genomic RNA. *J. Virol.* **2008**, *82*, 7047–7058. [CrossRef] [PubMed]

63. Malet, H.; Egloff, M.P.; Selisko, B.; Butcher, R.E.; Wright, P.J.; Roberts, M.; Gruez, A.; Sulzenbacher, G.; Vonrhein, C.; Bricogne, G.; *et al.* Crystal structure of the RNA polymerase domain of the West Nile virus non-structural protein 5. *J. Biol. Chem.* **2007**, *282*, 10678–10689. [CrossRef] [PubMed]

64. Potisopon, S.; Priet, S.; Collet, A.; Decroly, E.; Canard, B.; Selisko, B. The methyltransferase domain of dengue virus protein NS5 ensures efficient RNA synthesis initiation and elongation by the polymerase domain. *Nucleic Acids Res.* **2014**, *42*, 11642–11656. [CrossRef] [PubMed]

65. Teramoto, T.; Boonyasuppayakorn, S.; Handley, M.; Choi, K.H.; Padmanabhan, R. Substitution of NS5 N-terminal domain of dengue virus type 2 RNA with type 4 domain caused impaired replication and emergence of adaptive mutants with enhanced fitness. *J. Biol. Chem.* **2014**, *289*, 22385–22400. [CrossRef] [PubMed]

66. Bussetta, C.; Choi, K.H. Dengue virus nonstructural protein 5 adopts multiple conformations in solution. *Biochemistry* **2012**, *51*, 5921–5931. [CrossRef] [PubMed]

67. Takahashi, H.; Takahashi, C.; Moreland, N.J.; Chang, Y.T.; Sawasaki, T.; Ryo, A.; Vasudevan, S.G.; Suzuki, Y.; Yamamoto, N. Establishment of a robust dengue virus NS3-NS5 binding assay for identification of protein-protein interaction inhibitors. *Antivir. Res.* **2012**, *96*, 305–314. [CrossRef] [PubMed]

68. Lu, G.; Gong, P. Crystal structure of the full-length japanese encephalitis virus NS5 reveals a conserved methyltransferase-polymerase interface. *PLoS Pathog.* **2013**, *9*, e1003549. [CrossRef] [PubMed]

69. Zhao, Y.; Soh, T.S.; Zheng, J.; Chan, K.W.; Phoo, W.W.; Lee, C.C.; Tay, M.Y.; Swaminathan, K.; Cornvik, T.C.; Lim, S.P.; *et al.* A crystal structure of the dengue virus NS5 protein reveals a novel inter-domain interface essential for protein flexibility and virus replication. *PLoS Pathog.* **2015**, *11*, e1004682. [CrossRef] [PubMed]

70. Li, X.D.; Shan, C.; Deng, C.L.; Ye, H.Q.; Shi, P.Y.; Yuan, Z.M.; Gong, P.; Zhang, B. The interface between methyltransferase and polymerase of NS5 is essential for flavivirus replication. *PLoS Negl.*

Trop. Dis. **2014**, *8*, e2891. [CrossRef] [PubMed]

71. You, S.; Padmanabhan, R. A novel *in vitro* replication system for dengue virus. Initiation of RNA synthesis at the 3′-end of exogenous viral RNA templates requires 5′- and 3′-terminal complementary sequence motifs of the viral RNA. *J. Biol. Chem.* **1999**, *274*, 33714–33722. [CrossRef] [PubMed]

72. Nomaguchi, M.; Teramoto, T.; Yu, L.; Markoff, L.; Padmanabhan, R. Requirements for West Nile virus (−)- and (+)-strand subgenomic RNA synthesis *in vitro* by the viral RNA-dependent RNA polymerase expressed in escherichia coli. *J. Biol. Chem.* **2004**, *279*, 12141–12151. [CrossRef] [PubMed]

73. Dong, H.; Ray, D.; Ren, S.; Zhang, B.; Puig-Basagoiti, F.; Takagi, Y.; Ho, C.K.; Li, H.; Shi, P.Y. Distinct RNA elements confer specificity to flavivirus RNA cap methylation events. *J. Virol.* **2007**, *81*, 4412–4421. [CrossRef] [PubMed]

74. Dong, H.; Ren, S.; Zhang, B.; Zhou, Y.; Puig-Basagoiti, F.; Li, H.; Shi, P.Y. West Nile virus methyltransferase catalyzes two methylations of the viral RNA cap through a substrate-repositioning mechanism. *J. Virol.* **2008**, *82*, 4295–4307. [CrossRef] [PubMed]

10

APOBEC3 Interference during Replication of Viral Genomes

Luc Willems [1,2] **and Nicolas Albert Gillet** [1,2,*]

[1] Molecular and Cellular Epigenetics, Interdisciplinary Cluster for Applied Genoproteomics (GIGA) of University of Liège (ULg), B34, 1 avenue de L'Hôpital, Sart-Tilman Liège 4000, Belgium;
E-Mail: luc.willems@ulg.ac.be
[2] Molecular and Cellular Biology, Gembloux Agro-Bio Tech, University of Liège (ULg),
13 avenue Maréchal Juin, Gembloux 5030, Belgium

* Author to whom correspondence should be addressed; E-Mail: n.gillet@ulg.ac.be

Academic Editor: David Boehr

Abstract: Co-evolution of viruses and their hosts has reached a fragile and dynamic equilibrium that allows viral persistence, replication and transmission. In response, infected hosts have developed strategies of defense that counteract the deleterious effects of viral infections. In particular, single-strand DNA editing by Apolipoprotein B Editing Catalytic subunits proteins 3 (APOBEC3s) is a well-conserved mechanism of mammalian innate immunity that mutates and inactivates viral genomes. In this review, we describe the mechanisms of APOBEC3 editing during viral replication, the viral strategies that prevent APOBEC3 activity and the consequences of APOBEC3 modulation on viral fitness and host genome integrity. Understanding the mechanisms involved reveals new prospects for therapeutic intervention.

Keywords: viral replication; quasi-species; hypo-mutation; hyper-mutation; APOBEC3; cytidine deaminase

1. APOBEC3s Edit Single-Stranded DNA

The APOBEC3 enzymes are deaminases that edit single-stranded DNA (ssDNA) sequences by transforming deoxycytidine into deoxyuridine [1–3]. APOBEC3s are involved in the mechanisms of innate defense against exogenous viruses and endogenous retroelements [3]. The human genome codes for seven APOBEC3 genes clustered in tandem on chromosome 22 (namely A3A, A3B, A3C, A3DE, A3F, A3G, and A3H) and surrounded by the CBX6 and CBX7 genes. All APOBEC3 genes encode a single- or a double-zinc-coordinating-domain protein. Each zinc-domain belongs to one of the three distinct phylogenic clusters termed Z1, Z2 and Z3. The seven APOBEC3 genes arose via gene duplications and fusions of a key mammalian ancestor with a CBX6-Z1-Z2-Z3-CBX7 locus organization. Aside from mice and pigs, duplications of APOBEC3 genes have occurred independently in different lineages: humans and chimpanzees ($n = 7$), horses ($n = 6$), cats ($n = 4$), and sheep and cattle ($n = 3$) [4,5]. Read-through transcription, alternative splicing and internal transcription initiation may further extend the diversity of APOBEC3 proteins.

APOBEC3s are interferon-inducible genes [6] that are highly expressed in immune cells despite being present in almost all cell types [7,8]. The sub-cellular localization differs between the APOBEC3s isoforms: A3DE/A3F/A3G are excluded from chromatin throughout mitosis and become cytoplasmic during interphase, A3B is nuclear and A3A/A3C/A3H are cell-wide during interphase [9].

APOBEC3s exert an antiviral effect either dependently or independently of their deaminase activity. The deaminase activity involves the removal of the exocyclic amine group from deoxycytidine to form deoxyuridine. This process can generate different types of substitutions. First, DNA replication through deoxyuridine leads to the insertion of a deoxyadenosine, therefore causing a C to T transition. Alternatively, Rev1 translesion synthesis DNA polymerase can insert a C in front of an abasic site that is produced through uracil excision by uracil-DNA glycosylase (UNG2) leading to a C-to-G transversion [10]. In addition to inducing deleterious mutations in the viral genome, deamination of deoxycytidine can also initiate degradation of uracilated viral DNA via a UNG2-dependent pathway [11,12]. On the other hand, deaminase-independent inhibition requires binding of APOBEC3s to single-stranded DNA or RNA viral sequences at various steps of the replication cycle [13–23].

2. APOBEC3 Edition during Viral Replication Cycles

The mechanism of APOBEC3s inactivation is dependent on the type of virus and its mode of replication.

2.1. Retroviruses

Retroviruses are plus-strand single-stranded RNA viruses replicating via a DNA intermediate generated in the cytoplasm by reverse transcription. Human retroviruses notably include HIV (human immunodeficiency virus) and HTLV (human T-lymphotropic virus).

2.1.1. HIV-1

Historically, the first member of the APOBEC3 family was discovered in a groundbreaking study on HIV-1 [24]. A3G has indeed been shown to inhibit HIV infection and to be repressed by the viral Vif protein. Later on, a similar function was also attributed to other APOBEC3 proteins, namely A3DE, A3F and A3H [25–27]. Figure 1 illustrates the different mechanisms of HIV-1 inhibition by APOBEC3s. After binding of the HIV virion to the host cell membrane, the viral single-stranded RNA (ssRNA) genome is released into the cytoplasm and converted into double-stranded DNA (dsDNA) by reverse transcription. This dsDNA is then inserted into the host genome as an integrated provirus.

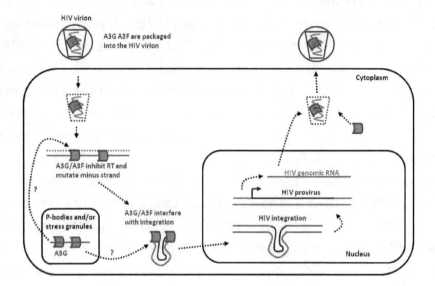

Figure 1. APOBEC3s interfere with several key steps of the HIV infectious cycle. After binding of the HIV virion to the cell membrane, the single-stranded RNA genome (**in blue**) is released into the cytoplasm together with APOBEC3G and 3F (**orange**). APOBEC3 proteins expressed by the host cell concentrate in P-bodies and stress granules. A3G and A3F inhibit reverse transcription, mutate viral DNA and perturb proviral integration into the host genome. In the absence of HIV Vif, A3G and A3F will be incorporated into the budding virions.

A3DE, A3F, A3G and A3H are expressed by CD4+ T cells upon HIV infection, are packaged into virions and lead to proviral DNA mutations [27]. A3G and A3F notably concentrate in cytoplasmic microdomains (non-membrane structures) called mRNA-processing bodies or P-bodies [28–30]. P-bodies are sites of RNA storage, translational repression and decay [31]. A3G exerts its anti-HIV effect mainly via its deaminase function inducing abundant and deleterious mutations within the HIV provirus, whereas A3F acts more preferentially through its deaminase-independent activity [32]. This deaminase-independent effect involves inhibition of reverse transcription priming and extension [14–17] and interference with proviral integration [21–23].

APOBEC3-induced mutations are almost always G-to-A transitions of the plus-strand genetic code. Moreover, the mutation load is not homogeneous along the HIV provirus but presents two highly polarized gradients, each peaking just 5′ to the central polypurine tract (cPPT) and 5′ to the LTR (long terminal repeat) proximal polypurine tract (3′PPT) [33]. As illustrated in Figure 2, this mutational signature is due to the mechanism of HIV reverse transcription. Binding of the human tRNALys3 to the primer binding sequence (PBS) initiates the minus strand DNA synthesis by the virus-encoded reverse transcriptase protein (RT). The RT-associated ribonuclease H activity (RNAse H) selectively degrades the RNA strand of the RNA:DNA hybrid leaving the nascent minus-strand DNA free to hybridize with the complementary sequence at the 3′ end of the viral genomic ssRNA. After minus strand transfer, the viral RNA is reverse-transcribed into DNA. Whilst DNA synthesis proceeds, the RNAse H function cleaves the RNA strand of the RNA:DNA. Two specific purine-rich sequences (polypurine tracts cPPT and 3′PPT) that are resistant to RNAse H remain annealed with the nascent minus strand DNA. The reverse transcriptase uses the PPTs as primers to synthesize the plus-strand DNA. Finally, another strand transfer allows the production of the 5′ end of the plus-strand DNA (reviewed in [34]). From this complex multistep process, it appears that only the minus strand can be single-stranded (light red in Figure 2). Thus, G to A mutations observed on the plus strand (dark red in Figure 2) originate from C-to-T mutations on the minus strand. The gradient of mutational load actually correlates with the time that the minus strand remains single-chain [33].

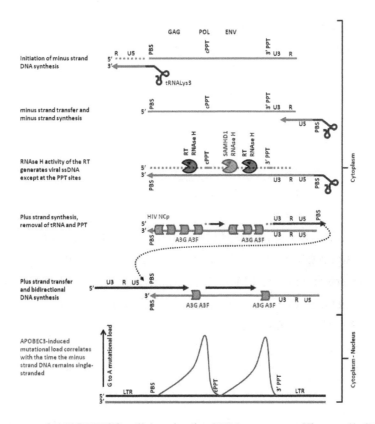

Figure 2. Hotspots of APOBEC3 editing in the HIV genome. Host cell tRNALys3 (**dark blue**) hybridizes to the primer binding sequence (PBS) of the single stranded plus-strand RNA genome (**light blue**) and initiates minus strand DNA synthesis (**light red**). After strand transfer, reverse transcription proceeds up to the PBS yielding minus-strand DNA. RNAse H then hydrolyses the RNA (**dotted light blue**) of the RNA:DNA hybrid leaving the minus-strand DNA single-stranded. APOBEC3 G and F (**orange**) have now access the ssDNA genome, deaminate deoxycytidine and inhibit plus strand DNA synthesis (**dark red**). RNAse H activity of SAMHD1 promotes exposure of the minus-strand DNA (red pacman) whereas HIV nucleocapsid (**green**) limits APOBEC3-edition. Deoxycytidine deamination of the minus strand generates G-to-A mutations on the plus strand. Since plus-strand DNA synthesis starts from the PolyPurine Tracts (cPPT, 3′PPT), ssDNA located distant to these sites will be accessible to APOBEC3 edition over a longer period of time. Therefore, the APOBEC3-related mutational load will also be higher (**brown curve**, schematic representation of the data from [33]).

To counteract inactivation, HIV-1 encodes the Vif protein that inhibits APOBEC3s. Vif prevents A3G, A3F and A3H from being packaged into the virion by recruitment to a cullin5-elonginB/C-Rbx2-CBFβ E3 ubiquitin ligase complex, resulting in their polyubiquitination and subsequent proteasomal degradation [35–37]. Other mechanisms can limit APOBEC3 access to the single-chain minus-strand DNA generated during reverse transcription. By stabilizing the viral core, the glycosylated Gag protein of the murine leukemia virus renders the reverse transcription complex resistant to APOBEC3 and to other cytosolic viral sensors [38]. The HIV nucleocapsid protein (NCp) is able to bind ssDNA in a sequence aspecific manner and prevents A3A from mutating genomic DNA during transient strand separation [39]. Degradation of the RNA strand from the RNA:DNA hybrid by the RNAse H activity of the reverse transcriptase contributes to expose the minus strand as a single-chain nucleic acid. Interestingly, the host factor SAMHD1 (sterile alpha motif and histidine-aspartic acid domain containing protein 1) restricts HIV via its RNAse H function, activity that may facilitate the access of the APOBEC3s to the transiently single-stranded minus strand [40,41].

2.1.2. HTLV-1

Another human retrovirus, human T-lymphotropic virus 1 (HTLV-1), is also a target of A3G [42,43]. As in HIV-1 infection, A3G induces G-to-A transitions on the plus strand via deamination of deoxycytidines on the minus strand. HTLV-1 proviruses contain A3G-related base substitutions, including non-sense mutations [43]. Because HTLV-1 proviral loads mainly result from clonal expansion of infected cells, non-sense mutations are stabilized and amplified by mitosis, provided that viral factors stimulating proliferation are functional [44–47]. Although HTLV-1 does not seem to encode for a Vif-like protein, the frequencies of G-to-A changes in HTLV-1 proviruses are low, likely due to the mode of replication of HTLV-1 by clonal expansion [43,48]. This phenotype has also been associated with the ability of the viral nucleocapsid to limit A3G encapsidation [49].

2.1.3. HERVs

Human endogenous retroviruses (HERV) are transposable elements which were evolutionary integrated into human lineage after infection of germline cells. HERVs are abundant in the human genome (about 8%) and exert important regulatory functions such as control of cellular gene transcription [50]. HERVs contain canonical retroviral *gag*, *pol* and *env* genes surrounded by two LTRs. Nevertheless, most HERVs are defective for replication because of inactivating mutations or deletions [51]. These mutations are likely associated with A3G activity because of a particular signature with a mutated C present in a 5'GC context instead of 5'TC for other APOBEC3s [52–54]. Interestingly, A3G is still able to inhibit a reconstituted functional form of HERV-K in cell culture [52].

2.1.4. Simian Foamy Virus

SFV (simian foamy virus) is a retrovirus that is widespread among non-human primates and can be transmitted to humans [55]. A3F and A3G target SFV genome *in vitro*, leading to G-to-A transitions on the plus strand [56]. SFV genomes found in humans also display G-to-A mutations [57–59]. SFV codes for the accessory protein Bet, limiting APOBEC3 action [60–63].

2.2. Retroelements

About half of the human genome is constituted by repetitive elements. Among them, non-LTR retroelements LINE-1 (long interspersed nuclear element-1), SINE (short interspersed nuclear elements) and Alu are capable of retrotransposition, *i.e.*, inserting a copy of themselves elsewhere in the genome. Since retrotranpositions can be harmful for genome integrity, these events are tightly controlled. In fact, only a small proportion of endogenous retroelements remains active in the germline cells because APOBEC3s protect the host genome from unscheduled retrotransposition (Figure 3). LINE-1 retrotransposition is initiated by transcription of a full-length LINE-1 RNA and translation of ORF1p and ORF2p. These two proteins associate with LINE-1 RNA to form the LINE-1 RiboNucleoProtein (L1 RNP) complex. Upon translocation of L1 RNP into the nucleus, LINE-1 is reverse transcribed and integrated into a new site of the host genome. A3C restricts LINE-1 retrotransposition in a deaminase-independent manner by redirecting and degrading the L1 RNP complex in P-bodies [20]. Within the nucleus, A3C also impairs LINE-1 minus strand DNA synthesis [20]. A3A prevents LINE-1 retrotransposition by deaminating the LINE-1 minus strand DNA [64]. Consistently, RNAse H treatment increases deamination of the LINE-1 minus strand [64].

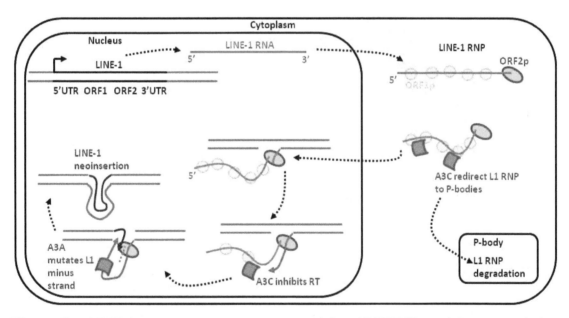

Figure 3. LINE-1 retrotransposons are targeted by APOBEC3s. After transcription, the LINE-1 mRNA is transported into the cytoplasm. After translation, the ORF1- and ORF2-encoded proteins associate with the LINE-1 RNA and form a ribonucleoprotein (RNP) complex. The LINE-1 RNP enters the nucleus, where the ORF2p endonuclease domain cleaves the chromosomal DNA. After cleavage, the 3′-hydroxyl is used by the LINE-1 reverse transcriptase to synthesize a cDNA of LINE-1. This target-site-primed reverse transcription typically results in the insertion of a 5′-truncated LINE-1 element into a new genomic location. Different APOBEC3s-dependent mechanisms control LINE-1 retrotransposition: (1) in the cytoplasm, A3C interacts with and redirects the L1-RNP into P-bodies for degradation; (2) in the nucleus, A3C inhibits reverse transcriptase processing while A3A mutates the minus strand LINE-1 DNA.

2.3. Hepadnaviruses

Since their genome is partially single-strand, hepadnaviruses, such as human hepatitis B virus (HBV), are susceptible to APOBEC3 editing. Except A3DE, all APOBEC3s are able to edit the HBV genome *in vitro,* A3A being the most efficient [65,66]. APOBEC3 editing of HBV DNA has also been validated *in vivo* [67,68]. Since both minus and plus strands are susceptible to APOBEC3 editing, the mutational signature is more complex than in retroviruses [65–67]. HBV viral particles contain a partially double-stranded circular DNA genome (relaxed circular DNA or rcDNA; Figure 4). After uncoating of the viral particle, the rcDNA migrates into the nucleus, where minus-strand DNA synthesis is completed to generate the covalently closed circular double-stranded DNA genome (cccDNA).

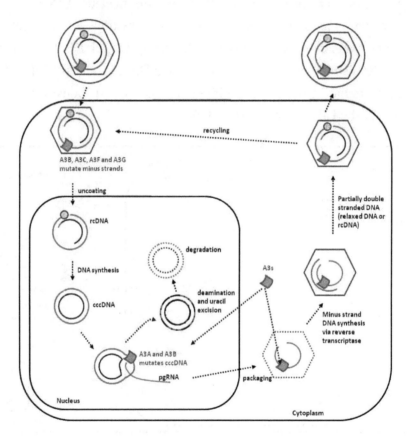

Figure 4. APOBEC3s interfere with several steps of the HBV replication cycle. The HBV viral particle contains a partially double-stranded DNA genome (relaxed circular DNA or rcDNA) that can be edited by A3G and A3F. Unlike HIV, HBV does not appear to encode Vif-like protein. Upon transfer into the nucleus, the plus strand of the rcDNA is replicated to form the covalently closed circular DNA genome (cccDNA). A3A and A3B deaminate the cccDNA genome leading to uracil excision and subsequent degradation.

In the nucleus, A3A and A3B deaminate HBV cccDNA (Figure 4). Since APOBEC3s require a ssDNA substrate, it is predicted that cccDNA melts during transcription. APOBEC3 deamination of deoxycytidine introduces deleterious mutations in the viral genome and initiates its catabolism via the uracil DNA glycosylase dependent pathway [12].

After transcription of the cccDNA, the pregenomic RNA (pgRNA) translocates into the cytoplasm and is reverse-transcribed into circular partially double-stranded DNA. This mechanism involves priming by the viral P protein, a strand transfer directed by DR1 annealing and degradation of the RNA template by the RNAse H activity of the reverse transcriptase (Figure 5, dotted light blue). The 5′ end of the pgRNA anneals with DR2, directs a second strand transfer and primes plus-strand DNA synthesis, yielding rcDNA. The minus strand DNA is deaminated proportionally to its exposure to APOBEC3s (Figure 5, orange curve) [68]. Since different subcellular compartments are involved (cytoplasm, nucleus, extracellular viral particles), multiple nuclear and cytoplasmic APOBEC3s (*i.e.*, A3A, A3B, A3C, A3F and A3G) edit the HBV genome [65].

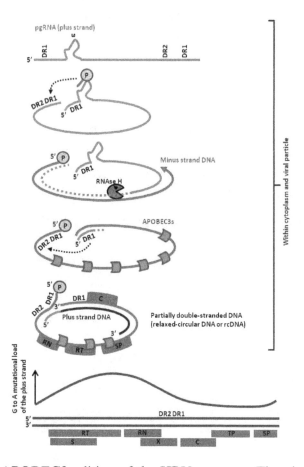

Figure 5. Profile of APOBEC3 editing of the HBV genome. The viral P protein initiates reverse transcription at the stem loop structure ε. The pregenomic RNA (pgRNA) contains two direct repeat sequences (DR1) at the 5′ and 3′ end of the viral genome, allowing strand transfer to the 5′ end of the viral genome. While synthesis of the minus-strand DNA proceeds, the RNAse H activity of the reverse transcriptase degrades the pgRNA except at the 5′ end. After a second strand transfer, the undigested pgRNA anneals with the direct repeat sequence DR2 and primes plus-strand gDNA synthesis, yielding relaxed circular DNA (rcDNA). Mutational load correlates with the time of exposure of ssDNA (**orange curve**, schematic representation of data extracted from reference [68]). Abbreviations within grey boxes read as follow: RT, Reverse Transcriptase; RN, RNAse; TP, Terminal Protein; SP, Spacer Domain; S, short surface gene; X, X gene, C, Core gene.

Compared to HIV, additional APOBEC3 proteins (A3A, A3B and A3C) target the HBV genome in the nucleus. Incorporation of HIV into chromatin instead of an episome for HBV may protect the provirus from APOBEC3s editing by a mechanism involving Tribbles 3 proteins [69].

2.4. Herpesviruses

Herpesviruses such as herpes simplex virus-1 (HSV-1) and Epstein-Barr virus (EBV) have a linear double-stranded DNA genome that is edited by APOBEC3 on both strands [70]. After infection, the HSV-1 capsid is transported to the nuclear pores and delivers the double-stranded linear DNA into the nucleus. After circularization of the viral genome, bidirectional DNA synthesis is initiated at the origins of replication [71,72]. This process requires DNA denaturation by the origin binding protein (UL9). The helicase/primase (UL5/UL8/UL52) and single-stranded DNA binding proteins (ICP8 coded by the UL29 gene) then associate with the origin of replication and recruit the DNA polymerase/UL42 complex (Figure 6). During DNA synthesis and transcription, nuclear APOBEC3s have access to single-stranded viral DNA. APOBEC3-edition of HSV-1 and EBV genomes is higher in the minus strand (G to A as opposed to C to T) [70]. It is hypothesized that, due to discontinued replication, the lagging strand exposes more viral ssDNA than the leading strand. HSV-1 and EBV encode orthologs of uracil-DNA glycosylases (UDG) that excise uridine at the replication fork. The HSV-1 UDG (UL2) binds to UL30, associates with the viral replisome and directs replication-coupled BER (base excision repair) to ensure genome integrity [73]. The viral UDG might therefore protect against APOBEC3 editing.

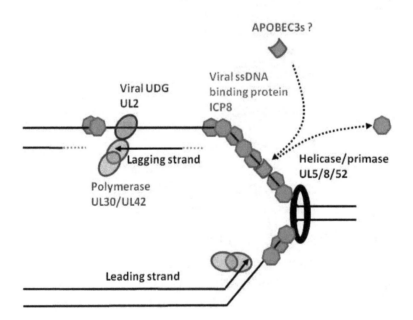

Figure 6. HSV-1 replication fork and hypothetical model of APOBEC3 editing. In the nucleus, replication of HSV-1 is initiated by the origin binding protein (UL9) that melts double-stranded DNA. The helicase/primase complex (UL5/UL8/UL52) unwinds and anneals RNA primers, allowing DNA replication by the UL30/UL42 complex. The viral protein ICP8 covers the transiently exposed single-stranded DNA and competes with the APOBEC3 binding. The viral UL2 is a uracil-DNA glycosylase (UDG) that favors replication-coupled base excision DNA repair.

2.5. Papillomavirus

Human papillomaviruses (HPVs) are circular double-stranded DNA viruses. A3A, A3C and A3H are able to deaminate both strands of the 8Kb viral genome [74–76]. APOBEC3-edited HPV DNA is found in benign and precancerous cervical lesions [74]. Replication of the HPV genome occurs in the nucleus and is primarily based on the host replication machinery. The HPV protein E1 recruits ssDNA-binding protein RPA (replication protein A) during replication to cover the transiently exposed viral ssDNA [77].

2.6. TT Virus

Transfusion-transmitted virus (TTV) is a non-enveloped virus causing a persistent and asymptomatic infection. Having a circular single-stranded DNA genome, TTV is a prototypical substrate of APOBEC3s and shows APOBEC3-related mutations [78].

Together, these data show that viruses are targeted by particular isoforms of APOBEC3 depending on their modes of replication (Table 1) and have developed strategies to dampen ssDNA edition. A3G and A3F are restricted to the cytoplasm whereas A3A, A3B and A3C preferentially act in the nucleus. Importantly, the mutational load is proportional to the duration of single-stranded DNA exposure to APOBEC3s.

Table 1. Summary of the anti-viral activity of the different APOBEC3 isoforms. * It has been recently shown that A3A can also edit RNA transcripts [79].

	Sub-cellular localization	Substrate edited	Retro viruses				Retro elements	Hepadna viruses	Herpes viruses
			HIV-1	HTLV-1	HERVs	SFV			
A3A	cell wide	single stranded DNA, RNA *					+	+	+
A3B	nuclear	single stranded DNA						+	
A3C	cell wide	single stranded DNA					+	+	+
A3DE	cytoplasmic	single stranded DNA	+						
A3F	cytoplasmic	single stranded DNA	+			+		+	
A3G	cytoplasmic	single stranded DNA	+	+	+	+		+	
A3H	cell wide	single stranded DNA	+					+	+

3. Therapeutic Strategies by Perturbation of the Viral Mutation Rate

Viral quasi-species refer to a population of distinct but closely related viral genomes that differ only by a limited number of mutations. The distribution of variants is dominated by a master sequence that displays the highest fitness within a given environment (Figure 7A). High mutation rates during viral replication are the driving force for quasi-species generation. Lethal mutations or inappropriate adaptation to the environmental conditions (e.g., anti-viral therapy, immune pressure) will clear unfit genomes. When conditions change, the fittest quasi-species may differ from the master sequence. Providing that the distribution contains an adequate variant, a new population will grow (Figure 7B). The rate of mutation and the selection pressure will dictate the wideness of the distribution. If the environmental changes are too drastic or the quasi-species distribution too narrow, the viral population will be unable to recover [80].

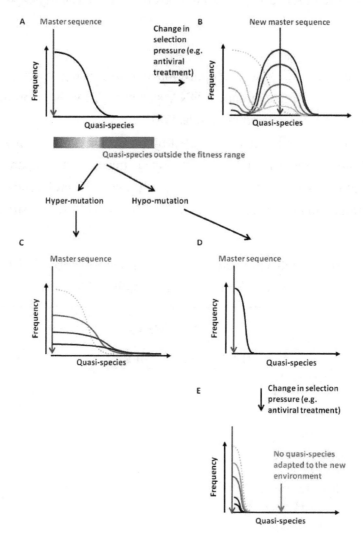

Figure 7. Antiviral strategy by hypo- or hyper-mutation. (**A**) Viral quasi-species refer to as a population of distinct but closely related viral genomes that only differ by a limited number of mutations. The frequency of these quasi-species spreads around a master sequence. The boundary of this population is dictated by the selection forces acting against the viral diversification. At equilibrium, quasi-species generated outside the fitness range will not persist. (**B**) If selection criteria are modified, the fittest sequence will change. If the original distribution contained this sequence, the population will first shrink and then re-grow around a new master sequence (from light to dark blue corresponding to the time evolution). (**C**) Excess of APOBEC3-directed mutations will affect fitness of the newly created quasi-species up to complete disappearance. (**D,E**) Hypo-mutation will restrict the range of quasi-species and limit adaptability to new environmental conditions.

Emergence of quasi-species is thus a major issue that limits antiviral therapy. Viral populations can indeed accommodate environmental changes due to improved immunity (vaccination) or pharmacological inhibition. It is possible to affect quasi-species adaptability by modulating the frequencies of mutation [81,82].

The first approach, referred to as lethal mutagenesis or the hyper-mutation strategy, aims to introduce an excess of mutations in the viral genomes. If the mutational load per viral genome is too high, a substantial proportion of the new viruses will be defective or inadequately adapted to their environment. Introducing mutations in viruses would therefore decrease viral load (Figure 7C). In principle, exogenous induction of APOBEC3 expression could achieve this goal. This strategy has recently been exemplified for HBV, where forced expression of A3A and A3B induced HBV cccDNA hypermutation with no detectable effect on genomic DNA [12]. Nevertheless, this approach raises serious safety issues because APOBEC3 mutations could also drive cancer development [83–85]. Indeed, A3A and A3B over-expression in yeast creates mutational clusters and genomic rearrangements similar to those observed in human cancers, the mutational burden being magnified by DNA strand breaks [86–89]. Because the processing of double-strand break repair transiently exposes single-stranded nucleic acids, DNA repair could provide a substrate for nuclear deaminases. What would, for example, happen if an HBV-infected liver cell is being forced to express deaminases and at the same time has to repair DNA strand breaks generated by reactive oxygen species produced during alcohol catabolism [90]?

Therefore, it would be safer to promote hyper-mutation by targeting the viral factors that inhibit APOBEC3s. In that respect, Vif inhibitors are being developed [91,92]. Inhibition of viral ssDNA-binding proteins (like HSV-1 ICP8) might lead to increased access for endogenously expressed APOBEC3s to the viral ssDNA (Figure 6). Promotion of RNAse H activity during retrotranscription might facilitate the binding of the APOBEC3s to the viral ssDNA (Figure 2). Because reverse transcription is thought to start within the virion, promotion of APOBEC3s loading in to the viral particle will increase editing (MLV glyco-Gag shields the reverse transcription complex from APOBEC3 and cytosolic sensors [38]). Alternatively, it would be possible to target viral DNA repair mechanisms (e.g., via inhibitors against the viral UDG UL2 of HSV-1, Figure 6). In these cases, safety issues are related to the emergence of sub-lethal APOBEC3-mutations and promotion of drug resistant quasi-species.

The reverse strategy would be to reduce mutation rate by inhibiting APOBEC3, thereby narrowing the quasi-species spectrum and limiting viral adaptability to new environmental conditions (Figure 7D,E). APOBEC3s inhibitors are currently being developed and evaluated [93,94]. This approach, which paradoxically targets a well-conserved mechanism of mammalian innate immunity, would preserve host genome integrity. Potential risks of this therapy pertain to adequate control of endogenous retroelements and opportunistic infections.

4. Conclusions

Single-strand DNA editing by APOBEC3 proteins is a very powerful mechanism of mammalian innate immunity that mutates and inactivates viral genomes. The outcome of infection is the result of a finely tuned balance between onset of mutations, generation of quasi-species and APOBEC inhibition by viral factors. Understanding the mechanisms involved reveals new prospects for therapeutic strategies that interfere with APOBEC3 deamination of cytosine residues in nascent viral DNA.

Acknowledgments

This work was supported by the "Fonds National de la Recherche Scientifique" (FNRS), the Télévie, the Interuniversity Attraction Poles (IAP) Program "Virus-host interplay at the early phases of infection" BELVIR initiated by the Belgian Science Policy Office, the Belgian Foundation against Cancer (FBC), the Sixth Research Framework Programme of the European Union (project "The role of infections in cancer" INCA LSHC-CT-2005-018704), the "Neoangio" excellence program and the "Partenariat Public Privé", PPP INCA, of the "Direction générale des Technologies, de la Recherche et de L'Energie/DG06" of the Walloon government, the "Action de Recherche Concertée Glyvir" (ARC) of the "Communauté française de Belgique", the "Centre anticancéreux près ULg" (CAC), the "Subside Fédéral de Soutien à la Recherche Synbiofor and Agricultureislife" projects of Gembloux Agrobiotech (GxABT), the "ULg Fonds Spéciaux pour la Recherche" and the "Plan Cancer" of the "Service Public Fédéral". We thank Sathya Neelature Sriramareddy, Srikanth Perike, Hélène Gazon, Alix de Brogniez, Alexandre Carpentier, Pierre-Yves Barez, Bernard Staumont and Malik Hamaidia for their careful reading and helpful comments.

Author Contributions

L.W. and N.A.G. wrote the paper.

References

1. Smith, H.C.; Bennett, R.P.; Kizilyer, A.; McDougall, W.M.; Prohaska, K.M. Functions and regulation of the APOBEC family of proteins. *Semin. Cell Dev. Biol.* **2012**, *23*, 258–268. [CrossRef] [PubMed]

2. Refsland, E.W.; Harris, R.S. The APOBEC3 family of retroelement restriction factors. *Curr. Top. Microbiol. Immunol.* **2013**, *371*, 1–27. [PubMed]

3. Vieira, V.C.; Soares, M.A. The role of cytidine deaminases on innate immune responses against human viral infections. *Biomed. Res. Int.* **2013**, *2013*, e683095. [CrossRef] [PubMed]

4. LaRue, R.S.; Andresdottir, V.; Blanchard, Y.; Conticello, S.G.; Derse, D.; Emerman, M.; Greene, W.C.; Jonsson, S.R.; Landau, N.R.; Lochelt, M.; *et al.* Guidelines for naming nonprimate APOBEC3 genes and proteins. *J. Virol.* **2009**, *83*, 494–497. [CrossRef] [PubMed]

5. Munk, C.; Willemsen, A.; Bravo, I.G. An ancient history of gene duplications, fusions and losses in the evolution of APOBEC3 mutators in mammals. *BMC Evol. Biol.* **2012**, *12*, e71. [CrossRef] [PubMed]

6. Mehta, H.V.; Jones, P.H.; Weiss, J.P.; Okeoma, C.M. IFN-alpha and lipopolysaccharide upregulate APOBEC3 mRNA through different signaling pathways. *J. Immunol.* **2012**, *189*, 4088–4103. [CrossRef] [PubMed]

7. Refsland, E.W.; Stenglein, M.D.; Shindo, K.; Albin, J.S.; Brown, W.L.; Harris, R.S. Quantitative profiling of the full APOBEC3 mRNA repertoire in lymphocytes and tissues: Implications for HIV-1 restriction. *Nucleic Acids Res.* **2010**, *38*, 4274–4284. [CrossRef] [PubMed]

8. Koning, F.A.; Newman, E.N.; Kim, E.Y.; Kunstman, K.J.; Wolinsky, S.M.; Malim, M.H. Defining

APOBEC3 expression patterns in human tissues and hematopoietic cell subsets. *J. Virol.* **2009**, *83*, 9474–9485. [CrossRef] [PubMed]

9. Lackey, L.; Law, E.K.; Brown, W.L.; Harris, R.S. Subcellular localization of the APOBEC3 proteins during mitosis and implications for genomic DNA deamination. *Cell Cycle* **2013**, *12*, 762–772. [CrossRef] [PubMed]

10. Krokan, H.E.; Saetrom, P.; Aas, P.A.; Pettersen, H.S.; Kavli, B.; Slupphaug, G. Error-free *versus* mutagenic processing of genomic uracil—Relevance to cancer. *DNA Repair* **2014**, *19*, 38–47. [CrossRef] [PubMed]

11. Weil, A.F.; Ghosh, D.; Zhou, Y.; Seiple, L.; McMahon, M.A.; Spivak, A.M.; Siliciano, R.F.; Stivers, J.T. Uracil DNA glycosylase initiates degradation of HIV-1 cDNA containing misincorporated dUTP and prevents viral integration. *Proc. Natl. Acad Sci. USA* **2013**, *110*, E448–E457. [CrossRef] [PubMed]

12. Lucifora, J.; Xia, Y.; Reisinger, F.; Zhang, K.; Stadler, D.; Cheng, X.; Sprinzl, M.F.; Koppensteiner, H.; Makowska, Z.; Volz, T.; *et al.* Specific and nonhepatotoxic degradation of nuclear hepatitis B virus cccDNA. *Science* **2014**, *343*, 1221–1228. [CrossRef] [PubMed]

13. Huthoff, H.; Autore, F.; Gallois-Montbrun, S.; Fraternali, F.; Malim, M.H. RNA-dependent oligomerization of APOBEC3G is required for restriction of HIV-1. *PLoS Pathog.* **2009**, *5*, e1000330. [CrossRef] [PubMed]

14. Guo, F.; Cen, S.; Niu, M.; Saadatmand, J.; Kleiman, L. Inhibition of tRNA(3)(Lys)—Primed reverse transcription by human APOBEC3G during human immunodeficiency virus type 1 replication. *J. Virol.* **2006**, *80*, 11710–11722. [CrossRef] [PubMed]

15. Anderson, J.L.; Hope, T.J. APOBEC3G restricts early HIV-1 replication in the cytoplasm of target cells. *Virology* **2008**, *375*, 1–12. [CrossRef] [PubMed]

16. Wang, X.; Ao, Z.; Chen, L.; Kobinger, G.; Peng, J.; Yao, X. The cellular antiviral protein APOBEC3G interacts with HIV-1 reverse transcriptase and inhibits its function during viral replication. *J. Virol.* **2012**, *86*, 3777–3786. [CrossRef] [PubMed]

17. Gillick, K.; Pollpeter, D.; Phalora, P.; Kim, E.Y.; Wolinsky, S.M.; Malim, M.H. Suppression of HIV-1 infection by APOBEC3 proteins in primary human CD4(+) T cells is associated with inhibition of processive reverse transcription as well as excessive cytidine deamination. *J. Virol.* **2013**, *87*, 1508–1517. [CrossRef] [PubMed]

18. Nguyen, D.H.; Gummuluru, S.; Hu, J. Deamination-independent inhibition of hepatitis B virus reverse transcription by APOBEC3G. *J. Virol.* **2007**, *81*, 4465–4472. [CrossRef] [PubMed]

19. Narvaiza, I.; Linfesty, D.C.; Greener, B.N.; Hakata, Y.; Pintel, D.J.; Logue, E.; Landau, N.R.; Weitzman, M.D. Deaminase-independent inhibition of parvoviruses by the APOBEC3A cytidine deaminase. *PLoS Pathog.* **2009**, *5*, e1000439. [CrossRef] [PubMed]

20. Horn, A.V.; Klawitter, S.; Held, U.; Berger, A.; Vasudevan, A.A.; Bock, A.; Hofmann, H.; Hanschmann, K.M.; Trosemeier, J.H.; Flory, E.; *et al.* Human LINE-1 restriction by APOBEC3C is deaminase independent and mediated by an ORF1p interaction that affects LINE reverse transcriptase activity. *Nucleic Acids Res.* **2014**, *42*, 396–416. [CrossRef] [PubMed]

21. Luo, K.; Wang, T.; Liu, B.; Tian, C.; Xiao, Z.; Kappes, J.; Yu, X.F. Cytidine deaminases

APOBEC3G and APOBEC3F interact with human immunodeficiency virus type 1 integrase and inhibit proviral DNA formation. *J. Virol.* **2007**, *81*, 7238–7248. [CrossRef] [PubMed]

22. Mbisa, J.L.; Barr, R.; Thomas, J.A.; Vandegraaff, N.; Dorweiler, I.J.; Svarovskaia, E.S.; Brown, W.L.; Mansky, L.M.; Gorelick, R.J.; Harris, R.S.; *et al.* Human immunodeficiency virus type 1 cDNAs produced in the presence of APOBEC3G exhibit defects in plus-strand DNA transfer and integration. *J. Virol.* **2007**, *81*, 7099–7110. [CrossRef] [PubMed]

23. Vetter, M.L.; D'Aquila, R.T. Cytoplasmic APOBEC3G restricts incoming Vif-positive human immunodeficiency virus type 1 and increases two-long terminal repeat circle formation in activated T-helper-subtype cells. *J. Virol.* **2009**, *83*, 8646–8654. [CrossRef] [PubMed]

24. Sheehy, A.M.; Gaddis, N.C.; Choi, J.D.; Malim, M.H. Isolation of a human gene that inhibits HIV-1 infection and is suppressed by the viral Vif protein. *Nature* **2002**, *418*, 646–650. [CrossRef] [PubMed]

25. Liddament, M.T.; Brown, W.L.; Schumacher, A.J.; Harris, R.S. APOBEC3F properties and hypermutation preferences indicate activity against HIV-1 *in vivo*. *Curr. Biol.* **2004**, *14*, 1385–1391. [CrossRef] [PubMed]

26. Wiegand, H.L.; Doehle, B.P.; Bogerd, H.P.; Cullen, B.R. A second human antiretroviral factor, APOBEC3F, is suppressed by the HIV-1 and HIV-2 Vif proteins. *EMBO J.* **2004**, *23*, 2451–2458. [CrossRef] [PubMed]

27. Hultquist, J.F.; Lengyel, J.A.; Refsland, E.W.; LaRue, R.S.; Lackey, L.; Brown, W.L.; Harris, R.S. Human and rhesus APOBEC3D, APOBEC3F, APOBEC3G, and APOBEC3H demonstrate a conserved capacity to restrict Vif-deficient HIV-1. *J. Virol.* **2011**, *85*, 11220–11234. [CrossRef] [PubMed]

28. Chiu, Y.L.; Witkowska, H.E.; Hall, S.C.; Santiago, M.; Soros, V.B.; Esnault, C.; Heidmann, T.; Greene, W.C. High-molecular-mass APOBEC3G complexes restrict Alu retrotransposition. *Proc. Natl. Acad. Sci. USA* **2006**, *103*, 15588–15593. [CrossRef] [PubMed]

29. Gallois-Montbrun, S.; Kramer, B.; Swanson, C.M.; Byers, H.; Lynham, S.; Ward, M.; Malim, M.H. Antiviral protein APOBEC3G localizes to ribonucleoprotein complexes found in P bodies and stress granules. *J. Virol.* **2007**, *81*, 2165–2178. [CrossRef] [PubMed]

30. Wichroski, M.J.; Robb, G.B.; Rana, T.M. Human retroviral host restriction factors APOBEC3G and APOBEC3F localize to mRNA processing bodies. *PLoS Pathog.* **2006**, *2*, e41. [CrossRef] [PubMed]

31. Phalora, P.K.; Sherer, N.M.; Wolinsky, S.M.; Swanson, C.M.; Malim, M.H. HIV-1 replication and APOBEC3 antiviral activity are not regulated by P bodies. *J. Virol.* **2012**, *86*, 11712–11724. [CrossRef] [PubMed]

32. Kobayashi, T.; Koizumi, Y.; Takeuchi, J.S.; Misawa, N.; Kimura, Y.; Morita, S.; Aihara, K.; Koyanagi, Y.; Iwami, S.; Sato, K. Quantification of deaminase activity-dependent and -independent restriction of HIV-1 replication mediated by APOBEC3F and APOBEC3G through experimental-mathematical investigation. *J. Virol.* **2014**, *88*, 5881–5887. [CrossRef] [PubMed]

33. Suspene, R.; Rusniok, C.; Vartanian, J.P.; Wain-Hobson, S. Twin gradients in APOBEC3 edited HIV-1 DNA reflect the dynamics of lentiviral replication. *Nucleic Acids Res.* **2006**, *34*, 4677–4684.

[CrossRef] [PubMed]

34. Esposito, F.; Corona, A.; Tramontano, E. HIV-1 Reverse Transcriptase Still Remains a New Drug Target: Structure, Function, Classical Inhibitors, and New Inhibitors with Innovative Mechanisms of Actions. *Mol. Biol. Int.* **2012**, *2012*, e586401. [CrossRef] [PubMed]

35. Mariani, R.; Chen, D.; Schrofelbauer, B.; Navarro, F.; Konig, R.; Bollman, B.; Munk, C.; Nymark-McMahon, H.; Landau, N.R. Species-specific exclusion of APOBEC3G from HIV-1 virions by Vif. *Cell* **2003**, *114*, 21–31. [CrossRef]

36. Jager, S.; Kim, D.Y.; Hultquist, J.F.; Shindo, K.; LaRue, R.S.; Kwon, E.; Li, M.; Anderson, B.D.; Yen, L.; Stanley, D.; *et al.* Vif hijacks CBF-beta to degrade APOBEC3G and promote HIV-1 infection. *Nature* **2012**, *481*, 371–375.

37. Zhao, K.; Du, J.; Rui, Y.; Zheng, W.; Kang, J.; Hou, J.; Wang, K.; Zhang, W.; Simon, V.A.; Yu, X.F. Evolutionarily conserved pressure for the existence of distinct G2/M cell cycle arrest and A3H inactivation functions in HIV-1 Vif. *Cell Cycle* **2015**, *14*, 838–847. [PubMed]

38. Stavrou, S.; Nitta, T.; Kotla, S.; Ha, D.; Nagashima, K.; Rein, A.R.; Fan, H.; Ross, S.R. Murine leukemia virus glycosylated Gag blocks apolipoprotein B editing complex 3 and cytosolic sensor access to the reverse transcription complex. *Proc. Natl. Acad. Sci. USA* **2013**, *110*, 9078–9083. [CrossRef] [PubMed]

39. Mitra, M.; Hercik, K.; Byeon, I.J.; Ahn, J.; Hill, S.; Hinchee-Rodriguez, K.; Singer, D.; Byeon, C.H.; Charlton, L.M.; Nam, G.; *et al.* Structural determinants of human APOBEC3A enzymatic and nucleic acid binding properties. *Nucleic Acids Res.* **2014**, *42*, 1095–1110. [CrossRef] [PubMed]

40. Beloglazova, N.; Flick, R.; Tchigvintsev, A.; Brown, G.; Popovic, A.; Nocek, B.; Yakunin, A.F. Nuclease activity of the human SAMHD1 protein implicated in the Aicardi-Goutieres syndrome and HIV-1 restriction. *J. Biol. Chem.* **2013**, *288*, 8101–8110. [CrossRef] [PubMed]

41. Ryoo, J.; Choi, J.; Oh, C.; Kim, S.; Seo, M.; Kim, S.Y.; Seo, D.; Kim, J.; White, T.E.; Brandariz-Nunez, A.; *et al.* The ribonuclease activity of SAMHD1 is required for HIV-1 restriction. *Nat. Med.* **2014**, *20*, 936–941. [CrossRef] [PubMed]

42. Mahieux, R.; Suspene, R.; Delebecque, F.; Henry, M.; Schwartz, O.; Wain-Hobson, S.; Vartanian, J.P. Extensive editing of a small fraction of human T-cell leukemia virus type 1 genomes by four APOBEC3 cytidine deaminases. *J. Gen. Virol.* **2005**, *86*, 2489–2494. [CrossRef] [PubMed]

43. Fan, J.; Ma, G.; Nosaka, K.; Tanabe, J.; Satou, Y.; Koito, A.; Wain-Hobson, S.; Vartanian, J.P.; Matsuoka, M. APOBEC3G generates nonsense mutations in human T-cell leukemia virus type 1 proviral genomes *in vivo*. *J. Virol.* **2010**, *84*, 7278–7287. [CrossRef] [PubMed]

44. Cavrois, M.; Wain-Hobson, S.; Gessain, A.; Plumelle, Y.; Wattel, E. Adult T-cell leukemia/lymphoma on a background of clonally expanding human T-cell leukemia virus type-1-positive cells. *Blood* **1996**, *88*, 4646–4650. [PubMed]

45. Etoh, K.; Tamiya, S.; Yamaguchi, K.; Okayama, A.; Tsubouchi, H.; Ideta, T.; Mueller, N.; Takatsuki, K.; Matsuoka, M. Persistent clonal proliferation of human T-lymphotropic virus type I-infected cells *in vivo*. *Cancer Res.* **1997**, *57*, 4862–4867. [PubMed]

46. Mortreux, F.; Gabet, A.S.; Wattel, E. Molecular and cellular aspects of HTLV-1 associated

leukemogenesis *in vivo*. *Leukemia* **2003**, *17*, 26–38. [CrossRef] [PubMed]

47. Gillet, N.A.; Malani, N.; Melamed, A.; Gormley, N.; Carter, R.; Bentley, D.; Berry, C.; Bushman, F.D.; Taylor, G.P.; Bangham, C.R. The host genomic environment of the provirus determines the abundance of HTLV-1-infected T-cell clones. *Blood* **2011**, *117*, 3113–3122. [CrossRef] [PubMed]

48. Yu, Q.; Konig, R.; Pillai, S.; Chiles, K.; Kearney, M.; Palmer, S.; Richman, D.; Coffin, J.M.; Landau, N.R. Single-strand specificity of APOBEC3G accounts for minus-strand deamination of the HIV genome. *Nat. Struct. Mol. Biol.* **2004**, *11*, 435–442. [CrossRef] [PubMed]

49. Derse, D.; Hill, S.A.; Princler, G.; Lloyd, P.; Heidecker, G. Resistance of human T cell leukemia virus type 1 to APOBEC3G restriction is mediated by elements in nucleocapsid. *Proc. Natl. Acad. Sci. USA* **2007**, *104*, 2915–2920. [CrossRef] [PubMed]

50. Medstrand, P.; van de Lagemaat, L.N.; Dunn, C.A.; Landry, J.R.; Svenback, D.; Mager, D.L. Impact of transposable elements on the evolution of mammalian gene regulation. *Cytogenet. Genome Res.* **2005**, *110*, 342–352. [CrossRef] [PubMed]

51. Stoye, J.P. Studies of endogenous retroviruses reveal a continuing evolutionary saga. *Nat. Rev. Microbiol.* **2012**, *10*, 395–406. [CrossRef] [PubMed]

52. Lee, Y.N.; Malim, M.H.; Bieniasz, P.D. Hypermutation of an ancient human retrovirus by APOBEC3G. *J. Virol.* **2008**, *82*, 8762–8770. [CrossRef] [PubMed]

53. Esnault, C.; Priet, S.; Ribet, D.; Heidmann, O.; Heidmann, T. Restriction by APOBEC3 proteins of endogenous retroviruses with an extracellular life cycle: *Ex vivo* effects and *in vivo* "traces" on the murine IAPE and human HERV-K elements. *Retrovirology* **2008**, *5*, e75. [CrossRef] [PubMed]

54. Anwar, F.; Davenport, M.P.; Ebrahimi, D. Footprint of APOBEC3 on the genome of human retroelements. *J. Virol.* **2013**, *87*, 8195–8204. [CrossRef] [PubMed]

55. Rua, R.; Gessain, A. Origin, evolution and innate immune control of simian foamy viruses in humans. *Curr. Opin. Virol.* **2015**, *10*, 47–55. [CrossRef] [PubMed]

56. Gartner, K.; Wiktorowicz, T.; Park, J.; Mergia, A.; Rethwilm, A.; Scheller, C. Accuracy estimation of foamy virus genome copying. *Retrovirology* **2009**, *6*, e32. [CrossRef] [PubMed]

57. Delebecque, F.; Suspene, R.; Calattini, S.; Casartelli, N.; Saib, A.; Froment, A.; Wain-Hobson, S.; Gessain, A.; Vartanian, J.P.; Schwartz, O. Restriction of foamy viruses by APOBEC cytidine deaminases. *J. Virol.* **2006**, *80*, 605–614. [CrossRef] [PubMed]

58. Rua, R.; Betsem, E.; Gessain, A. Viral latency in blood and saliva of simian foamy virus-infected humans. *PLoS ONE* **2013**, *8*, e77072. [CrossRef] [PubMed]

59. Matsen, F.A.T.; Small, C.T.; Soliven, K.; Engel, G.A.; Feeroz, M.M.; Wang, X.; Craig, K.L.; Hasan, M.K.; Emerman, M.; Linial, M.L.; *et al.* A novel bayesian method for detection of APOBEC3-mediated hypermutation and its application to zoonotic transmission of simian foamy viruses. *PLoS Comput. Biol.* **2014**, *10*, e1003493. [CrossRef] [PubMed]

60. Russell, R.A.; Wiegand, H.L.; Moore, M.D.; Schafer, A.; McClure, M.O.; Cullen, B.R. Foamy virus Bet proteins function as novel inhibitors of the APOBEC3 family of innate antiretroviral defense factors. *J. Virol.* **2005**, *79*, 8724–8731. [CrossRef] [PubMed]

61. Lochelt, M.; Romen, F.; Bastone, P.; Muckenfuss, H.; Kirchner, N.; Kim, Y.B.; Truyen, U.;

Rosler, U.; Battenberg, M.; Saib, A.; *et al.* The antiretroviral activity of APOBEC3 is inhibited by the foamy virus accessory Bet protein. *Proc. Natl. Acad. Sci. USA* **2005**, *102*, 7982–7987. [CrossRef] [PubMed]

62. Perkovic, M.; Schmidt, S.; Marino, D.; Russell, R.A.; Stauch, B.; Hofmann, H.; Kopietz, F.; Kloke, B.P.; Zielonka, J.; Strover, H.; *et al.* Species-specific inhibition of APOBEC3C by the prototype foamy virus protein bet. *J. Biol. Chem.* **2009**, *284*, 5819–5826. [CrossRef] [PubMed]

63. Jaguva Vasudevan, A.A.; Perkovic, M.; Bulliard, Y.; Cichutek, K.; Trono, D.; Haussinger, D.; Munk, C. Prototype foamy virus Bet impairs the dimerization and cytosolic solubility of human APOBEC3G. *J. Virol.* **2013**, *87*, 9030–9040. [CrossRef] [PubMed]

64. Richardson, S.R.; Narvaiza, I.; Planegger, R.A.; Weitzman, M.D.; Moran, J.V. APOBEC3A deaminates transiently exposed single-strand DNA during LINE-1 retrotransposition. *eLife* **2014**, *3*, e02008. [CrossRef] [PubMed]

65. Suspene, R.; Guetard, D.; Henry, M.; Sommer, P.; Wain-Hobson, S.; Vartanian, J.P. Extensive editing of both hepatitis B virus DNA strands by APOBEC3 cytidine deaminases *in vitro* and *in vivo*. *Proc. Natl. Acad. Sci. USA* **2005**, *102*, 8321–8326. [CrossRef] [PubMed]

66. Henry, M.; Guetard, D.; Suspene, R.; Rusniok, C.; Wain-Hobson, S.; Vartanian, J.P. Genetic editing of HBV DNA by monodomain human APOBEC3 cytidine deaminases and the recombinant nature of APOBEC3G. *PLoS ONE* **2009**, *4*, e4277. [CrossRef] [PubMed]

67. Vartanian, J.P.; Henry, M.; Marchio, A.; Suspene, R.; Aynaud, M.M.; Guetard, D.; Cervantes-Gonzalez, M.; Battiston, C.; Mazzaferro, V.; Pineau, P.; *et al.* Massive APOBEC3 editing of hepatitis B viral DNA in cirrhosis. *PLoS Pathog.* **2010**, *6*, e1000928. [CrossRef] [PubMed]

68. Beggel, B.; Munk, C.; Daumer, M.; Hauck, K.; Haussinger, D.; Lengauer, T.; Erhardt, A. Full genome ultra-deep pyrosequencing associates G-to-A hypermutation of the hepatitis B virus genome with the natural progression of hepatitis B. *J. Viral. Hepat.* **2013**, *20*, 882–889. [CrossRef] [PubMed]

69. Aynaud, M.M.; Suspene, R.; Vidalain, P.O.; Mussil, B.; Guetard, D.; Tangy, F.; Wain-Hobson, S.; Vartanian, J.P. Human Tribbles 3 protects nuclear DNA from cytidine deamination by APOBEC3A. *J. Biol. Chem.* **2012**, *287*, 39182–39192. [CrossRef] [PubMed]

70. Suspene, R.; Aynaud, M.M.; Koch, S.; Pasdeloup, D.; Labetoulle, M.; Gaertner, B.; Vartanian, J.P.; Meyerhans, A.; Wain-Hobson, S. Genetic editing of herpes simplex virus 1 and Epstein-Barr herpesvirus genomes by human APOBEC3 cytidine deaminases in culture and *in vivo*. *J. Virol.* **2011**, *85*, 7594–7602. [CrossRef] [PubMed]

71. Muylaert, I.; Tang, K.W.; Elias, P. Replication and recombination of herpes simplex virus DNA. *J. Biol. Chem.* **2011**, *286*, 15619–15624. [CrossRef] [PubMed]

72. Weller, S.K.; Coen, D.M. Herpes simplex viruses: Mechanisms of DNA replication. *Cold Spring Harb. Perspect. Biol.* **2012**, *4*, a013011. [CrossRef] [PubMed]

73. Bogani, F.; Corredeira, I.; Fernandez, V.; Sattler, U.; Rutvisuttinunt, W.; Defais, M.; Boehmer, P.E. Association between the herpes simplex virus-1 DNA polymerase and uracil DNA glycosylase. *J. Biol. Chem.* **2010**, *285*, 27664–27672. [CrossRef] [PubMed]

74. Vartanian, J.P.; Guetard, D.; Henry, M.; Wain-Hobson, S. Evidence for editing of human papillomavirus DNA by APOBEC3 in benign and precancerous lesions. *Science* **2008**, *320*, 230–233. [CrossRef] [PubMed]

75. Wang, Z.; Wakae, K.; Kitamura, K.; Aoyama, S.; Liu, G.; Koura, M.; Monjurul, A.M.; Kukimoto, I.; Muramatsu, M. APOBEC3 deaminases induce hypermutation in human papillomavirus 16 DNA upon beta interferon stimulation. *J. Virol.* **2014**, *88*, 1308–1317. [CrossRef] [PubMed]

76. Warren, C.J.; Xu, T.; Guo, K.; Griffin, L.M.; Westrich, J.A.; Lee, D.; Lambert, P.F.; Santiago, M.L.; Pyeon, D. APOBEC3A functions as a restriction factor of human papillomavirus. *J. Virol.* **2015**, *89*, 688–702. [CrossRef] [PubMed]

77. Loo, Y.M.; Melendy, T. Recruitment of replication protein A by the papillomavirus E1 protein and modulation by single-stranded DNA. *J. Virol.* **2004**, *78*, 1605–1615. [CrossRef] [PubMed]

78. Tsuge, M.; Noguchi, C.; Akiyama, R.; Matsushita, M.; Kunihiro, K.; Tanaka, S.; Abe, H.; Mitsui, F.; Kitamura, S.; Hatakeyama, T.; *et al.* G to A hypermutation of TT virus. *Virus Res.* **2010**, *149*, 211–216. [CrossRef] [PubMed]

79. Sharma, S.; Patnaik, S.K.; Thomas Taggart, R.; Kannisto, E.D.; Enriquez, S.M.; Gollnick, P.; Baysal, B.E. APOBEC3A cytidine deaminase induces RNA editing in monocytes and macrophages. *Nat. Commun.* **2015**, *6*, 6881. [CrossRef] [PubMed]

80. Domingo, E.; Sheldon, J.; Perales, C. Viral quasispecies evolution. *Microbiol. Mol. Biol. Rev.* **2012**, *76*, 159–216. [CrossRef] [PubMed]

81. Hultquist, J.F.; Harris, R.S. Leveraging APOBEC3 proteins to alter the HIV mutation rate and combat AIDS. *Future Virol.* **2009**, *4*, 605. [CrossRef] [PubMed]

82. Harris, R.S. Enhancing immunity to HIV through APOBEC. *Nat. Biotechnol.* **2008**, *26*, 1089–1090. [CrossRef] [PubMed]

83. Alexandrov, L.B.; Nik-Zainal, S.; Wedge, D.C.; Aparicio, S.A.; Behjati, S.; Biankin, A.V.; Bignell, G.R.; Bolli, N.; Borg, A.; Borresen-Dale, A.L.; *et al.* Signatures of mutational processes in human cancer. *Nature* **2013**, *500*, 415–421. [CrossRef] [PubMed]

84. Burns, M.B.; Temiz, N.A.; Harris, R.S. Evidence for APOBEC3B mutagenesis in multiple human cancers. *Nat. Genet.* **2013**, *45*, 977–983. [CrossRef] [PubMed]

85. Roberts, S.A.; Lawrence, M.S.; Klimczak, L.J.; Grimm, S.A.; Fargo, D.; Stojanov, P.; Kiezun, A.; Kryukov, G.V.; Carter, S.L.; Saksena, G.; *et al.* An APOBEC cytidine deaminase mutagenesis pattern is widespread in human cancers. *Nat. Genet.* **2013**, *45*, 970–976. [CrossRef] [PubMed]

86. Taylor, B.J.; Nik-Zainal, S.; Wu, Y.L.; Stebbings, L.A.; Raine, K.; Campbell, P.J.; Rada, C.; Stratton, M.R.; Neuberger, M.S. DNA deaminases induce break-associated mutation showers with implication of APOBEC3B and 3A in breast cancer kataegis. *eLife* **2013**, *2*, e00534. [CrossRef] [PubMed]

87. Lada, A.G.; Dhar, A.; Boissy, R.J.; Hirano, M.; Rubel, A.A.; Rogozin, I.B.; Pavlov, Y.I. AID/APOBEC cytosine deaminase induces genome-wide kataegis. *Biol. Direct.* **2012**, *7*, e47. [CrossRef] [PubMed]

88. Lada, A.G.; Stepchenkova, E.I.; Waisertreiger, I.S.; Noskov, V.N.; Dhar, A.; Eudy, J.D.;

Boissy, R.J.; Hirano, M.; Rogozin, I.B.; Pavlov, Y.I. Genome-wide mutation avalanches induced in diploid yeast cells by a base analog or an APOBEC deaminase. *PLoS Genet.* **2013**, *9*, e1003736. [CrossRef] [PubMed]

89. Roberts, S.A.; Sterling, J.; Thompson, C.; Harris, S.; Mav, D.; Shah, R.; Klimczak, L.J.; Kryukov, G.V.; Malc, E.; Mieczkowski, P.A.; *et al.* Clustered mutations in yeast and in human cancers can arise from damaged long single-strand DNA regions. *Mol. Cell* **2012**, *46*, 424–435. [CrossRef] [PubMed]

90. Seitz, H.K.; Stickel, F. Molecular mechanisms of alcohol-mediated carcinogenesis. *Nat. Rev. Cancer* **2007**, *7*, 599–612. [CrossRef] [PubMed]

91. Zuo, T.; Liu, D.; Lv, W.; Wang, X.; Wang, J.; Lv, M.; Huang, W.; Wu, J.; Zhang, H.; Jin, H.; *et al.* Small-molecule inhibition of human immunodeficiency virus type 1 replication by targeting the interaction between Vif and ElonginC. *J. Virol.* **2012**, *86*, 5497–5507. [CrossRef] [PubMed]

92. Pery, E.; Sheehy, A.; Nebane, N.M.; Brazier, A.J.; Misra, V.; Rajendran, K.S.; Buhrlage, S.J.; Mankowski, M.K.; Rasmussen, L.; White, E.L.; *et al.* Identification of a Novel HIV-1 Inhibitor Targeting Vif-dependent Degradation of Human APOBEC3G. *J. Biol. Chem.* **2015**, *290*, 10504–10517. [CrossRef] [PubMed]

93. Li, M.; Shandilya, S.M.; Carpenter, M.A.; Rathore, A.; Brown, W.L.; Perkins, A.L.; Harki, D.A.; Solberg, J.; Hook, D.J.; Pandey, K.K.; *et al.* First-in-class small molecule inhibitors of the single-strand DNA cytosine deaminase APOBEC3G. *ACS Chem. Biol.* **2012**, *7*, 506–517. [CrossRef] [PubMed]

94. Olson, M.E.; Li, M.; Harris, R.S.; Harki, D.A. Small-molecule APOBEC3G DNA cytosine deaminase inhibitors based on a 4-amino-1,2,4-triazole-3-thiol scaffold. *ChemMedChem* **2013**, *8*, 112–117. [CrossRef] [PubMed]

11

Translational Control of the HIV Unspliced Genomic RNA

Bárbara Rojas-Araya [1], Théophile Ohlmann [2,3,4,5,6],* and Ricardo Soto-Rifo [1],*

[1] Molecular and Cellular Virology Laboratory, Program of Virology, Institute of Biomedical Sciences, Faculty of Medicine, University of Chile, Independencia 834100, Santiago, Chile;
E-Mail: barbara.rojas.araya@gmail.com

[2] CIRI, International Center for Infectiology Research, Université de Lyon, Lyon 69007, France

[3] Inserm, U1111, Lyon 69007, France

[4] Ecole Normale Supérieure de Lyon, Lyon 69007, France

[5] Université Lyon 1, Centre International de Recherche en Infectiologie, Lyon 69007, France

[6] CNRS, UMR5308, Lyon 69007, France

* Authors to whom correspondence should be addressed; E-Mails: tohlmann@ens-lyon.fr (T.O.); rsotorifo@med.uchile.cl (R.S.-R.)

Academic Editor: David Boehr

Abstract: Post-transcriptional control in both HIV-1 and HIV-2 is a highly regulated process that commences in the nucleus of the host infected cell and finishes by the expression of viral proteins in the cytoplasm. Expression of the unspliced genomic RNA is particularly controlled at the level of RNA splicing, export, and translation. It appears increasingly obvious that all these steps are interconnected and they result in the building of a viral ribonucleoprotein complex (RNP) that must be efficiently translated in the cytosolic compartment. This review summarizes our knowledge about the genesis, localization, and expression of this viral RNP.

Keywords: HIV-1; Rev; nuclear export; DDX3; cap-dependent translation; IRES

1. Introduction

Human Immunodeficiency virus type-1 (HIV-1) and type-2 (HIV-2) belong to the Lentivirus genus of the *Retroviridae* family and are the etiological agents of the Acquired Immunodeficiency Syndrome (AIDS) in humans [1]. Both viruses primarily infect cells of the immune system that express the CD4 receptor and one of the chemokine receptors CCR5 or CXCR4 that act as co-receptors for viral entry. The HIV replication cycle begins with the interactions between the surface glycoprotein gp120 with CD4 and one of the co-receptors in a process that induces conformational changes allowing insertion of the viral transmembrane protein gp41 in the host cell membrane to trigger fusion of both membranes and entry of the viral capsid into the host cell cytoplasm. Then, the positive single stranded RNA genome is converted into double stranded DNA by the virally encoded reverse transcriptase, which is located in the capsid. In association with viral and cellular proteins, viral DNA forms the so-called pre-integration complex (PIC), which is imported to the host cell nucleus in an active process orchestrated by the viral proteins capsid and integrase [2]. The latter then catalyzes integration of viral DNA into the host cell genome to establish what is known as the proviral state. Once integrated, the provirus can remain latent or undergo efficient gene expression in order to continue with late steps of the replication cycle. The full-length unspliced genomic RNA (hence referred as unspliced mRNA) has a dual function as it is both used as mRNA for the synthesis of Gag and Gag-Pol precursors and the genome that is incorporated into the viral particles. The structural protein Gag drives both packaging of the genomic RNA and assembly of newly synthesized viral particles, which will be maturated by the viral protease allowing initiation of a new replication cycle.

HIV gene expression relies on the host for transcription, RNA processing, nuclear export and translation, a series of complex processes that are assisted by at least, two major viral regulators namely Tat and Rev. HIV transcription relies both on the promoter sequences present in the viral 5′ long-terminal repeat (5′-LTR) region and the *trans*-activator viral protein Tat, which acts together with host cellular proteins including the RNA polymerase II and the pTEFb transcription factor [3–7]. Transcription from the provirus results in expression of the full-length unspliced mRNA, which is 9-kb long and encodes structural and enzymatic proteins (Gag and Gag-Pol). However, the presence of multiple splice donor and acceptor sites within the full-length mRNA supports alternative splicing which results in the generation of a complex pattern of viral mRNAs harboring the open reading frames of Vif, Vpr, Vpu/Env, Tat, Rev and Nef, which differ in their 5′ untranslated regions (5′-UTR) [8]. These transcripts are both incompletely (4-kb) and completely spliced (2-kb) and are used for expression of all remaining viral proteins. Several of these completely spliced transcripts coding for Tat, Rev, and Nef are produced during the early steps of infection [9–11]. Later on, the full-length unspliced mRNA together with further different 4-kb transcripts coding for Env/Vpu, Vif, Vpr, and Tat are then generated, exported and translated in the cytoplasm [9–12]. All these RNA processing events generate different viral mRNP complexes that will differ in the routes used to reach the host translational machinery. As such, while completely spliced transcripts are exported by the canonical nuclear export pathway, the unspliced and the 4-kb incompletely spliced transcripts require the binding of the virally encoded protein Rev to the *cis*-acting RNA element called the Rev responsive element (RRE) present in all of these intron-containing transcripts; this allows their export through the CRM1 pathway [13–20] (Figure 1).

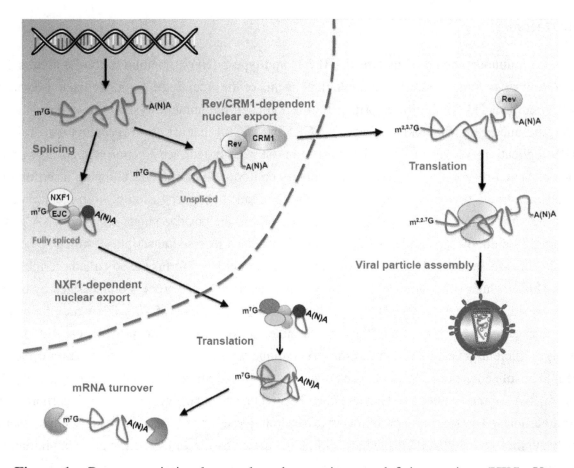

Figure 1. Post-transcriptional control on human immunodeficiency virus (HIV). Upon transcription, the capped and polyadenylated full-length genomic RNA is used as a template for the host mRNA processing machinery in order to generate fully spliced and partially spliced transcripts (partially spliced transcripts have been omitted for simplicity). In the nucleus, fully spliced transcripts form a classical messenger ribonucleoprotein complex (mRNP) together with host proteins such as the exon junction complex (EJC) and the mRNA export factor NXF1. In the cytoplasm, fully spliced mRNAs recruit the host translational apparatus for protein synthesis and later they are degraded by the mRNA turnover machinery. In the presence of the viral protein Rev, the unspliced genomic RNA (and partially spliced mRNAs) reaches the cytoplasm through the CRM1-dependent pathway avoiding the host cell surveillance mechanisms. During this journey to the cytoplasm, the unspliced genomic RNA forms a unique mRNP that favors its association with the host translational machinery. In contrast to the fully spliced transcripts, the unspliced genomic RNA does not undergo turnover as it is incorporated into viral particles.

As mentioned above, in addition to its nuclear function as a pre-mRNA template for the generation of the 2 and 4-kb transcripts, the 9-kb full-length unspliced mRNA plays two additional roles in the cytoplasm by serving both as a mRNA for viral protein production and as the packaged genome (Figure 1). In order to combine these different functions, the unspliced mRNA needs to overcome several structural and functional constraints that could affect cellular post-transcriptional events such as nuclear export and translation. In this review, we focus on how the virus has evolved to combine the building of a complex and specific mRNP on its mRNAs ensuring proper viral gene expression.

2. Reaching the Cytoplasm Avoiding Surveillance Mechanisms

HIV-1 transcripts are synthesized by the RNA polymerase II and, consequently, are capped and polyadenylated by the host machinery [21–24]. As described above, many different viral mRNA species can be found in infected cells with at least four different 5′- and eight different 3′-splice sites being used during pre-mRNA processing [9–11]. However, the cellular splicing machinery must be inefficient in the usage of viral splice sites in order to ensure that appropriate pools of each subset of viral mRNAs can be produced in the nucleus [25–27]. The vast majority of cellular mRNAs are usually spliced to completion and thus all introns are removed during splicing [28]. This is the case for the viral 2-kb transcripts that are completely processed and can be exported to the cytoplasm through the nuclear export factor NXF1 [29–31]. However, nuclear export of mRNAs that harbor functional introns is quite unusual and they are often retained in the nucleus by the interaction with splicing factors until they are either spliced to completion or degraded [32–34]. In addition, viral intron-containing transcripts cannot be exported through the NXF1-dependent pathway due to surveillance mechanism ruled by, amongst others, the cellular protein Tpr [35–37]. As such, the 4-kb incompletely spliced and the 9-kb unspliced transcripts are retained and degraded in the host cell nucleus unless the viral protein Rev is present [25,38].

Rev is synthesized from a 2-kb completely spliced transcript and is essential for virus replication [39]. Although the step of the replication cycle in which Rev activity is the most important has been the subject of some controversy [39–41], there is no doubt that synthesis of the viral structural proteins Gag and Gag-Pol from the unspliced mRNA is dramatically reduced in the absence of Rev. Rev is a phosphoprotein of approximately 18-kDa that constantly shuttles between the nucleus and the cytoplasm but accumulates in the nucleus [39]. The N-terminal domain of the protein contains an arginine-rich motif that serves both as a nuclear localization signal (NLS) and as an RNA-binding domain (RBD) [39,42–50]. While the NLS allows recognition and nuclear import of Rev by Importin-β, the RBD allows the interaction with the Rev Responsive Element (RRE) which is present exclusively in the incompletely spliced and unspliced viral transcripts as it is located within the *env* gene [19,39,51]. The arginine-rich sequence is flanked from both sides by less defined sequences required for oligomerization [39,49,50]. The C-terminal domain contains the leucine-rich nuclear export signal (NES) that allows the interaction and nuclear export of the Rev-RRE complex with the karyopherin CRM1 (Chromosome maintenance-1) bound to Ran-GTP [16,52–55]. Recent structural studies have revealed that once bound to the RRE, the Rev protein oligomerizes in order to promote nuclear export [49,56] while CRM1 forms a dimer that favors nuclear export of the Rev-RRE complex [57]. Moreover, it was recently shown that Rev can interact with the nuclear cap-binding complex (CBC) component CBP80 and block NXF1 recruitment in order to specifically enter the nuclear export pathway through CRM1 [58]. In addition to CRM1-RanGTP and CBP80, Rev recruits several host proteins including eIF5A, hRIP, DDX3, DDX1, and Sam68 to promote nuclear export [59]. Thus, by using this alternative pathway, the viral protein Rev ensures the cytoplasmic accumulation of intron-containing transcripts and avoids NXF1-associated quality control mechanisms. This explains that despite the presence of introns, viral transcripts that do not undergo complete splicing are not substrates for non-sense mediated decay (NMD) [60,61].

After completion of their journey from the nucleus and through the nuclear pores, the viral transcripts must compete with cellular mRNAs in the cytoplasm to recruit the host translational machinery. In mammals, ribosome recruitment onto the mRNA occurs by two main mechanisms: the cap-dependent and the internal ribosome entry sites (IRES)-driven mechanisms [62,63] and HIV-1 has evolved strategies to use both [64].

3. An Overview on mRNA Translation Initiation in Eukaryotes

The vast majority of cellular mRNAs recruit ribosomes through a cap-dependent translation initiation mechanism. This process sequentially involves: (i) formation of a 43S pre-initiation complex; (ii) cap structure recognition and loading of the 43S pre-initiation complex onto the mRNA; (iii) ribosomal scanning of the 5′-UTR; (iv) initiation codon recognition and (v) joining of the 60S ribosomal subunit [62]. The 43S pre-initiation complex is composed of a recycled 40S small ribosomal subunit, an eIF2-GTP-tRNAi ternary complex (TC), eIF3, eIF1, eIF1A and probably eIF5 [62]. At the 5′ end of the mRNA, the eIF4F holoenzyme binds to the cap-structure and unwinds local RNA structures assisted by eIF4B or eIF4H creating the landing pad for the 43S pre-initiation complex. The eIF4F multimeric complex is composed of the cap-binding protein eIF4E, the RNA helicase eIF4A, and the scaffold protein eIF4G [65,66]. eIF4E exhibits high affinity for the cap structure and interacts with eIF4G to mediate cap-dependent translation initiation by promoting assembly of eIF4F onto the capped mRNA. The DEAD-box protein eIF4A is an RNA helicase with ATP-dependent RNA unwinding activity [67,68]. Although the intrinsic helicase activity of eIF4A is weak, its inclusion into the eIF4F complex together with the binding of eIF4B and the related factor eIF4H strongly stimulates its enzymatic activity [69]. As mentioned above, the eIF4G scaffold protein associates with eIF4E and eIF4A to form the eIF4F holoenzyme that binds to the 5′ end of capped mRNAs [70–73]. By further interacting with eIF3, eIF4G promotes attachment of the 43S pre-initiation complex onto the transcript to allow formation of a 48S pre-initiation complex [72,74–78]. Once attached, this complex immediately starts scanning in a 5′ to 3′ direction from the cap structure until it reaches an initiation codon, which often corresponds to the first AUG codon [79–81]. The ribosomal scanning model proposes that the translation initiation complex unwinds secondary structures present in the 5′-UTR and moves in the 5′ to 3′ direction in an ATP-dependent manner [82,83]. Thus, in addition to their role in 43S pre-initiation complex attachment, eIF4G, eIF4A, eIF4B (or eIF4H) also assist the scanning process [83,84]. Although the RNA helicase eIF4A and its associated factors eIF4B/eIF4H can support the unwinding process of the scanning pre-initiation complexes, it has been recently shown that additional RNA helicases can also be recruited [85]. As such, the related DExH box protein 29 (DHX29) binds the 40S small ribosomal subunits while RNA helicase A binds selected mRNAs and both are required for efficient scanning of mRNAs containing highly structured 5′-UTRs [86–88].

An alternative model of translation initiation has been described for mRNAs that harbor specific RNA sequences termed Internal Ribosome Entry Sites (IRES). These sequences are generally present in the 5′-UTR of the mRNA whose function is to recruit ribosomes for translation initiation in a cap-independent manner. IRES elements were first discovered in viral RNA genomes more than 25 years ago with the studies of picornavirus translation [89,90] and have now been characterized in many viral mRNAs including HCV, Pestiviruses, and Retroviruses [63]. Although IRES elements have also been described in near 100 cellular mRNAs their existence remains controversial mainly by the lack of essential controls discarding cryptic promoters and/or alternative splicing during the characterization process [91].

IRES elements promote the direct binding of the 43S pre-initiation complex and associated factors to the mRNA. However, the precise mechanism of IRES-mediated translation initiation is not completely understood. Although a classification of IRES elements by structural criteria is not possible due to the lack of any conserved sequence, viral IRES elements can be grouped based on a mechanistic and functional point of view involving: (i) the way by which the 43S pre-initiation complexes is recruited, e.g., whether it is assisted or not by eIFs; and (ii) the site where the 43S pre-initiation complex is positioned onto the mRNA, which can be close to the initiation codon or if it involves an additional step of scanning. Moreover, IRES elements can also be characterized by the requirement of diverse cellular accessory proteins denominated IRES *trans*-acting factors (ITAFs) for proper function.

IRES elements allow the selective translation of viral mRNAs under conditions in which global host translation is compromised. When faced by several stresses (such as viral infections, hypoxia, or heat shock) or particular cellular conditions (such as mitosis or apoptosis), Eukaryotic cells often respond by reducing the global rates of translation [92]. However, a significant fraction of cellular mRNAs was shown to remain associated to polysomes [93] and several of these are IRES-containing transcripts. This shows that the presence of the IRES element allows mRNAs to be translated under unfavorable conditions in which cap-dependent translation is slowed down or arrested [94–98].

4. Recruiting the Host Translational Machinery onto the Unspliced HIV Genomic RNA

The unspliced mRNA harbors a long (5′-UTR) organized in several RNA structures involved in many steps of the replication cycle [3,99–103]. Given the structure and complexity of the 5'-UTR, the mechanism by which translation initiation takes place on the HIV-1 genomic RNA has been the subject of debate for several years [104]. Indeed, it was initially shown that sequences derived from the 5′-UTR were inhibitory for translation [105–109]. Particularly, cell-free *in vitro* translation assays and *ex vivo* experiments using reporter genes suggested that the presence and folding of the TAR RNA motif, which is located at the very 5′ end of the viral transcripts, exerted a negative effect on protein synthesis both by impeding ribosome recruitment and by activating the kinase PKR [105–108,110,111]. However, despite this incompatibility with ribosome recruitment by a cap-dependent ribosomal scanning mechanism, an IRES-driven mechanism on HIV-1 transcripts was rapidly discarded indicating that cap-dependent translation initiation was the major mechanism for ribosome recruitment [112].

5. Identification of a Cell Cycle-Dependent IRES

As mentioned above, initial attempts to identify sequences within viral 5′-UTR supporting IRES activity failed and the cap-dependent ribosomal scanning was proposed as the only mechanism to drive Gag synthesis [112]. However, a more detailed study revealed that an IRES element was indeed present within the 5′-UTR of the HIV-1 unspliced mRNA [113]. This IRES element was mapped to nucleotides 104 to 336 where it spans the primer-binding site (PBS), the dimerization site (DIS), the major splice donor (SD) and RNA motifs that are critical for encapsidation [113]. Interestingly, this IRES element was shown to be activated during the G2/M phase of the cell cycle [113]. This peculiarity not only explained why previous studies failed to detect IRES activity but also emphasized the physiological relevance that the use of an alternative mechanism of ribosome recruitment could have during viral replication. Indeed, during HIV-1 and other lentiviral infections, the viral protein Vpr induces a cell cycle arrest at the G2/M phase [114–117]. Although the G2/M phase is characterized by a strong inhibition of cap-dependent protein synthesis [95,118,119], HIV-1 viral gene expression was shown to continue during this phase of the cell cycle [120–122]. Although IRES elements were demonstrated to be able to drive efficiently protein synthesis in G2/M [96], other authors have proposed another alternative in which the translation of the HIV-1 unspliced mRNA was rather conducted by a eIF4E-independent, CBC-driven, cap-dependent mechanism during the G2/M arrest induced by Vpr [122]. Nevertheless, the ability of the 5′-UTR of HIV-1 transcripts to drive IRES-driven translation has now been evidenced by several groups in different experimental contexts and on different HIV-1 prototype strains [123–130]. Therefore, it is conceivable that the HIV-1 genomic RNA can use both strategies depending on some physiological conditions that remain to be found. Moreover, similar to poliovirus, the HIV-1 and HIV-2 proteases were shown to process translation initiation factors eIF4GI and PABP *in vitro* and *ex vivo* leading to the inhibition of cap-dependent ribosomal scanning with modest impact on viral unspliced mRNA translation [70,131–134]. However, processing of eIF4GI and PABP during viral infection was rather modest and occurred late during infection [131] and thus, the significance of these events in the course of viral replication remains to be demonstrated.

Protein synthesis from the HIV-1 and HIV-2 unspliced mRNAs presents an additional layer of complexity as IRES elements have also been characterized within the Gag coding region [64,135]. By using the HIV-1 Gag ORF lacking the viral 5′-UTR it was shown that this region was able to drive synthesis of full-length p55 Gag and a novel 40-kDa N-terminally truncated isoform of Gag (p40) initiated at an internal *in frame* AUG codon [136]. The presence of IRES elements downstream to the authentic initiation codon and the synthesis of N-terminally truncated isoforms of Gag were also characterized in other related lentiviruses such as HIV-2, SIV, and FIV indicating that the conservation of the mechanism is a common feature of the genus and could be important for replication [135,137–139]. In HIV-1 and HIV-2, these Gag isoforms are incorporated into viral particles despite the lack of a myristoylation site at their N-terminus, probably by protein-protein interactions with the full length Gag polyprotein; this suggests a role for these truncated isoforms in the replication cycle [136,138]. Although the molecular mechanisms controlling this process in the HIV-1 unspliced mRNA are not completely understood, an *in vitro* study revealed that the different modes of ribosome recruitment have different levels of requirements for eIF4F [140]. In the case of the HIV-2, it was shown that three IRES elements within the Gag coding region were able to directly recruit three independent 43S pre-initiation complexes [138,141–143].

6. Translation by a Cap-Dependent Mechanism

More recently, by using synthetic constructs Berkhout and co-workers demonstrated that cap-dependent ribosomal scanning occurs throughout the 5′-UTR of the HIV-1 unspliced mRNA [144]. Using similar approaches, other groups including ours demonstrated that the cap-dependent mechanism of translation initiation occurs both *in vitro* and *ex vivo* [109,121]. The ability of the 43S pre-initiation complex to scan through the highly structured 5′-UTR could be explained by the recruitment of the helicase RHA, which was shown to promote polysome association of the unspliced mRNA by interacting with a post-transcriptional control element (PCE) located at the 5'-UTR [88,145]. Although it is thought that RHA helicase activity contributes to the unwinding of secondary structures during ribosomal scanning, the involvement of other RNA helicases such as DHX29 has not yet been investigated and thus, cannot be discarded.

While the involvement of RHA shed light on how the 43S pre-initiation complex moves along the highly structured viral 5'-UTR, it was still unclear how the cap-structure could be recognized by the eIF4F complex in the presence of the TAR structure. Indeed, the 5' end cap moiety of all HIV-1 transcripts is base-paired and embedded within the basis of the stem of the TAR RNA motif and thus, is likely to be inaccessible for the binding of the eIF4F complex and the ribosomal 43S subunit. Surprisingly, although the presence of TAR was shown to strongly interfere with translation initiation in the rabbit reticulocytes lysate (RRL), this was not the case in constructs expressed in living cells [109]. These data suggested that some specific host factor(s) that are absent or in limited concentration in the RRL can be used to overcome the structural constraint imposed by TAR. A likely candidate was found amongst one of the Rev cofactors namely the RNA helicase DDX3 [146]. DDX3 belongs to the DEAD-box family of proteins whose prototype member is the initiation factor eIF4A [147]. DEAD-box proteins are ATP-dependent RNA helicases that play pleiotropic functions within the cell by participating in all steps of RNA metabolism [148]. These proteins are thought to participate in RNA:RNA and RNA:protein remodeling or to act as RNA clamps for the assembly of large macromolecular complexes [148]. DDX3 was first proposed to be a host factor involved in Rev-dependent nuclear export [146]. By using a full-length reporter proviral DNA and viral infection, we were also able to show that DDX3 was required for translation of the unspliced genomic RNA both in HeLa and T-cells and this function required the ATP binding and ATPase activity of the enzyme [149,150]. We also reported that the molecular target for DDX3 was actually the TAR RNA motif, an observation recently validated by another group [151]. Interestingly, we observed that DDX3 was required to unwind TAR in cells and this functional interaction was necessary when the latter was at its original location (e.g., at the 5' end of the HIV-1 transcript) but the dependence in DDX3 was abolished when the TAR motif was preceded by an unstructured spacer sequence [149]. These data suggested that DDX3 binds and unwinds TAR during a pre-translation initiation step that is necessary to remodel secondary structures in order to render the cap moiety accessible to the eIF4F holenzyme and the 43S complex [149] (Figure 2). In agreement with this, we could also show that DDX3 was bound to, at least, two additional and specific sites within the 5'-UTR of the HIV-1 genomic RNA [149]. These sites were located exclusively on RNA single stranded regions and could correspond to loading platforms for DDX3 as had been previously suggested [148]. In addition, an interaction between DDX3 and translation initiation factors eIF4GI and PABPC1 was also evidenced by biochemical assays as well as confocal microscopy [149]. Interestingly, we observed that the complex formed between the unspliced HIV-1 mRNA, DDX3, eIF4GI, and PABPC1 was assembled in localized cytoplasmic granules that resembled but were different from stress granules as they lacked the eIF4F components eIF4E, eIF4A, eIF4B, and the CBC component CBP80 [150] (Figure 2).

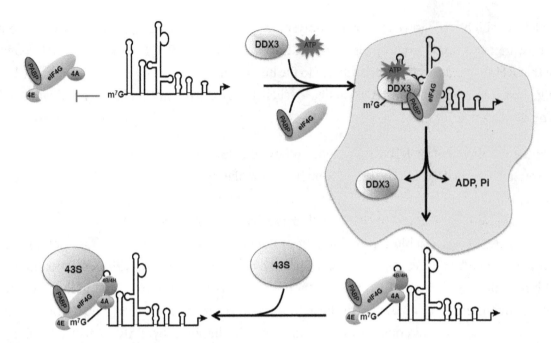

Figure 2. DDX3-mediated translation of the HIV-1 genomic RNA. In the absence of DDX3, the TAR RNA motif impedes binding of the eIF4F holoenzyme to the cap. Thus, DDX3 binds the viral 5'-UTR to nucleate formation of a pre-translation initiation complex that involves ATP-dependent unwinding of TAR and specific recruitment of translation initiation factors eIF4GI and PABP (and probably other unidentified cellular proteins). TAR unwinding renders the cap accessible for eIF4F binding and subsequent recruitment of the 43S pre-initiation complex. It is possible that such a pre-translation initiation step driven by DDX3 occurs compartmentalized in RNA granules (in yellow).

Another intriguing feature of the unspliced HIV-1 mRNA that could also influence its translation is the presence of a trimethylguanosine (TMG) cap structure [152]. A few years ago, the peroxisome proliferator-activated receptor-interacting protein with methyltransferase domain (PIMT) (the human homolog of the yeast cap hypermethylase TGS1) was shown to interact with Rev and this resulted in the hypermethylation of the 5' cap structure of the HIV-1 unspliced mRNA [152]. It has been known for quite some time that TMG-capped mRNAs present reduced translational rates *in vitro* [153]. However, in the case of the HIV-1 genomic RNA, the latter is efficiently used for viral protein production and trimethylation of its cap was shown to be required in this process although the molecular mechanism underlying was not elucidated [152]. In light of our data showing the presence of a pre-initiation complex composed of DDX3/eIF4G/PABP, but lacking any of the major cap binding proteins CBC and eIF4E [150], a further investigation into the role of a TMG cap would be interesting. Indeed, as the affinity of the CBC and eIF4E for the TMG cap is largely reduced compared to the classical m^7G monomethylated cap [154], this could explain the exclusion of both eIF4E and CBC from this pre-initiation complex and could suggest that other TMG-bound cellular proteins may be recruited for initiation of HIV-1 unspliced mRNA translation.

7. Assembly of Unspliced mRNA-Containing Granules

Cellular mRNAs are in a dynamic equilibrium between polysomes and cytoplasmic granules such as stress granules and p-bodies [155,156]. While stress granules are sites of triage for mRNAs stalled in translation initiation as a response to cellular stress, p-bodies are sites intimately related to the mRNA decay machinery [155,156]. Both structures and/or some of their components have been shown to play pivotal roles during replication of several viruses and thus, it is not surprising that viruses have evolved different strategies to manipulate the assembly/disassembly of mRNA granules [157–159].

Although it was first proposed that HIV-1 translation could be negatively regulated by some components of p-bodies including APOBEC3G [160,161], there is new evidence showing that HIV-1 replication induces the disassembly of p-bodies [162]. Moreover, it was also recently shown that APOBEC3G activity on HIV-1 replication was independent of p-bodies [163] and that p-bodies components such as DDX6 and Argonaute 2 were rather involved in viral particle assembly independent of RNA packaging [164]. Therefore, further work is necessary to clarify the role of p-bodies in HIV-1 unspliced mRNA metabolism.

Interestingly, HIV-1 and HIV-2 have evolved completely opposite strategies to modulate and control the assembly of stress granules. As such, it was shown that HIV-1 has the ability to interfere with stress granule assembly induced by different types of stresses [162,165]. Indeed, the authors showed that the HIV-1 Gag protein has the ability to interfere with stress granules assembly through a direct interaction with eEF2 and G3BP1, two key factors required for assembly of these cytoplasmic structures [165]. Thus, it is possible that by doing so, the HIV-1 unspliced mRNA promotes the assembly of a pre-initiation complex with DDX3 and subset of eIFs in order to enter in translation initiation and associate with polysomes [150] (Figure 3A). Then, the HIV-1 unspliced mRNA is assembled into a Staufen1-dependent mRNP, which also contains the viral protein Gag and is required for RNA packaging [162] (Figure 3A).

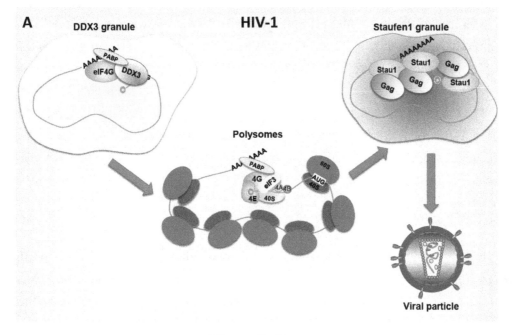

Figure 3. *Cont·*

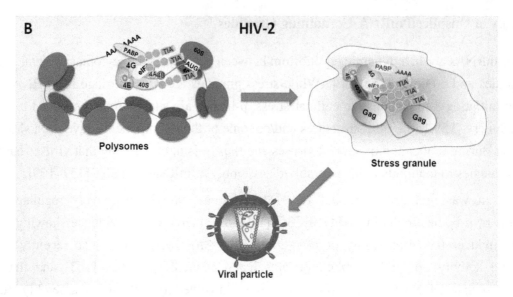

Figure 3. RNA granules assembled during HIV replication. (**A**) Polysome association of HIV-1 unspliced mRNA requires its previous assembly in DDX3-dependent granules together with eIF4GI, PABPC1 and probably other, yet, unidentified cellular proteins. Once translated, unspliced mRNA associates with the dsRNA-binding protein Staufen1 and the viral protein Gag in order to form another specific RNA granule (Staufen1 granule), which is required for viral particle assembly. This dynamic assembly of different RNA granules allows HIV-1 to coordinate genomic RNA translation and packaging; (**B**) The HIV-2 unspliced mRNA recruits the stress granule assembly factor TIAR to form a specific viral mRNP that accumulates in stress granules in the absence of active translation. The viral protein Gag also accumulates in stress granules suggesting that the transition from translation to RNA packaging could occur in these structures.

In sharp contrast with what was described for HIV-1, we showed that HIV-2 replication induces the spontaneous assembly of stress granules [166] (Figure 3B). Moreover, we observed that HIV-2 unspliced mRNA was directly associated with the stress granule assembly factor TIAR in order to form a specific viral mRNP [166] (Figure 3B). We have previously shown that ribosome recruitment onto the HIV-2 genomic RNA is very inefficient due to a strong interference imposed by the highly structured TAR RNA motif [109]. Thus, stress granules could serve as sites of storage for the viral genome while threshold levels of Gag required for RNA packaging are produced. Interestingly, the HIV-2 Gag polyprotein was also observed in stress granules indicating that the transition from translation to RNA packaging may occur in these structures [166] (Figure 3B).

8. Viral Proteins Promoting Translation

Some of the virally encoded proteins, namely Tat, Rev, and Gag have been involved in the control of viral mRNA translation (Figure 4). Initial studies carried out in the RRL and *Xenopus leavis* oocytes revealed that Tat was involved in the control of translation notably by counteracting the deleterious activation of PKR [107,167,168]. Indeed, secondary RNA structures constituting the TAR motif at the 5′-UTR were shown to activate the protein kinase R (PKR) leading to inhibition of translation [169,170].

Once activated, PKR phosphorylates the α subunit of eIF2 resulting in a global inhibition of translation initiation [171]. However, Tat binding to TAR and/or PKR could prevent activation of the kinase the phosphorylation of eIF2α [172]. In addition, it was shown that Tat is able to stimulate translation both *in vitro* and in living cells [129]. Moreover, binding of Tat to the 5′-UTR of the unspliced mRNA could stimulate the programmed-1 ribosomal frameshift [173]. More recently, Tat was shown to interact with DDX3 and remain associated to polysomes together with the unspliced mRNA further indicating its role in viral mRNA translation [151].

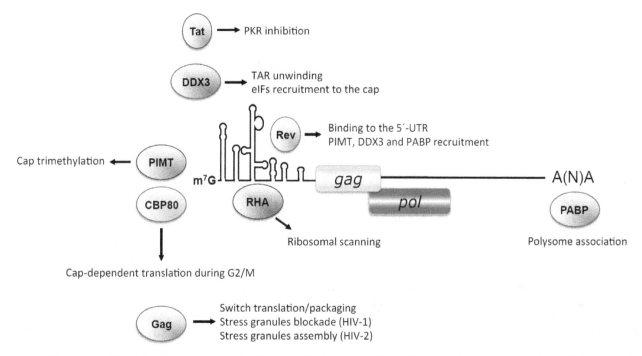

Figure 4. Translational control by host and viral proteins. Schematic representation of the panel of viral (Rev, Tat, Gag) and cellular (DDX3, PIMT, CBP80, RHA, and PABP) proteins required to assist translation initiation from the HIV-1 and HIV-2 unspliced mRNA.

Another viral protein, Rev, was demonstrated to be required for association of the incompletely spliced mRNAs *vif, vpr, env,* and *vpu* into polysomes [174]. By using a Gag expression vector lacking the RRE, it was also shown that polysome association of the resulting gag mRNA was deficient either in the presence or absence of Rev, suggesting that the Rev-RRE interaction and not the presence of Rev *per se* is critical for ribosome recruitment [175]. Such a function of Rev in translation could be explained by an enhanced recruitment of PABPC1 to Rev-dependent mRNAs [176] or by direct binding to the loop-A of stem-loop 1 located within the packaging signal [177].

Finally the Gag protein was shown to modulate its own translation by exerting a bimodal effect depending on its concentration [143,166,178]. As such, it was shown that HIV-1 and HIV-2 Gag stimulate translation at low concentrations to then inhibit protein synthesis. In the case of HIV-1, stimulation required the matrix domain while inhibition was dependent on the binding of nucleocapsid domain to the packaging signal [178]. In the case of HIV-2, we also observed a bimodal effect of Gag on translation with stimulation at low concentration and inhibition and higher concentrations [166]. Interestingly, we observed that such an effect of Gag on translation was concordant with the subcellular localization of the unspliced mRNA [166]. As such, we observed that both the unspliced mRNA and Gag localized diffusely in the cytoplasm at low concentrations of Gag while both components were assembled in stress granules at high concentrations of the viral protein [166].

9. Concluding Remarks

Post-transcriptional control of HIV-1 viral gene expression is regulated from the nucleus to the cytoplasm and involves many host and viral proteins throughout this process. It is amazing to realize that almost every step, from splicing, export, and translation, has its own regulatory pathway, which often differs from that used by cellular mRNAs. This results in the constitution of a unique viral RNP that reaches the cytoplasm to be translated by a spectrum of different mechanisms juggling with cap-dependent and cap-independent mechanisms of initiation. The great diversity of these means of expression confers to the virus several selective advantages such as preventing degradation of the unspliced mRNA by surveillance mechanisms in the nucleus and allowing selective translation under conditions that are not favorable for host gene expression. Evolution of diverse mechanisms for gene expression also allows the conciliation of the presence of multiple RNA structures in the 5'-UTR, that are required for genome replication, with the need of an efficient mechanism for viral protein synthesis. For instance, the TAR structure at the 5' end of the mRNA represents an essential element for transcription that would be severely inhibitory for ribosome binding and scanning unless it can be counteracted by the recruitment of the host RNA helicase DDX3 to assist pre-initiation complex formation. An interesting, promising new direction concerns the recent identification of compartmentalized cytoplasmic foci containing HIV-1 and HIV-2 viral RNPs. Although, the function of these foci are not fully characterized, they may serve as sites of storage to ensure an equilibrium between unspliced mRNA translation and its packaging into assembly virions. Interestingly, use of these cytoplasmic foci seems to be radically different between the two closely related human immunodeficiency viruses. A better understanding of this process may shed light on our understanding of the replication cycle of these two relatives. Such a specific and complex control of post-transcription gene expression in lentiviruses can point to new directions in the treatment of disease. As such, the targeting of essential host factors that are required for viral replication, such as DDX3 for instance, could bring new therapeutical approaches. Above all, due to great diversity of their strategies developed to express their genome, lentiviruses represent good paradigms for the studies on the control of post-transcriptional gene expression.

Acknowledgments

Research at the RSR laboratory is funded by the Comisión Nacional de Investigación en Ciencia y Tecnología (Conicyt) through the Fondecyt Initiation into Research Program (No 11121339) and the International Cooperation Program (DRI USA2013-0005). Research in the TO laboratory is financed by the ANRS.

References

1. Killian, M.S.; Levy, J.A. HIV/AIDS: 30 Years of progress and future challenges. *Eur. J. Immunol.* **2011**, *41*, 3401–3411. [CrossRef] [PubMed]

2. Matreyek, K.A.; Engelman, A. Viral and cellular requirements for the nuclear entry of retroviral preintegration nucleoprotein complexes. *Viruses* **2013**, *5*, 2483–2511. [CrossRef] [PubMed]

3. Arya, S.K. Human and simian immunodeficiency retroviruses: Activation and differential transactivation of gene expression. *AIDS Res. Hum. Retrovir.* **1988**, *4*, 175–186. [CrossRef] [PubMed]

4. Berkhout, B.; Jeang, K.T. Trans activation of human immunodeficiency virus type 1 is sequence specific for both the single-stranded bulge and loop of the trans-acting-responsive hairpin: A quantitative analysis. *J. Virol.* **1989**, *63*, 5501–5504. [PubMed]

5. Cullen, B.R. Trans-activation of human immunodeficiency virus occurs via a bimodal mechanism. *Cell* **1986**, *46*, 973–982. [CrossRef]

6. Gatignol, A. Transcription of HIV: Tat and cellular chromatin. *Adv. Pharmacol.* **2007**, *55*, 137–159. [PubMed]

7. Jones, K.A. HIV trans-activation and transcription control mechanisms. *New Biol.* **1989**, *1*, 127–135. [PubMed]

8. Tazi, J.; Bakkour, N.; Marchand, V.; Ayadi, L.; Aboufirassi, A.; Branlant, C. Alternative splicing: Regulation of HIV-1 multiplication as a target for therapeutic action. *FEBS J.* **2010**, *277*, 867–876. [CrossRef] [PubMed]

9. Purcell, D.F.; Martin, M.A. Alternative splicing of human immunodeficiency virus type 1 mRNA modulates viral protein expression, replication, and infectivity. *J. Virol.* **1993**, *67*, 6365–6378. [PubMed]

10. Schwartz, S.; Felber, B.K.; Benko, D.M.; Fenyo, E.M.; Pavlakis, G.N. Cloning and functional analysis of multiply spliced mRNA species of human immunodeficiency virus type 1. *J. Virol.* **1990**, *64*, 2519–2529. [PubMed]

11. Schwartz, S.; Felber, B.K.; Pavlakis, G.N. Expression of human immunodeficiency virus type 1 vif and vpr mRNAs is Rev-dependent and regulated by splicing. *Virology* **1991**, *183*, 677–686. [CrossRef]

12. Guerrero, S.; Batisse, J.; Libre, C.; Bernacchi, S.; Marquet, R.; Paillart, J.C. HIV-1 replication and the cellular eukaryotic translation apparatus. *Viruses* **2015**, *7*, 199–218. [CrossRef] [PubMed]

13. Bogerd, H.P.; Echarri, A.; Ross, T.M.; Cullen, B.R. Inhibition of human immunodeficiency virus Rev and human T-cell leukemia virus Rex function, but not Mason-Pfizer monkey virus constitutive transport element activity, by a mutant human nucleoporin targeted to Crm1. *J. Virol.* **1998**, *72*, 8627–8635. [PubMed]

14. Dillon, P.J.; Nelbock, P.; Perkins, A.; Rosen, C.A. Function of the human immunodeficiency virus types 1 and 2 Rev proteins is dependent on their ability to interact with a structured region present in env gene mRNA. *J. Virol.* **1990**, *64*, 4428–4437. [PubMed]

15. Emerman, M.; Vazeux, R.; Peden, K. The rev gene product of the human immunodeficiency virus affects envelope-specific RNA localization. *Cell* **1989**, *57*, 1155–1165. [CrossRef]

16. Fischer, U.; Huber, J.; Boelens, W.C.; Mattaj, I.W.; Luhrmann, R. The HIV-1 Rev activation domain is a nuclear export signal that accesses an export pathway used by specific cellular RNAs. *Cell* **1995**, *82*, 475–483. [CrossRef]

17. Le, S.Y.; Malim, M.H.; Cullen, B.R.; Maizel, J.V. A highly conserved RNA folding

region coincident with the Rev response element of primate immunodeficiency viruses. *Nucleic Acids Res.* **1990**, *18*, 1613–1623. [CrossRef] [PubMed]

18. Lewis, N.; Williams, J.; Rekosh, D.; Hammarskjold, M.L. Identification of a cis-acting element in human immunodeficiency virus type 2 (HIV-2) that is responsive to the HIV-1 rev and human T-cell leukemia virus types I and II rex proteins. *J. Virol.* **1990**, *64*, 1690–1697. [PubMed]

19. Malim, M.H.; Hauber, J.; Le, S.Y.; Maizel, J.V.; Cullen, B.R. The HIV-1 Rev *trans*-activator acts through a structured target sequence to activate nuclear export of unspliced viral mRNA. *Nature* **1989**, *338*, 254–257. [CrossRef] [PubMed]

20. Neville, M.; Stutz, F.; Lee, L.; Davis, L.I.; Rosbash, M. The importin-β family member Crm1p bridges the interaction between Rev and the nuclear pore complex during nuclear export. *Curr. Biol.* **1997**, *7*, 767–775. [CrossRef]

21. Chiu, Y.L.; Coronel, E.; Ho, C.K.; Shuman, S.; Rana, T.M. HIV-1 Tat protein interacts with mammalian capping enzyme and stimulates capping of TAR RNA. *J. Biol. Chem.* **2001**, *276*, 12959–12966. [CrossRef] [PubMed]

22. Chiu, Y.L.; Ho, C.K.; Saha, N.; Schwer, B.; Shuman, S.; Rana, T.M. Tat stimulates cotranscriptional capping of HIV mRNA. *Mol. Cell* **2002**, *10*, 585–597. [CrossRef]

23. Shatkin, A.J. Capping of eucaryotic mRNAs. *Cell* **1976**, *9*, 645–653. [CrossRef]

24. Karn, J.; Stoltzfus, C.M. Transcriptional and posttranscriptional regulation of HIV-1 gene expression. *Cold Spring Harb. Perspect. Med.* **2012**, *2*. [CrossRef] [PubMed]

25. Chang, D.D.; Sharp, P.A. Regulation by HIV Rev depends upon recognition of splice sites. *Cell* **1989**, *59*, 789–795. [CrossRef]

26. Lu, X.B.; Heimer, J.; Rekosh, D.; Hammarskjold, M.L. U1 small nuclear RNA plays a direct role in the formation of a rev-regulated human immunodeficiency virus *env* mRNA that remains unspliced. *Proc. Natl. Acad. Sci. USA* **1990**, *87*, 7598–7602. [CrossRef] [PubMed]

27. Stutz, F.; Rosbash, M. A functional interaction between Rev and yeast pre-mRNA is related to splicing complex formation. *EMBO J.* **1994**, *13*, 4096–4104. [PubMed]

28. Han, J.; Xiong, J.; Wang, D.; Fu, X.D. Pre-mRNA splicing: Where and when in the nucleus. *Trends Cell Biol.* **2011**, *21*, 336–343. [CrossRef] [PubMed]

29. Cullen, B.R. Nuclear mRNA export: Insights from virology. *Trends Biochem. Sci.* **2003**, *28*, 419–424. [CrossRef]

30. Cullen, B.R. Nuclear RNA export. *J. Cell Sci.* **2003**, *116*, 587–597. [CrossRef] [PubMed]

31. Gruter, P.; Tabernero, C.; von Kobbe, C.; Schmitt, C.; Saavedra, C.; Bachi, A.; Wilm, M.; Felber, B.K.; Izaurralde, E. TAP, the human homolog of Mex67p, mediates CTE-dependent RNA export from the nucleus. *Mol. Cell* **1998**, *1*, 649–659. [CrossRef]

32. Legrain, P.; Rosbash, M. Some *cis*- and *trans*-acting mutants for splicing target pre-mRNA to the cytoplasm. *Cell* **1989**, *57*, 573–583. [CrossRef]

33. Nakielny, S.; Dreyfuss, G. Nuclear export of proteins and RNAs. *Curr. Opin. Cell Biol.* **1997**, *9*, 420–429. [CrossRef]

34. Stutz, F.; Izaurralde, E. The interplay of nuclear mRNP assembly, mRNA surveillance and export. *Trends Cell Biol.* **2003**, *13*, 319–327. [CrossRef]

35. Coyle, J.H.; Bor, Y.C.; Rekosh, D.; Hammarskjold, M.L. The Tpr protein regulates export of mRNAs with retained introns that traffic through the Nxf1 pathway. *RNA* **2011**, *17*, 1344–1356. [CrossRef] [PubMed]

36. Rajanala, K.; Nandicoori, V.K. Localization of nucleoporin Tpr to the nuclear pore complex is essential for Tpr mediated regulation of the export of unspliced RNA. *PLoS ONE* **2012**, *7*, e29921. [CrossRef] [PubMed]

37. Porrua, O.; Libri, D. RNA quality control in the nucleus: The Angels' share of RNA. *Biochim. Biophys. Acta* **2013**, *1829*, 604–611. [CrossRef] [PubMed]

38. Felber, B.K.; Hadzopoulou-Cladaras, M.; Cladaras, C.; Copeland, T.; Pavlakis, G.N. Rev protein of human immunodeficiency virus type 1 affects the stability and transport of the viral mRNA. *Proc. Natl. Acad. Sci. USA* **1989**, *86*, 1495–1499. [CrossRef] [PubMed]

39. Pollard, V.W.; Malim, M.H. The HIV-1 Rev protein. *Annu. Rev. Microbiol.* **1998**, *52*, 491–532. [CrossRef] [PubMed]

40. Groom, H.C.; Anderson, E.C.; Lever, A.M. Rev: Beyond nuclear export. *J. Gen. Virol.* **2009**, *90*, 1303–1318. [CrossRef] [PubMed]

41. Blissenbach, M.; Grewe, B.; Hoffmann, B.; Brandt, S.; Uberla, K. Nuclear RNA export and packaging functions of HIV-1 Rev revisited. *J. Virol.* **2010**, *84*, 6598–6604. [CrossRef] [PubMed]

42. Berger, J.; Aepinus, C.; Dobrovnik, M.; Fleckenstein, B.; Hauber, J.; Bohnlein, E. Mutational analysis of functional domains in the HIV-1 Rev trans-regulatory protein. *Virology* **1991**, *183*, 630–635. [CrossRef]

43. Bohnlein, E.; Berger, J.; Hauber, J. Functional mapping of the human immunodeficiency virus type 1 Rev RNA binding domain: New insights into the domain structure of Rev and Rex. *J. Virol.* **1991**, *65*, 7051–7055. [PubMed]

44. Daly, T.J.; Cook, K.S.; Gray, G.S.; Maione, T.E.; Rusche, J.R. Specific binding of HIV-1 recombinant Rev protein to the Rev-responsive element *in vitro*. *Nature* **1989**, *342*, 816–819. [CrossRef] [PubMed]

45. Hope, T.J.; McDonald, D.; Huang, X.J.; Low, J.; Parslow, T.G. Mutational analysis of the human immunodeficiency virus type 1 Rev transactivator: Essential residues near the amino terminus. *J. Virol.* **1990**, *64*, 5360–5366. [PubMed]

46. Kubota, S.; Siomi, H.; Satoh, T.; Endo, S.; Maki, M.; Hatanaka, M. Functional similarity of HIV-I [rev] and HTLV-I [rex] proteins: Identification of a new nucleolar-targeting signal in [rev] protein. *Biochem. Biophys. Res. Commun.* **1989**, *162*, 963–970. [CrossRef]

47. Malim, M.H.; Bohnlein, S.; Hauber, J.; Cullen, B.R. Functional dissection of the HIV-1 Rev *trans*-activator—Derivation of a *trans*-dominant repressor of Rev function. *Cell* **1989**, *58*, 205–214. [CrossRef]

48. Perkins, A.; Cochrane, A.W.; Ruben, S.M.; Rosen, C.A. Structural and functional characterization of the human immunodeficiency virus rev protein. *J. Acquir. Immune Defic. Syndr.* **1989**, *2*, 256–263. [PubMed]

49. Daugherty, M.D.; Booth, D.S.; Jayaraman, B.; Cheng, Y.; Frankel, A.D. HIV Rev response element (RRE) directs assembly of the Rev homooligomer into discrete asymmetric complexes.

Proc. Natl. Acad. Sci. USA **2010**, *107*, 12481–12486. [CrossRef] [PubMed]

50. DiMattia, M.A.; Watts, N.R.; Stahl, S.J.; Rader, C.; Wingfield, P.T.; Stuart, D.I.; Steven, A.C.; Grimes, J.M. Implications of the HIV-1 Rev dimer structure at 3.2 Å resolution for multimeric binding to the Rev response element. *Proc. Natl. Acad. Sci. USA* **2010**, *107*, 5810–5814. [CrossRef] [PubMed]

51. Cochrane, A.W.; Chen, C.H.; Rosen, C.A. Specific interaction of the human immunodeficiency virus Rev protein with a structured region in the env mRNA. *Proc. Natl. Acad. Sci. USA* **1990**, *87*, 1198–1202. [CrossRef] [PubMed]

52. Malim, M.H.; McCarn, D.F.; Tiley, L.S.; Cullen, B.R. Mutational definition of the human immunodeficiency virus type 1 Rev activation domain. *J. Virol.* **1991**, *65*, 4248–4254. [PubMed]

53. Venkatesh, L.K.; Chinnadurai, G. Mutants in a conserved region near the carboxy-terminus of HIV-1 Rev identify functionally important residues and exhibit a dominant negative phenotype. *Virology* **1990**, *178*, 327–330. [CrossRef]

54. Weichselbraun, I.; Farrington, G.K.; Rusche, J.R.; Bohnlein, E.; Hauber, J. Definition of the human immunodeficiency virus type 1 Rev and human T-cell leukemia virus type I Rex protein activation domain by functional exchange. *J. Virol.* **1992**, *66*, 2583–2587. [PubMed]

55. Fornerod, M.; Ohno, M.; Yoshida, M.; Mattaj, I.W. CRM1 is an export receptor for leucine-rich nuclear export signals. *Cell* **1997**, *90*, 1051–1060. [CrossRef]

56. Daugherty, M.D.; Liu, B.; Frankel, A.D. Structural basis for cooperative RNA binding and export complex assembly by HIV Rev. *Nat. Struct. Mol. Biol.* **2010**, *17*, 1337–1342. [CrossRef] [PubMed]

57. Booth, D.S.; Cheng, Y.; Frankel, A.D. The export receptor Crm1 forms a dimer to promote nuclear export of HIV RNA. *eLife* **2014**, *3*. [CrossRef] [PubMed]

58. Taniguchi, I.; Mabuchi, N.; Ohno, M. HIV-1 Rev protein specifies the viral RNA export pathway by suppressing TAP/NXF1 recruitment. *Nucleic Acids Res.* **2014**, *42*, 6645–6658. [CrossRef] [PubMed]

59. Suhasini, M.; Reddy, T.R. Cellular proteins and HIV-1 Rev function. *Curr. HIV Res.* **2009**, *7*, 91–100. [CrossRef] [PubMed]

60. Bohne, J.; Wodrich, H.; Krausslich, H.G. Splicing of human immunodeficiency virus RNA is position-dependent suggesting sequential removal of introns from the 5′ end. *Nucleic Acids Res.* **2005**, *33*, 825–837. [CrossRef] [PubMed]

61. Ajamian, L.; Abrahamyan, L.; Milev, M.; Ivanov, P.V.; Kulozik, A.E.; Gehring, N.H.; Mouland, A.J. Unexpected roles for UPF1 in HIV-1 RNA metabolism and translation. *RNA* **2008**, *14*, 914–927. [CrossRef] [PubMed]

62. Jackson, R.J.; Hellen, C.U.; Pestova, T.V. The mechanism of eukaryotic translation initiation and principles of its regulation. *Nat. Rev. Mol. Cell Biol.* **2010**, *11*, 113–127. [CrossRef] [PubMed]

63. Balvay, L.; Soto Rifo, R.; Ricci, E.P.; Decimo, D.; Ohlmann, T. Structural and functional diversity of viral IRESes. *Biochim. Biophys. Acta* **2009**, *1789*, 542–557. [CrossRef] [PubMed]

64. De Breyne, S.; Soto-Rifo, R.; Lopez-Lastra, M.; Ohlmann, T. Translation initiation is driven by different mechanisms on the HIV-1 and HIV-2 genomic RNAs. *Virus Res.* **2013**, *171*, 366–381. [CrossRef] [PubMed]

65. Grifo, J.A.; Tahara, S.M.; Morgan, M.A.; Shatkin, A.J.; Merrick, W.C. New initiation factor activity required for globin mRNA translation. *J. Biol. Chem.* **1983**, *258*, 5804–5810. [PubMed]

66. Prevot, D.; Darlix, J.L.; Ohlmann, T. Conducting the initiation of protein synthesis: The role of eIF4G. *Biol. Cell* **2003**, *95*, 141–156. [CrossRef]

67. Pause, A.; Methot, N.; Svitkin, Y.; Merrick, W.C.; Sonenberg, N. Dominant negative mutants of mammalian translation initiation factor eIF-4A define a critical role for eIF-4F in cap-dependent and cap-independent initiation of translation. *EMBO J.* **1994**, *13*, 1205–1215. [PubMed]

68. Pause, A.; Sonenberg, N. Mutational analysis of a DEAD box RNA helicase: The mammalian translation initiation factor eIF-4A. *EMBO J.* **1992**, *11*, 2643–2654. [PubMed]

69. Rogers, G.W., Jr.; Richter, N.J.; Lima, W.F.; Merrick, W.C. Modulation of the helicase activity of eIF4A by eIF4B, eIF4H, and eIF4F. *J. Biol. Chem.* **2001**, *276*, 30914–30922. [CrossRef] [PubMed]

70. Prevot, D.; Decimo, D.; Herbreteau, C.H.; Roux, F.; Garin, J.; Darlix, J.L.; Ohlmann, T. Characterization of a novel RNA-binding region of eIF4GI critical for ribosomal scanning. *EMBO J.* **2003**, *22*, 1909–1921. [CrossRef] [PubMed]

71. Korneeva, N.L.; Lamphear, B.J.; Hennigan, F.L.; Merrick, W.C.; Rhoads, R.E. Characterization of the two eIF4A-binding sites on human eIF4G-1. *J. Biol. Chem.* **2001**, *276*, 2872–2879. [CrossRef] [PubMed]

72. Lamphear, B.J.; Kirchweger, R.; Skern, T.; Rhoads, R.E. Mapping of functional domains in eukaryotic protein synthesis initiation factor 4G (eIF4G) with picornaviral proteases. Implications for cap-dependent and cap-independent translational initiation. *J. Biol. Chem.* **1995**, *270*, 21975–21983. [CrossRef] [PubMed]

73. Mader, S.; Lee, H.; Pause, A.; Sonenberg, N. The translation initiation factor eIF-4E binds to a common motif shared by the translation factor eIF-4 gamma and the translational repressors 4E-binding proteins. *Mol. Cell. Biol.* **1995**, *15*, 4990–4997. [PubMed]

74. Imataka, H.; Sonenberg, N. Human eukaryotic translation initiation factor 4G (eIF4G) possesses two separate and independent binding sites for eIF4A. *Mol. Cell. Biol.* **1997**, *17*, 6940–6947. [PubMed]

75. Korneeva, N.L.; Lamphear, B.J.; Hennigan, F.L.; Rhoads, R.E. Mutually cooperative binding of eukaryotic translation initiation factor (eIF) 3 and eIF4A to human eIF4G-1. *J. Biol. Chem.* **2000**, *275*, 41369–41376. [CrossRef] [PubMed]

76. Ohlmann, T.; Rau, M.; Pain, V.M.; Morley, S.J. The C-terminal domain of eukaryotic protein synthesis initiation factor (eIF) 4G is sufficient to support cap-independent translation in the absence of eIF4E. *EMBO J.* **1996**, *15*, 1371–1382. [PubMed]

77. Rau, M.; Ohlmann, T.; Morley, S.J.; Pain, V.M. A reevaluation of the cap-binding protein, eIF4E, as a rate-limiting factor for initiation of translation in reticulocyte lysate. *J. Biol. Chem.* **1996**, *271*, 8983–8990. [CrossRef] [PubMed]

78. Safer, B.; Kemper, W.; Jagus, R. Identification of a 48S preinitiation complex in reticulocyte lysate. *J. Biol. Chem.* **1978**, *253*, 3384–3386. [PubMed]

79. Kozak, M. Adherence to the first-AUG rule when a second AUG codon follows closely upon the first. *Proc. Natl. Acad. Sci. USA* **1995**, *92*, 2662–2666. [CrossRef] [PubMed]

80. Kozak, M. Recognition of AUG and alternative initiator codons is augmented by G in position +4 but is not generally affected by the nucleotides in positions +5 and +6. *EMBO J.* **1997**, *16*, 2482–2492. [CrossRef] [PubMed]

81. Wegrzyn, J.L.; Drudge, T.M.; Valafar, F.; Hook, V. Bioinformatic analyses of mammalian 5′-UTR sequence properties of mRNAs predicts alternative translation initiation sites. *BMC Bioinform.* **2008**, *9*. [CrossRef] [PubMed]

82. Jackson, R.J. The ATP requirement for initiation of eukaryotic translation varies according to the mRNA species. *Eur. J. Biochem.* **1991**, *200*, 285–294. [CrossRef] [PubMed]

83. Pestova, T.V.; Kolupaeva, V.G. The roles of individual eukaryotic translation initiation factors in ribosomal scanning and initiation codon selection. *Genes Dev.* **2002**, *16*, 2906–2922. [CrossRef] [PubMed]

84. Poyry, T.A.; Kaminski, A.; Jackson, R.J. What determines whether mammalian ribosomes resume scanning after translation of a short upstream open reading frame? *Genes Dev.* **2004**, *18*, 62–75. [CrossRef] [PubMed]

85. Parsyan, A.; Svitkin, Y.; Shahbazian, D.; Gkogkas, C.; Lasko, P.; Merrick, W.C.; Sonenberg, N. mRNA helicases: The tacticians of translational control. *Nat. Rev. Mol. Cell Biol.* **2011**, *12*, 235–245. [CrossRef] [PubMed]

86. Parsyan, A.; Shahbazian, D.; Martineau, Y.; Petroulakis, E.; Alain, T.; Larsson, O.; Mathonnet, G.; Tettweiler, G.; Hellen, C.U.; Pestova, T.V.; *et al.* The helicase protein DHX29 promotes translation initiation, cell proliferation, and tumorigenesis. *Proc. Natl. Acad. Sci. USA* **2009**, *106*, 22217–22222. [CrossRef] [PubMed]

87. Pisareva, V.P.; Pisarev, A.V.; Komar, A.A.; Hellen, C.U.; Pestova, T.V. Translation initiation on mammalian mRNAs with structured 5′ UTRs requires DExH-box protein DHX29. *Cell* **2008**, *135*, 1237–1250. [CrossRef] [PubMed]

88. Hartman, T.R.; Qian, S.; Bolinger, C.; Fernandez, S.; Schoenberg, D.R.; Boris-Lawrie, K. RNA helicase A is necessary for translation of selected messenger RNAs. *Nat. Struct. Mol. Biol.* **2006**, *13*, 509–516. [CrossRef] [PubMed]

89. Jang, S.K.; Krausslich, H.G.; Nicklin, M.J.; Duke, G.M.; Palmenberg, A.C.; Wimmer, E. A segment of the 5′ nontranslated region of encephalomyocarditis virus RNA directs internal entry of ribosomes during *in vitro* translation. *J. Virol.* **1988**, *62*, 2636–2643. [PubMed]

90. Pelletier, J.; Sonenberg, N. Internal initiation of translation of eukaryotic mRNA directed by a sequence derived from poliovirus RNA. *Nature* **1988**, *334*, 320–325. [CrossRef] [PubMed]

91. Jackson, R.J. The current status of vertebrate cellular mRNA IRESs. *Cold Spring Harb. Perspect. Biol.* **2013**, *5*. [CrossRef] [PubMed]

92. Yamasaki, S.; Anderson, P. Reprogramming mRNA translation during stress. *Curr. Opin. Cell Biol.* **2008**, *20*, 222–226. [CrossRef] [PubMed]

93. Johannes, G.; Carter, M.S.; Eisen, M.B.; Brown, P.O.; Sarnow, P. Identification of eukaryotic mRNAs that are translated at reduced cap binding complex eIF4F concentrations using a cDNA microarray. *Proc. Natl. Acad. Sci. USA* **1999**, *96*, 13118–13123. [CrossRef] [PubMed]

94. Holcik, M.; Sonenberg, N.; Korneluk, R.G. Internal ribosome initiation of translation and the control of cell death. *Trends Genet.* **2000**, *16*, 469–473. [CrossRef]

95. Pyronnet, S.; Dostie, J.; Sonenberg, N. Suppression of cap-dependent translation in mitosis. *Genes Dev.* **2001**, *15*, 2083–2093. [CrossRef] [PubMed]

96. Pyronnet, S.; Pradayrol, L.; Sonenberg, N. A cell cycle-dependent internal ribosome entry site. *Mol. Cell* **2000**, *5*, 607–616. [CrossRef]

97. Spriggs, K.A.; Stoneley, M.; Bushell, M.; Willis, A.E. Re-programming of translation following cell stress allows IRES-mediated translation to predominate. *Biol. Cell* **2008**, *100*, 27–38. [CrossRef] [PubMed]

98. Stoneley, M.; Willis, A.E. Cellular internal ribosome entry segments: Structures, trans-acting factors and regulation of gene expression. *Oncogene* **2004**, *23*, 3200–3207. [CrossRef] [PubMed]

99. Baudin, F.; Marquet, R.; Isel, C.; Darlix, J.L.; Ehresmann, B.; Ehresmann, C. Functional sites in the 5' region of human immunodeficiency virus type 1 RNA form defined structural domains. *J. Mol. Biol.* **1993**, *229*, 382–397. [CrossRef] [PubMed]

100. Berkhout, B. Structure and function of the human immunodeficiency virus leader RNA. *Prog. Nucleic Acid Res. Mol. Biol.* **1996**, *54*, 1–34. [PubMed]

101. Paillart, J.C.; Dettenhofer, M.; Yu, X.F.; Ehresmann, C.; Ehresmann, B.; Marquet, R. First snapshots of the HIV-1 RNA structure in infected cells and in virions. *J. Biol. Chem.* **2004**, *279*, 48397–48403. [CrossRef] [PubMed]

102. Rosen, C.A.; Sodroski, J.G.; Haseltine, W.A. The location of cis-acting regulatory sequences in the human T cell lymphotropic virus type III (HTLV-III/LAV) long terminal repeat. *Cell* **1985**, *41*, 813–823. [CrossRef]

103. Watts, J.M.; Dang, K.K.; Gorelick, R.J.; Leonard, C.W.; Bess, J.W., Jr.; Swanstrom, R.; Burch, C.L.; Weeks, K.M. Architecture and secondary structure of an entire HIV-1 RNA genome. *Nature* **2009**, *460*, 711–716. [CrossRef] [PubMed]

104. Yilmaz, A.; Bolinger, C.; Boris-Lawrie, K. Retrovirus translation initiation: Issues and hypotheses derived from study of HIV-1. *Curr. HIV Res.* **2006**, *4*, 131–139. [CrossRef] [PubMed]

105. Geballe, A.P.; Gray, M.K. Variable inhibition of cell-free translation by HIV-1 transcript leader sequences. *Nucleic Acids Res.* **1992**, *20*, 4291–4297. [CrossRef] [PubMed]

106. Parkin, N.T.; Cohen, E.A.; Darveau, A.; Rosen, C.; Haseltine, W.; Sonenberg, N. Mutational analysis of the 5' non-coding region of human immunodeficiency virus type 1: Effects of secondary structure on translation. *EMBO J.* **1988**, *7*, 2831–2837. [PubMed]

107. SenGupta, D.N.; Berkhout, B.; Gatignol, A.; Zhou, A.M.; Silverman, R.H. Direct evidence for translational regulation by leader RNA and Tat protein of human immunodeficiency virus type 1. *Proc. Natl. Acad. Sci. USA* **1990**, *87*, 7492–7496. [CrossRef] [PubMed]

108. Svitkin, Y.V.; Pause, A.; Sonenberg, N. La autoantigen alleviates translational repression by the 5' leader sequence of the human immunodeficiency virus type 1 mRNA. *J. Virol.* **1994**, *68*, 7001–7007. [PubMed]

109. Soto-Rifo, R.; Limousin, T.; Rubilar, P.S.; Ricci, E.P.; Decimo, D.; Moncorge, O.; Trabaud, M.A.; André, P.; Cimarelli, A.; Ohlmann, T.; *et al.* Different effects of the TAR structure on HIV-1 and HIV-2 genomic RNA translation. *Nucleic Acids Res.* **2012**, *40*, 2653–2667. [CrossRef] [PubMed]

110. Dorin, D.; Bonnet, M.C.; Bannwarth, S.; Gatignol, A.; Meurs, E.F.; Vaquero, C. The TAR RNA-binding protein, TRBP, stimulates the expression of TAR-containing RNAs *in vitro* and

in vivo independently of its ability to inhibit the dsRNA-dependent kinase PKR. *J. Biol. Chem.* **2003**, *278*, 4440–4448. [CrossRef] [PubMed]

111. Dugre-Brisson, S.; Elvira, G.; Boulay, K.; Chatel-Chaix, L.; Mouland, A.J.; DesGroseillers, L. Interaction of Staufen1 with the 5′ end of mRNA facilitates translation of these RNAs. *Nucleic Acids Res.* **2005**, *33*, 4797–4812. [CrossRef] [PubMed]

112. Miele, G.; Mouland, A.; Harrison, G.P.; Cohen, E.; Lever, A.M. The human immunodeficiency virus type 1 5′ packaging signal structure affects translation but does not function as an internal ribosome entry site structure. *J. Virol.* **1996**, *70*, 944–951. [PubMed]

113. Brasey, A.; Lopez-Lastra, M.; Ohlmann, T.; Beerens, N.; Berkhout, B.; Darlix, J.L.; Sonenberg, N. The leader of human immunodeficiency virus type 1 genomic RNA harbors an internal ribosome entry segment that is active during the G2/M phase of the cell cycle. *J. Virol.* **2003**, *77*, 3939–3949. [CrossRef] [PubMed]

114. Andersen, J.L.; Planelles, V. The role of Vpr in HIV-1 pathogenesis. *Curr. HIV Res.* **2005**, *3*, 43–51. [CrossRef] [PubMed]

115. Elder, R.T.; Benko, Z.; Zhao, Y. HIV-1 VPR modulates cell cycle G2/M transition through an alternative cellular mechanism other than the classic mitotic checkpoints. *Front. Biosci.* **2002**, *7*, d349–d357. [CrossRef] [PubMed]

116. Gemeniano, M.C.; Sawai, E.T.; Sparger, E.E. Feline immunodeficiency virus Orf-A localizes to the nucleus and induces cell cycle arrest. *Virology* **2004**, *325*, 167–174. [CrossRef] [PubMed]

117. He, J.; Choe, S.; Walker, R.; di Marzio, P.; Morgan, D.O.; Landau, N.R. Human immunodeficiency virus type 1 viral protein R (Vpr) arrests cells in the G2 phase of the cell cycle by inhibiting p34cdc2 activity. *J. Virol.* **1995**, *69*, 6705–6711. [PubMed]

118. Fan, H.; Penman, S. Regulation of protein synthesis in mammalian cells. II. Inhibition of protein synthesis at the level of initiation during mitosis. *J. Mol. Biol.* **1970**, *50*, 655–670. [CrossRef]

119. Tarnowka, M.A.; Baglioni, C. Regulation of protein synthesis in mitotic HeLa cells. *J. Cell. Physiol.* **1979**, *99*, 359–367. [CrossRef] [PubMed]

120. Goh, W.C.; Rogel, M.E.; Kinsey, C.M.; Michael, S.F.; Fultz, P.N.; Nowak, M.A.; Hahn, B.H.; Emerman, M. HIV-1 Vpr increases viral expression by manipulation of the cell cycle: A mechanism for selection of Vpr *in vivo*. *Nat. Med.* **1998**, *4*, 65–71. [CrossRef] [PubMed]

121. Monette, A.; Valiente-Echeverria, F.; Rivero, M.; Cohen, E.A.; Lopez-Lastra, M.; Mouland, A.J. Dual mechanisms of translation initiation of the full-length HIV-1 mRNA contribute to gag synthesis. *PLoS ONE* **2013**, *8*, e68108. [CrossRef] [PubMed]

122. Sharma, A.; Yilmaz, A.; Marsh, K.; Cochrane, A.; Boris-Lawrie, K. Thriving under stress: Selective translation of HIV-1 structural protein mRNA during Vpr-mediated impairment of eif4e translation activity. *PLoS Pathog.* **2012**, *8*, e1002612. [CrossRef] [PubMed]

123. Amorim, R.; Costa, S.M.; Cavaleiro, N.P.; da Silva, E.E.; da Costa, L.J. HIV-1 transcripts use IRES-initiation under conditions where Cap-dependent translation is restricted by poliovirus 2A protease. *PLoS ONE* **2014**, *9*, e88619. [CrossRef] [PubMed]

124. Plank, T.D.; Whitehurst, J.T.; Kieft, J.S. Cell type specificity and structural determinants of IRES activity from the 5′ leaders of different HIV-1 transcripts. *Nucleic Acids Res.* **2013**, *41*,

6698–6714. [CrossRef] [PubMed]

125. Vallejos, M.; Deforges, J.; Plank, T.D.; Letelier, A.; Ramdohr, P.; Abraham, C.G.; Valiente-Echeverría, F.; Kieft, J.S.; Sargueil, B.; López-Lastra, M.; *et al.* Activity of the human immunodeficiency virus type 1 cell cycle-dependent internal ribosomal entry site is modulated by IRES *trans*-acting factors. *Nucleic Acids Res.* **2011**, *39*, 6186–6200. [CrossRef] [PubMed]

126. Vallejos, M.; Carvajal, F.; Pino, K.; Navarrete, C.; Ferres, M.; Huidobro-Toro, J.P.; Sargueil, B.; López-Lastra, M. Functional and structural analysis of the internal ribosome entry site present in the mRNA of natural variants of the HIV-1. *PLoS ONE* **2012**, *7*, e35031. [CrossRef] [PubMed]

127. Valiente-Echeverria, F.; Vallejos, M.; Monette, A.; Pino, K.; Letelier, A.; Huidobro-Toro, J.P.; Mouland, A.J.; López-Lastra, M. A cis-acting element present within the Gag open reading frame negatively impacts on the activity of the HIV-1 IRES. *PLoS ONE* **2013**, *8*, e56962. [CrossRef] [PubMed]

128. Gendron, K.; Ferbeyre, G.; Heveker, N.; Brakier-Gingras, L. The activity of the HIV-1 IRES is stimulated by oxidative stress and controlled by a negative regulatory element. *Nucleic Acids Res.* **2011**, *39*, 902–912. [CrossRef] [PubMed]

129. Charnay, N.; Ivanyi-Nagy, R.; Soto-Rifo, R.; Ohlmann, T.; Lopez-Lastra, M.; Darlix, J.L. Mechanism of HIV-1 Tat RNA translation and its activation by the Tat protein. *Retrovirology* **2009**, *6*. [CrossRef] [PubMed]

130. Rivas-Aravena, A.; Ramdohr, P.; Vallejos, M.; Valiente-Echeverria, F.; Dormoy-Raclet, V.; Rodriguez, F.; Pino, K.; Holzmann, C.; Huidobro-Toro, J.P.; Gallouzi, I.E.; *et al.* The Elav-like protein HuR exerts translational control of viral internal ribosome entry sites. *Virology* **2009**, *392*, 178–185. [CrossRef] [PubMed]

131. Alvarez, E.; Menendez-Arias, L.; Carrasco, L. The eukaryotic translation initiation factor 4GI is cleaved by different retroviral proteases. *J. Virol.* **2003**, *77*, 12392–12400. [CrossRef] [PubMed]

132. Castello, A.; Franco, D.; Moral-Lopez, P.; Berlanga, J.J.; Alvarez, E.; Wimmer, E.; Carrasco, L. HIV-1 protease inhibits Cap- and poly(A)-dependent translation upon eIF4GI and PABP cleavage. *PLoS ONE* **2009**, *4*, e7997. [CrossRef] [PubMed]

133. Ohlmann, T.; Prevot, D.; Decimo, D.; Roux, F.; Garin, J.; Morley, S.J.; Darlix, J.L. *In vitro* cleavage of eIF4GI but not eIF4GII by HIV-1 protease and its effects on translation in the rabbit reticulocyte lysate system. *J. Mol. Biol.* **2002**, *318*, 9–20. [CrossRef]

134. Ventoso, I.; Blanco, R.; Perales, C.; Carrasco, L. HIV-1 protease cleaves eukaryotic initiation factor 4G and inhibits cap-dependent translation. *Proc. Natl. Acad. Sci. USA* **2001**, *98*, 12966–12971. [CrossRef] [PubMed]

135. Chamond, N.; Locker, N.; Sargueil, B. The different pathways of HIV genomic RNA translation. *Biochem. Soc. Trans.* **2010**, *38*, 1548–1552. [CrossRef] [PubMed]

136. Buck, C.B.; Shen, X.; Egan, M.A.; Pierson, T.C.; Walker, C.M.; Siliciano, R.F. The human immunodeficiency virus type 1 gag gene encodes an internal ribosome entry site. *J. Virol.* **2001**, *75*, 181–191. [CrossRef] [PubMed]

137. Camerini, V.; Decimo, D.; Balvay, L.; Pistello, M.; Bendinelli, M.; Darlix, J.L.; Ohlmann, T. A dormant internal ribosome entry site controls translation of feline immunodeficiency virus. *J. Virol.* **2008**, *82*, 3574–3583. [CrossRef] [PubMed]

138. Herbreteau, C.H.; Weill, L.; Decimo, D.; Prevot, D.; Darlix, J.L.; Sargueil, B.; Ohlmann, T. HIV-2 genomic RNA contains a novel type of IRES located downstream of its initiation codon. *Nat. Struct. Mol. Biol.* **2005**, *12*, 1001–1007. [CrossRef] [PubMed]

139. Nicholson, M.G.; Rue, S.M.; Clements, J.E.; Barber, S.A. An internal ribosome entry site promotes translation of a novel SIV Pr55(Gag) isoform. *Virology* **2006**, *349*, 325–334. [CrossRef] [PubMed]

140. De Breyne, S.; Chamond, N.; Decimo, D.; Trabaud, M.A.; Andre, P.; Sargueil, B.; Ohlmann, T. *In vitro* studies reveal that different modes of initiation on HIV-1 mRNA have different levels of requirement for eukaryotic initiation factor 4F. *FEBS J.* **2012**, *279*, 3098–3111. [CrossRef] [PubMed]

141. Weill, L.; James, L.; Ulryck, N.; Chamond, N.; Herbreteau, C.H.; Ohlmann, T.; Sargueil, B. A new type of IRES within gag coding region recruits three initiation complexes on HIV-2 genomic RNA. *Nucleic Acids Res.* **2010**, *38*, 1367–1381. [CrossRef] [PubMed]

142. Locker, N.; Chamond, N.; Sargueil, B. A conserved structure within the HIV gag open reading frame that controls translation initiation directly recruits the 40S subunit and eIF3. *Nucleic Acids Res.* **2011**, *39*, 2367–2377. [CrossRef] [PubMed]

143. Ricci, E.P.; Herbreteau, C.H.; Decimo, D.; Schaupp, A.; Datta, S.A.; Rein, A.; Darlix, J.L.; Ohlmann, T. *In vitro* expression of the HIV-2 genomic RNA is controlled by three distinct internal ribosome entry segments that are regulated by the HIV protease and the Gag polyprotein. *RNA* **2008**, *14*, 1443–1455. [CrossRef] [PubMed]

144. Berkhout, B.; Arts, K.; Abbink, T.E. Ribosomal scanning on the 5′-untranslated region of the human immunodeficiency virus RNA genome. *Nucleic Acids Res.* **2011**, *39*, 5232–5244. [CrossRef] [PubMed]

145. Bolinger, C.; Sharma, A.; Singh, D.; Yu, L.; Boris-Lawrie, K. RNA helicase A modulates translation of HIV-1 and infectivity of progeny virions. *Nucleic Acids Res.* **2010**, *38*, 1686–1696. [CrossRef] [PubMed]

146. Yedavalli, V.S.; Neuveut, C.; Chi, Y.H.; Kleiman, L.; Jeang, K.T. Requirement of DDX3 DEAD box RNA helicase for HIV-1 Rev-RRE export function. *Cell* **2004**, *119*, 381–392. [CrossRef] [PubMed]

147. Soto-Rifo, R.; Ohlmann, T. The role of the DEAD-box RNA helicase DDX3 in mRNA metabolism. *Wiley Interdiscip. Rev. RNA* **2013**, *4*, 369–385. [CrossRef] [PubMed]

148. Linder, P.; Jankowsky, E. From unwinding to clamping—The DEAD box RNA helicase family. *Nat. Rev. Mol. Cell Biol.* **2011**, *12*, 505–516. [CrossRef] [PubMed]

149. Soto-Rifo, R.; Rubilar, P.S.; Limousin, T.; de Breyne, S.; Decimo, D.; Ohlmann, T. DEAD-box protein DDX3 associates with eIF4F to promote translation of selected mRNAs. *EMBO J.* **2012**, *31*, 3745–3756. [CrossRef] [PubMed]

150. Soto-Rifo, R.; Rubilar, P.S.; Ohlmann, T. The DEAD-box helicase DDX3 substitutes for the cap-binding protein eIF4E to promote compartmentalized translation initiation of the HIV-1 genomic RNA. *Nucleic Acids Res.* **2013**, *41*, 6286–6299. [CrossRef] [PubMed]

151. Lai, M.C.; Wang, S.W.; Cheng, L.; Tarn, W.Y.; Tsai, S.J.; Sun, H.S. Human DDX3 interacts with the HIV-1 Tat protein to facilitate viral mRNA translation. *PLoS ONE* **2013**, *8*, e68665.

[CrossRef] [PubMed]

152. Yedavalli, V.S.; Jeang, K.T. Trimethylguanosine capping selectively promotes expression of Rev-dependent HIV-1 RNAs. *Proc. Natl. Acad. Sci. USA* **2010**, *107*, 14787–14792. [CrossRef] [PubMed]

153. Darzynkiewicz, E.; Stepinski, J.; Ekiel, I.; Jin, Y.; Haber, D.; Sijuwade, T.; Tahara, S.M. β-globin mRNAs capped with m^7G, $m_2^{2.7}(2)G$ or $m_3^{2.2.7}G$ differ in intrinsic translation efficiency. *Nucleic Acids Res.* **1988**, *16*, 8953–8962. [CrossRef] [PubMed]

154. Worch, R.; Niedzwiecka, A.; Stepinski, J.; Mazza, C.; Jankowska-Anyszka, M.; Darzynkiewicz, E.; Cusack, S.; Stolarski, R. Specificity of recognition of mRNA 5′ cap by human nuclear cap-binding complex. *RNA* **2005**, *11*, 1355–1363. [CrossRef] [PubMed]

155. Decker, C.J.; Parker, R. P-bodies and stress granules: Possible roles in the control of translation and mRNA degradation. *Cold Spring Harb. Perspect. Biol.* **2012**, *4*. [CrossRef] [PubMed]

156. Stoecklin, G.; Kedersha, N. Relationship of GW/P-bodies with stress granules. *Adv. Exp. Med. Biol.* **2013**, *768*, 197–211. [PubMed]

157. Lloyd, R.E. Regulation of stress granules and P-bodies during RNA virus infection. *Wiley Interdiscip. Rev. RNA* **2013**, *4*, 317–331. [CrossRef] [PubMed]

158. Reineke, L.C.; Lloyd, R.E. Diversion of stress granules and P-bodies during viral infection. *Virology* **2013**, *436*, 255–267. [CrossRef] [PubMed]

159. Valiente-Echeverria, F.; Melnychuk, L.; Mouland, A.J. Viral modulation of stress granules. *Virus Res.* **2012**, *169*, 430–437. [CrossRef] [PubMed]

160. Nathans, R.; Chu, C.Y.; Serquina, A.K.; Lu, C.C.; Cao, H.; Rana, T.M. Cellular microRNA and P bodies modulate host-HIV-1 interactions. *Mol. Cell* **2009**, *34*, 696–709. [CrossRef] [PubMed]

161. Chable-Bessia, C.; Meziane, O.; Latreille, D.; Triboulet, R.; Zamborlini, A.; Wagschal, A.; Jacquet, J.M.; Reynes, J.; Levy, Y.; Saib, A.; *et al.* Suppression of HIV-1 replication by microRNA effectors. *Retrovirology* **2009**, *6*. [CrossRef] [PubMed]

162. Abrahamyan, L.G.; Chatel-Chaix, L.; Ajamian, L.; Milev, M.P.; Monette, A.; Clement, J.F.; Song, R.; Lehmann, M.; DesGroseillers, L.; Laughrea, M.; *et al.* Novel Staufen1 ribonucleoproteins prevent formation of stress granules but favour encapsidation of HIV-1 genomic RNA. *J. Cell Sci.* **2010**, *123*, 369–383. [CrossRef] [PubMed]

163. Phalora, P.K.; Sherer, N.M.; Wolinsky, S.M.; Swanson, C.M.; Malim, M.H. HIV-1 replication and APOBEC3 antiviral activity are not regulated by P bodies. *J. Virol.* **2012**, *86*, 11712–11724. [CrossRef] [PubMed]

164. Reed, J.C.; Molter, B.; Geary, C.D.; McNevin, J.; McElrath, J.; Giri, S.; Klein, K.C.; Lingappa, J.R. HIV-1 Gag co-opts a cellular complex containing DDX6, a helicase that facilitates capsid assembly. *J. Cell Biol.* **2012**, *198*, 439–456. [CrossRef] [PubMed]

165. Valiente-Echeverria, F.; Melnychuk, L.; Vyboh, K.; Ajamian, L.; Gallouzi, I.E.; Bernard, N.; Mouland, A.J. eEF2 and Ras-GAP SH3 domain-binding protein (G3BP1) modulate stress granule assembly during HIV-1 infection. *Nat. Commun.* **2014**, *5*. [CrossRef] [PubMed]

166. Soto-Rifo, R.; Valiente-Echeverria, F.; Rubilar, P.S.; Garcia-de-Gracia, F.; Ricci, E.P.; Limousin, T.; Décimo, D.; Mouland, A.J.; Ohlmann, T. HIV-2 genomic RNA accumulates in stress granules in the absence of active translation. *Nucleic Acids Res.* **2014**, *42*, 12861–12875.

[CrossRef] [PubMed]

167. Braddock, M.; Thorburn, A.M.; Chambers, A.; Elliott, G.D.; Anderson, G.J.; Kingsman, A.J.; Kingsman, S.M. A nuclear translational block imposed by the HIV-1 U3 region is relieved by the Tat-TAR interaction. *Cell* **1990**, *62*, 1123–1133. [CrossRef]

168. Braddock, M.; Powell, R.; Blanchard, A.D.; Kingsman, A.J.; Kingsman, S.M. HIV-1 TAR RNA-binding proteins control TAT activation of translation in Xenopus oocytes. *FASEB J.* **1993**, *7*, 214–222. [PubMed]

169. SenGupta, D.N.; Silverman, R.H. Activation of interferon-regulated, dsRNA-dependent enzymes by human immunodeficiency virus-1 leader RNA. *Nucleic Acids Res.* **1989**, *17*, 969–978. [CrossRef] [PubMed]

170. Edery, I.; Petryshyn, R.; Sonenberg, N. Activation of double-stranded RNA-dependent kinase (DSL) by the TAR region of HIV-1 mRNA: A novel translational control mechanism. *Cell* **1989**, *56*, 303–312. [CrossRef]

171. Williams, B.R. Signal integration via PKR. *Sci. STKE* **2001**, *2001*. [CrossRef] [PubMed]

172. Clerzius, G.; Gelinas, J.F.; Gatignol, A. Multiple levels of PKR inhibition during HIV-1 replication. *Rev. Med. Virol.* **2011**, *21*, 42–53. [CrossRef] [PubMed]

173. Charbonneau, J.; Gendron, K.; Ferbeyre, G.; Brakier-Gingras, L. The 5′ UTR of HIV-1 full-length mRNA and the Tat viral protein modulate the programmed-1 ribosomal frameshift that generates HIV-1 enzymes. *RNA* **2012**, *18*, 519–529. [CrossRef] [PubMed]

174. Arrigo, S.J.; Chen, I.S. Rev is necessary for translation but not cytoplasmic accumulation of HIV-1 vif, vpr, and env/vpu 2 RNAs. *Genes Dev.* **1991**, *5*, 808–819. [CrossRef] [PubMed]

175. Kimura, T.; Hashimoto, I.; Nishikawa, M.; Fujisawa, J.I. A role for Rev in the association of HIV-1 gag mRNA with cytoskeletal beta-actin and viral protein expression. *Biochimie* **1996**, *78*, 1075–1080. [CrossRef]

176. Campbell, L.H.; Borg, K.T.; Haines, J.K.; Moon, R.T.; Schoenberg, D.R.; Arrigo, S.J. Human immunodeficiency virus type 1 Rev is required *in vivo* for binding of poly(A)-binding protein to Rev-dependent RNAs. *J. Virol.* **1994**, *68*, 5433–5438. [PubMed]

177. Groom, H.C.; Anderson, E.C.; Dangerfield, J.A.; Lever, A.M. Rev regulates translation of human immunodeficiency virus type 1 RNAs. *J. Gen. Virol.* **2009**, *90*, 1141–1147. [CrossRef] [PubMed]

178. Anderson, E.C.; Lever, A.M. Human immunodeficiency virus type 1 Gag polyprotein modulates its own translation. *J. Virol.* **2006**, *80*, 10478–10486. [CrossRef] [PubMed]

12

Both ERK1 and ERK2 are Required for Enterovirus 71 (EV71) Efficient Replication

Meng Zhu, Hao Duan, Meng Gao, Hao Zhang and Yihong Peng *

Department of Microbiology, School of Basic Medical Sciences, Peking University Health Science Center, 38 Xueyuan Road, Beijing 100191, China; E-Mails: mengzhu1984@gmail.com (M.Z.); dight119@163.com (H.D.); gm0323@126.com (M.G.); hao_zhang@bjmu.edu.cn (H.Z.)

* Author to whom correspondence should be addressed; E-Mail: ypeng78@bjmu.edu.cn

Academic Editor: David Boehr

Abstract: It has been demonstrated that MEK1, one of the two MEK isoforms in Raf-MEK-ERK1/2 pathway, is essential for successful EV71 propagation. However, the distinct function of ERK1 and ERK2 isoforms, the downstream kinases of MEKs, remains unclear in EV71 replication. In this study, specific ERK siRNAs and selective inhibitor U0126 were applied. Silencing specific ERK did not significantly impact on the EV71-caused biphasic activation of the other ERK isoform, suggesting the EV71-induced activations of ERK1 and ERK2 were non-discriminative and independent to one another. Knockdown of either ERK1 or ERK2 markedly impaired progeny EV71 propagation (both by more than 90%), progeny viral RNA amplification (either by about 30% to 40%) and protein synthesis (both by around 70%), indicating both ERK1 and ERK2 were critical and not interchangeable to EV71 propagation. Moreover, suppression of EV71 replication by inhibiting both early and late phases of ERK1/2 activation showed no significant difference from that of only blocking the late phase, supporting the late phase activation was more importantly responsible for EV71 life cycle. Taken together, this study for the first time identified both ERK1 and ERK2 were required for EV71 efficient replication and further verified the important role of MEK1-ERK1/2 in EV71 replication.

Keywords: EV71; virus replication; ERK1; ERK2

---

1. Introduction

Enterovirus 71 (EV71) is a non-enveloped, positive single-stranded RNA virus belonging to the enterovirus A species of the genus *Enterovirus*, family *Picornaviridae* [1]. The EV71 genome, about 7.5 kb in length, consists of a single open reading frame encoding seven nonstructural proteins (2A, 2B, 2C, 3A, 3B, 3C, and 3D) and four structural proteins (VP1, VP2, VP3, and VP4) [2]. As one of the major causative agents leading to large outbreaks of hand, foot and mouth disease (HFMD) worldwide, especially in the Asia-Pacific area recently, EV71 has become the most dangerous neurotropic enterovirus after the control of poliovirus [3,4]. There is currently no effective vaccine or specific therapy that has been applied to prevent or treat EV71-caused severe HFMD, due to insufficient understanding of the molecular mechanisms of EV71 replication and host response to EV71 infection.

It is universally acknowledged that successful viral replication is reliant on many functioning components of cellular metabolism and a prerequisite for all the following pathogenic consequences in host cells. Accumulated data show that virus takes advantage of various signaling cascades for its life cycle, among which studies done by us and others have demonstrated that the extracellular signal regulated kinase (ERK) pathway is essential for EV71 and other viruses replication, inhibition of this signaling pathway has been found to severely impair EV71 and other variety of viruses production [5–11].

Cellular ERK signaling pathway, one of the three major mitogen-activated protein kinase (MAPK) cascades, which consists of three tiered serine/threonine kinases of Raf, MEK and ERK, plays an important role in regulating cell physiological functions [12–14] as well as many pathologic processes, including brain injury, cancer, diabetes, infectious diseases and inflammation *etc.* [15–18]. The two isoforms of ERKs, ERK1 and ERK2 (also referring to ERK1/2), are considered to be the only downstream substrates of MEK (including MEK1 and MEK2, also referring to MEK1/2) to date [19].

Therefore, ERK1 and ERK2, undertaking the upstream signals from MEK1/2 and in turn activating variety of their downstream substrates, are key players in ERK pathway. They share 85% similarity at the amino acid level [20] and yet it is still controversial whether the individual ERK isoform plays a distinctive role(s). Some studies suggest that ERK1 and ERK2 are interchangeable [21–23].

However, considerable evidences indicate that they might act differentially [24–27]. Thus, it is still an open question, which needs to be further explored as to whether roles are unique or preferred to one or the other ERK isoform in the physiological and/or pathological processes.

Our previous work have proved that MEK1 and MEK2 play differential roles, and MEK1, rather than MEK2, is critical to promote EV71 efficient replication [6], highlighting that MEK1 and MEK2 could exert distinct effects on the replication of EV71. However, as the downstream kinases of MEK1, the specific contributions of ERK1 and ERK2 to EV71 replication have not been addressed yet. The objective of the present study is to determine the role(s) of individual ERK isoform on the life cycle of EV71. In addition, here we showed that either ERK1 or ERK2 were both required and not functionally redundant for EV71 efficient replication.

2. Materials and Methods

2.1. Cells Culture and Virus Preparation

Rhabdomyosarcoma (RD) cell line was obtained from The National Institute for the Control of Pharmaceutical and Biological Products. Cells were cultured in Dulbecco's modified Eagle's medium (DMEM, GIBCO) supplemented with 10% fetal bovine serum (FBS, Gibco) at 37 °C in an atmosphere of 5% CO_2.

The titer of Enterovirus 71 (EV71-BC08 stain) was determined by titration in RD cells and stored at −80 °C until use [28].

2.2. Inhibitor against ERK Pathway and Antibodies

U0126 (Pierce, Thermo Scientific, Waltham, MA, USA), the inhibitor of ERK pathway, was dissolved in DMSO at the stocking concentration of 2 mM. Antibodies were purchased from Cell Signaling Technology (Danvers, MA, USA) (CST, anti-ERK1/2, anti-phospho-ERK1/2), Abcam (anti-EV71 VP1 and anti-VP3/4), Santa Cruz (anti-β-actin).

2.3. siRNAs and Transfection

siRNAs targeting human ERK1 (siERK1) and ERK2 (siERK2) were synthesized from Genepharma Co., Ltd. (Shanghai, China). The sequences, coming from Christopher A. Dimitri's paper [29], are showing as follows:

ERK1 siRNA (siERK1): 5'-CCCUGACCCGUCUAAUAUAdTdT-3' (sense),
5'-UAUAUUAGACGGGUCAGGGdAdG-3' (antisense);
ERK2 siRNA (siERK2): 5'-CAUGGUAGUCACUAACAUAdTdT-3'(sense),
5'-UAUGUUAGUGACUACCAUGdAdT-3' (antisense).

In addition, the negative control siRNAs (siNC) were purchased from Genepharma Co., Ltd. siERK1 and siERK2 (siERK1+2) were used together to knock down both ERK1 and ERK2.

Lipofectamine 2000 (Invitrogen) was used according to the manufacturer's instructions for siRNA transfection. RD cells were grown to 60% confluency in 6 or 12 well plates before transfection. RD cells were then transfected with siRNAs at the indicated concentrations.

2.4. Morphological Analysis

RD cells infected with EV71 were examined at every 8 h intervals post infection (p.i.) for the cytopathic effect (CPE) with phase-contrast microscopy.

2.5. Real Time Quantitative PCR (qPCR)

Total and intracellular viral RNAs were prepared for relative qPCR by Trizol reagent (Invitrogen). Then, according to the manufacturer's instructions of the ReverAid First strand cDNA synthesis kit (Thermo Scientific), 11 μL of viral RNAs and 1.5 μg of total RNAs were reversed transcribed into cDNA. Then 1 μL of cDNA was amplified with forward and reverse primers for EV71 VP1 gene and GAPDH control using LightCycler DNA Master SYBR Green I kit (Roche Diagnostics Corporation,

Basel, Switzerland). The forward and reverse EV71 VP1 gene primers were: 5'- GCA GCC CAA AAG AAC TTC AC-3' and 5'- ATT TCA GCA GCT TGG AGT GC-3', respectively. The forward and reverse GAPDH primers were: 5'-TGTTCCAATATGATTCCACCC-3' and 5'- CTTCTCCATGGTGCGTGAAGA-3', respectively. The reactions were performed with the Roche Light Cycler 480 system under the following conditions: Initial denaturation step at 95 °C for 10 min, followed by 40 cycles of 30 sat 94 °C, at 55 °C for 30 s and at 72 °C for 30 s. The CT value was normalized to that of GAPDH. All samples were run in triplicate.

Intracellular EV71 virions for absolute qPCR were prepared as described in Dr. Mingliang He's paper [30]. A quantitative standard curve was achieved as described in our previous paper [6]. Quantified results were extrapolated from the standard curve with all samples being run in triplicate.

2.6. Western Blot Analysis

Western blots were performed as described in our previous study [9]. Cells were harvested at indicated time points and lysed for 1 h in lysis buffer (Santa Cruz) containing complete protease inhibitors (Roche Applied Science). Total protein concentration was determined by the Bicinchoninic Acid Protein Assay Kit (Pierce, Thermo Scientific) after obtaining cell extracts by centrifugation at 13,000 rpm and 4 °C. Before transferred to PVDF membranes (Millipore), proteins were resolved on the sodium dodecyl sulfatesulfate polyacrylamide gel electrophoresis (SDS-PAGE). Then the membranes were blocked in 5% non-fat-dry-milk solution for 1 h at room temperature and then blotted with specific primary antibodies over night at 4 °C, following incubated with horseradish peroxidase antibodies for 1 h at room temperature. The immunoreactive bands were developed with SuperSignal West Femto Maximum Sensitivity Substrate (Pierce, Thermo Scientific) or enhanced chemiluminescent substrate (ECL), followed by autoradiography.

2.7. Statistics

All the curves and diagrams were made by using the Graph Pad Prism 5 Program (GraphPad). Data were shown as the mean ± standard deviation (SD) and analyzed by Student's t-test. $p < 0.05$ was considered statistically significant.

3. Results

3.1. Specific Knockdown of ERK Isoform Did Not Affect the Activation of the Other Isoform Induced by EV71

Biphasic activation of ERK1/2 caused by EV71 infection, including early transient and late sustained phases, has been demonstrated in our previous study [6]. To further identify the activation status of specific ERK isoform (ERK1 or ERK2) in viral infection, the phosph-ERK1 (pERK1) and phosph-ERK2 (pERK2) under EV71 infection were determined separately. As shown in Figure 1, a biphasic activation of ERK2 induced by EV71 was observed in RD cells pre-treated with ERK1 siRNA (designated as siERK1) compared with that of respective uninfected groups pre-transfected with negative control siRNA (designated as siNC).

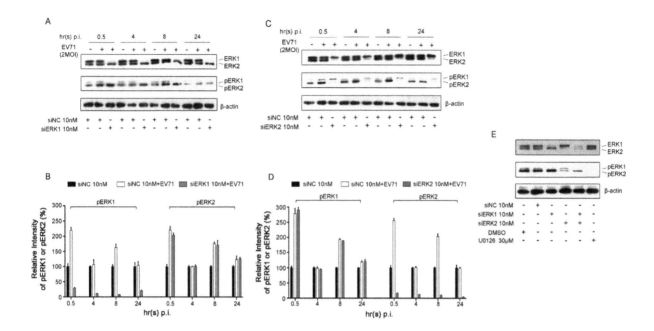

Figure 1. EV71-induced activation of ERK1/2 in Rhabdomyosarcoma (RD) cells. (**A**) RD cells were infected with EV71 (MOI = 2) at 36 h post-transfection with ERK1 siRNA (siERK1). Cells transfected with negative control siRNA (siNC) were used as controls. Cell lysates were collected at the indicated time points post infection (p.i.). ERK1/2, pERK1/2 and β-actin were detected by Western blot analysis, respectively; (**B**) The intensity of pERK1 or pERK2 normalized to that of corresponding ERK was determined by densitometric scanning based on the results from panel A; (**C**) The experiment was performed as described in panel A except that RD cells were transfected with ERK2 siRNA (siERK2); (**D**) Based on the results from panel C, the intensity was determined as described in panel B; (**E**) Cell lysates were collected from RD cells treated with siERK1, siERK2 and both (siERK1+2) at 36 h after transfection and cells treated with U0126 at 12 h after addition. ERK1/2 and pERK1/2 were blotted with specific antibodies. Experiments were repeated three times.

No compensatory activation of ERK2 was found when specific knockdown of ERK1. Similarly, the activation of ERK1 was not impacted when specifically silencing of ERK2 with ERK2 siRNA (designated as siERK2) in RD cells infected with EV71 (Figure 1C, D). These data indicated that the activation of ERK1 and ERK2 did not affect one another in the presence of EV71 infection in RD cells.

Notably, siERK1 and siERK2 were confirmed very efficient in knocking down corresponding ERK protein in quiescent RD cells collected 36 h post transfection and detected by Western blot analysis, which correlated with a strong decrease in ERK phosphorylation (Figure 1E). Moreover, treatment of U0126, a specific inhibitor of ERK activation, decreased more than 99% of ERK1/2 activation at 12 h after addition. No significant cytotoxicity of siERK1, siERK2, siERK1+2 and U0126 to the proliferation and survival of RD cells were observed in the current study (Figure S1).

3.2. Depletion of Individual ERK Isoform Resulted in a Similar Reduction of EV71 Proliferation

It has been demonstrated that MEK1 and MEK2 play a different role in EV71 replication [6]. To further elucidate the roles of downstream kinases of MEKs, ERK1 and ERK2 in EV71 replication, the effects of distinct knockdown of ERK1 or ERK2, or both on progeny viral titers were investigated. A clear reduction of progeny EV71 titers by about 90% were obtained in cells pre-treated with siERK1, siERK2 and siERK1+2, respectively, when compared with that of siNC control (Figure 2A).

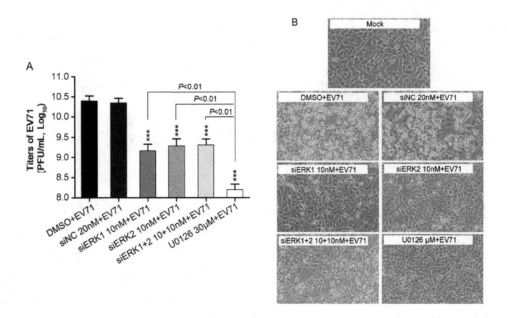

Figure 2. EV71 titers and CPE in RD cells treatment with distinct siERK (s) or U0126. (**A**) RD cells, pre-transfected with siERK1, siERK2 or siERK1+2 for 36 h, respectively, or treated with U0126 1 h prior to infection, were infected with EV71 at an MOI of 2. Cells pre-treated with DMSO or siNC were done as parallels. At 24 h p.i., both supernatants and cell lysates were collected and applied for determining the total titers of EV71 by titration; (**B**) Experiments were performed as described in panel A except CPE was examined every 8 h and images were taken at 24 h p.i.. Each result represents the average of three independent experiments and is shown as the means ± standard deviations (SD). *** $p < 0.001$, *versus* corresponding controls by Student's *t*-test.

No obvious difference of progeny viral titers was observed either knocking down of both ERK1 and ERK2, or each, whereas a stronger reduction of progeny viral titers was found when treating cells with U0126. In addition, EV71-induced cytopathic effect (CPE) was examined under phase-contrast microscopy. At 24 h p.i., pre-treatment with siERK1, siERK2, siERK1+2 or U0126 remarkably suppressed the morphological changes caused by EV71 infection (Figure 2B), as compared to corresponding controls, which was consistent with the viral titers in corresponding groups. Taken together, these results indicated ERK1 and ERK2 were required, but not functionally redundant for EV71 proliferation.

3.3. Distinct Knockdown of ERK Isoform Reduced Viral Genomic RNAs

To further identify the effects of ERK1 and ERK2 on EV71 replication cycle, VP1 gene among total viral RNAs extracted from siERK(s)- and U0126-treated RD cells was quantified by relative quantitative PCR (qPCR). Viral VP1 gene was significantly suppressed by about 30% to 40% at 4 h, 8 h and 12 h p.i. by siERK1, siERK2, siERK1+2 and U0126 when compared with corresponding controls in RD cells infected with EV71 (Figure 3).

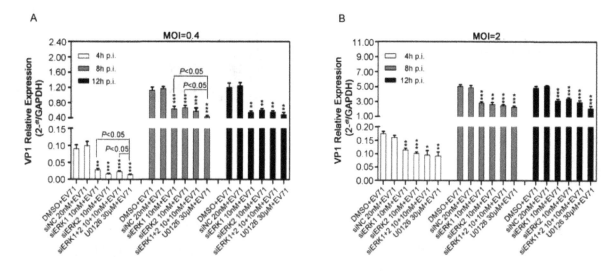

Figure 3. Effects of ERK1 and ERK2 on viral RNA synthesis in RD cells. RD cells pre-treated with siERK(s) or U0126 were infected with EV71 at the MOI of 0.4 (panel **A**) or 2 (panel **B**). Then at the indicated time points p.i., both supernatants and cell lysates were collected, viral VP1 gene among total viral genomes was quantified by relative qPCR. Data were the means of three independent experiments and error bars were denoted the SD. ** $p < 0.01$ *** $p < 0.001$, *versus* respective controls by Student's *t*-test.

In addition, U0126 caused a stronger reduction of viral genomic RNA than those of siERK1 and siERK1+2 at 4 and 8 h p.i., when RD cells was infected with 0.4 MOI of EV71 (Figure 3A). It should be noted that amounts of VP1 gene remained almost unchanged in each group from 8 h p.i. to 12 h p.i., indicating that the life cycle of EV71 might be less than 8 h. All these results revealed that both ERK1 and ERK2 might play important roles in viral RNA synthesis.

3.4. Disruption of Either ERK1 or ERK2 Resulted in a Reduction of EV71 Protein

To further determine the impact of specific ERK isoform on EV71 protein, viral structural proteins VP1 and VP3/4 were detected by Western blot analysis in EV71-infected RD cells pre-transfected with siERK1,or siERK2, or both, respectively. As shown in Figure 4A, ERK1, or ERK2, or ERK1/2 expression were almost diminished by siERK1, or ERK2, or siERK1+2, which resulted in the activation inhibition of ERK1, ERK2, or ERK1/2 by about 80% to 90%. The silencing effect decreased the expression of VP1 and VP3/4 by around 70%, when compared with that of siNC control (Figure 4B).

It seemed that siERK1 had a better inhibitory effect on VP1 and VP3/4 but no significance difference compared with that of siERK2 and siERK1+2 (Figure 4B). In addition, U0126 reduced more than 99%

of viral VP1 and VP3/4 expression which was significantly reduction than that of siERK(s)-treated groups. Our results suggested that both ERK1 and ERK2 might be crucial for EV71 protein production.

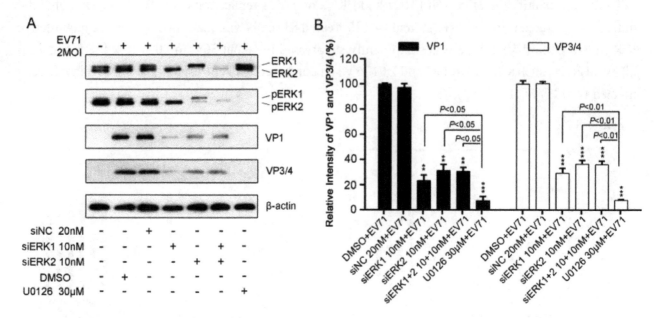

Figure 4. Impacts of ERK1 and ERK2 on protein production of EV71 in RD cells. (**A**). RD cells pre-treated with siERK(s) or U0126 were infected with 2 MOI of EV71. At 12 h p.i., total protein harvested was blotted with antibodies specific to VP1 and VP3/4. β-actin was used as the loading control. The relative intensities of VP1 (panel **B**) and VP3/4 (panel **C**) normalized to β-actin were presented by the percentage of the intensity of respective DMSO+EV71 group (100%). Data represent the average of three independent experiments and is shown as the means ± SD. ** $p < 0.01$ *** $p < 0.001$, *versus* corresponding control groups by Student's *t*-test.

3.5. Inhibiting both Early and Late Phases of ERK1/2 Activation Showed No Significant Difference from Blocking only the Late One for EV71 Replication

Since depletion of ERK1 or ERK2 both impaired virus replication severely, a time-of-drug addition assay was performed next in EV71-infected RD cells at the indicated time points to further specify the roles of the two phases of ERK1/2 activation in the viral life cycle. VP1 gene representing intracellular viral RNAs or intracellular virions was determined.

As shown in Figure 5A, disturbing activation of both ERK1 and ERK2 by treating cells with U0126 at 1 h before infection (−1), 0 h (0), 1 h(1) and 4 h (4) p.i. resulted in about 30% of suppression of intercellular viral RNAs. The similar results of intercellular virions were also obtained in the time-of-drug addition assay (Figure 5B). These data suggested that the late phase of ERK1/2 activated by EV71 might be more important for EV71 life cycle which may include viral RNA replication and/or translation.

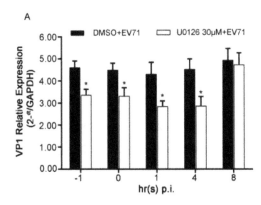 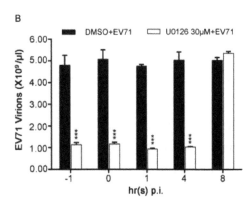

Figure 5. Effects of early ERK1/2 activation on EV71 infection in RD cells. (**A**) RD cells infected with EV71 at the MOI of 2 was treated with 30 μM of U0126 at the indicated time points p.i.. At 12 h p.i., cells were collected and viral VP1 gene among intracellular viral RNAs was quantified by relative qPCR; (**B**) The experiments were performed as described in panel A except 100 μg/mL of RNase A was used to eliminate naked viral RNAs before RNA was extracted. VP1 gene among intracellular EV71 virions was quantified by absolute qPCR. Data shown were the means ± SD ($n = 3$). * $p < 0.05$ ** $p < 0.01$ *** $p < 0.001$, *versus* respective controls by Student's *t*-test.

4. Discussion

Viruses hijack components of host's metabolic machinery for productive replication. The significance of activation of ERK pathway, as one of the specific responses to infection of varieties of viruses including EV71 [5–11,31,32], has been broadly reported. Our previous study has established the key role of MEK1 in the activation of ERK pathway induced by EV71 and proposed that MEK1-ERK1/2 signaling pathway acts as a central "hub" in EV71 replication [6]. However, it is still unclear if ERK1 and ERK2, the only known downstream kinases of MEK1 to date, play distinct role(s) in the life cycle of EV71. In fact, there are many studies focusing on distinction of ERK1 and ERK2 in several other areas rather than in virus replication. Meanwhile, the conclusions of these studies yet remain controversial. Although ERK1 and ERK2 are inclined to be thought of interchangeably [23], increasing studies have provided evidences for differential roles of ERK1 and ERK2 in cell movement [24], embryonic mice viability [33–35], and pathophysiology of central neural system *etc.* [36]. Our present work investigated distinction(s) of the individual involvement of ERK1 and ERK2 in EV71 replication.

Consistent with previous studies [6], EV71 infection caused a non-discriminative biphasic activation of ERK1/2 in RD cells. Some literatures reported that depletion of ERK1 induced higher activation of ERK2 in MEFs [34] and removal of ERK2 led to increased activation of ERK1 in NIH3T3 cells [37]. However, in our study, single silencing specific ERK isoform did not lead any significant effect on the biphasic activation of the other isoform in cells infected with or without EV71 (Figure 1), indicating that the activation of ERK1 and ERK2 might be various under different circumstances, although the mechanism is elusive.

Moreover, the present study shows both ERK1 and ERK2 are essential and not functionally redundant for EV71 replication due to the fact that individually silencing specific ERK isoform impaired EV71 propagation significantly, resulting in a marked suppression of viral progeny titer and EV71-induced

CPE (Figure 2) as well as a clear reduction of viral RNA and protein synthesis (Figures 3 and 4). U0126 showed stronger inhibitory effect on virus replication than ERK siRNA(s) did, probably because U0126 was able to block the activation of ERK pathway more efficiently. Interestingly, double knockdown of ERK1/2 by siRNAs suppressed virus propagation, viral RNAs and proteins to a similar level compared with those of single knockdown groups, indicating that ERK1 and ERK2 probably work together in EV71 life cycle in RD cells. In addition, it is reasonable to hypothesize that activated ERK1 and ERK2 act as a functional ensemble rather than playing their roles separately in EV71 replication. This is the first to report that ERK1 and ERK2 are both required as key factors in EV71 efficient propagation, which is quite different from distinctive or interchangeable functions of ERK1 and ERK2 in several other areas [21–24,33,34]. Combined with our previous study on MEK1 [6], the upstream kinase of ERK1/2, these findings further verified the critical role of MEK1-ERK1/2 pathway in EV71 life cycle.

In addition, according to the time-of-drug addition assay (Figure 5), it appeared that virus successful biosynthesis process was more closely related to the late phase of ERK1/2 activation induced by EV71 than the early one. This result was consistent with and further confirmed previous studies in which the early phase of ERK1/2 activation is not involved in viral propagation directly whereas might be the consequence of the virus-receptor binding [6,11].

In conclusion, this study provided evidence that ERK1 and ERK2 are not interchangeable in the EV71 life cycle and indicated they might act their functions as a whole. The findings deepened our understanding in the roles played by MEK1-ERK1/2 signaling pathway in virus life cycle and shed light on possibilities to offer flexible choices to selectively target the one between ERK1 and ERK2 in future anti-viral strategies. Further studies will be needed to specify the effect(s) and mechanism(s) of ERK1/2 on specific EV71 replication process(es).

5. Conclusions

Overall, the present study identified the effects of specific ERK isoform on EV71 propagation as well as on the viral genomic RNAs and proteins. Our data revealed that ERK1 and ERK2 are both required and not interchangeable for EV71 life cycle. The activations of ERK1 and ERK2 induced by EV71 were non-discriminative and independent to each another. Besides, the late phase of ERK1/2 activation under infection might be more responsible for the replication of EV71.

Acknowledgments

This work was supported by the National Natural Science Foundation of China (NSFC grant 81371816).

Author Contributions

M.Z. and Y.P. conceived and designed the experiments; M.Z. and H.D. performed the experiments; M.Z. and M.G. analyzed the data; H.Z. contributed reagents/materials/analysis tools; M.Z. wrote and Y.P. revised the manuscript, respectively.

References

1. The Picornavirus Pages: Enterovirus A. Available online: http://www.picornaviridae.com/enterovirus/ev-a/ev-a.htm (accessed on 12 February 2015).
2. Lin, J.Y.; Chen, T.C.; Weng, K.F.; Chang, S.C.; Chen, L.L.; Shih, S.R. Viral and host proteins involved in picornavirus life cycle. *J. Biomed. Sci.* **2009**, *16*, e103.
3. Yi, L.; Lu, J.; Kung, H.F.; He, M.L. The virology and developments toward control of human enterovirus 71. *Crit. Rev. Microbiol.* **2011**, *37*, 313–327.
4. Wang, S.M.; Liu, C.C. Update of enterovirus 71 infection: epidemiology, pathogenesis and vaccine. *Expert Rev. Anti Infect. Ther.* **2014**, *12*, 447–456.
5. Tung, W.H.; Hsieh, H.L.; Lee, I.T.; Yang, C.M. Enterovirus 71 modulates a COX-2/PGE2/cAMP-dependent viral replication in human neuroblastoma cells: role of the c-Src/EGFR/p42/p44 MAPK/CREB signaling pathway. *J. Cell Biochem.* **2011**, *112*, 559–570.
6. Wang, B.; Zhang, H.; Zhu, M.; Luo, Z.; Peng, Y. MEK1-ERKs signal cascade is required for the replication of Enterovirus 71 (EV71). *Antivir. Res.* **2012**, *93*, 110–117.
7. Cai, Y.; Liu, Y.; Zhang, X. Suppression of coronavirus replication by inhibition of the MEK signaling pathway. *J. Virol.* **2007**, *81*, 446–456.
8. Pleschka, S.; Wolff, T.; Ehrhardt, C.; Hobom, G.; Planz, O.; Rapp, U.R.; Ludwig, S. Influenza virus propagation is impaired by inhibition of the Raf/MEK/ERK signalling cascade. *Nat. Cell Biol.* **2001**, *3*, 301–305.
9. Zhang, H.; Feng, H.; Luo, L.; Zhou, Q.; Luo, Z.; Peng, Y. Distinct effects of knocking down MEK1 and MEK2 on replication of herpes simplex virus type 2. *Virus Res.* **2010**, *150*, 22–27.
10. Panteva, M.; Korkaya, H.; Jameel, S. Hepatitis viruses and the MAPK pathway: Is this a survival strategy? *Virus Res.* **2003**, *92*, 131–140.
11. Luo, H.; Yanagawa, B.; Zhang, J.; Luo, Z.; Zhang, M.; Esfandiarei, M.; Carthy, C.; Wilson, J.E.; Yang, D.; McManus, B.M. Coxsackievirus B3 replication is reduced by inhibition of the extracellular signal-regulated kinase (ERK) signaling pathway. *J. Virol.* **2002**, *76*, 3365–3373.
12. Murphy, L.O.; Blenis, J. MAPK signal specificity: The right place at the right time. *Trends Biochem. Sci.* **2006**, *31*, 268–275.
13. McCubrey, J.A.; Steelman, L.S.; Chappell, W.H.; Abrams, S.L.; Wong, E.W.; Chang, F.; Lehmann, B.; Terrian, D.M.; Milella, M.; Tafuri, A.; *et al.* Roles of the Raf/MEK/ERK pathway in cell growth, malignant transformation and drug resistance. *Biochim. Biophys. Acta* **2007**, *1773*, 1263–1284.
14. Zhang, W.; Liu, H.T. MAPK signal pathways in the regulation of cell proliferation in mammalian cells. *Cell Res.* **2002**, *12*, 9–18.
15. Kim, E.K.; Choi, E.J. Pathological roles of MAPK signaling pathways in human diseases. *Biochim. Biophys. Acta* **2010**, *1802*, 396–405.
16. Deschenes-Simard, X.; Kottakis, F.; Meloche, S.; Ferbeyre, G. ERKs in cancer: Friends or foes? *Cancer Res.* **2014**, *74*, 412–419.
17. Arthur, J.S.; Ley, S.C. Mitogen-activated protein kinases in innate immunity. *Nat. Rev. Immunol.* **2013**, *13*, 679–692.
18. Tanti, J.F.; Jager, J. Cellular mechanisms of insulin resistance: Role of stress-regulated serine

kinases and insulin receptor substrates (IRS) serine phosphorylation. *Curr. Opin. Pharmacol.* **2009**, *9*, 753–762.

19. Chambard, J.C.; Lefloch, R.; Pouyssegur, J.; Lenormand, P. ERK implication in cell cycle regulation. *Biochim. Biophys. Acta* **2007**, *1773*, 1299–1310.

20. Boulton, T.G.; Nye, S.H.; Robbins, D.J.; Ip, N.Y.; Radziejewska, E.; Morgenbesser, S.D.; DePinho, R.A.; Panayotatos, N.; Cobb, M.H.; Yancopoulos, G.D. ERKs: A family of protein-serine/threonine kinases that are activated and tyrosine phosphorylated in response to insulin and NGF. *Cell* **1991**, *65*, 663–675.

21. Roskoski, R., Jr. ERK1/2 MAP kinases: structure, function, and regulation. *Pharmacol. Res.* **2012**, *66*, 105–143.

22. Lefloch, R.; Pouyssegur, J.; Lenormand, P. Total ERK1/2 activity regulates cell proliferation. *Cell Cycle* **2009**, *8*, 705–711.

23. Yoon, S.; Seger, R. The extracellular signal-regulated kinase: Multiple substrates regulate diverse cellular functions. *Growth Factors* **2006**, *24*, 21–44.

24. Krens, S.F.; He, S.; Lamers, G.E.; Meijer, A.H.; Bakkers, J.; Schmidt, T.; Spaink, H.P.; Snaar-Jagalska, B.E. Distinct functions for ERK1 and ERK2 in cell migration processes during zebrafish gastrulation. *Dev. Biol.* **2008**, *319*, 370–383.

25. Vantaggiato, C.; Formentini, I.; Bondanza, A.; Bonini, C.; Naldini, L.; Brambilla, R. ERK1 and ERK2 mitogen-activated protein kinases affect Ras-dependent cell signaling differentially. *J. Biol.* **2006**, *5*, e14.

26. Lloyd, A.C. Distinct functions for ERKs? *J. Biol.* **2006**, *5*, e13.

27. Cargnello, M.; Roux, P.P. Activation and function of the MAPKs and their substrates, the MAPK-activated protein kinases. *Microbiol. Mol. Biol. Rev.* **2011**, *75*, 50–83.

28. Wang, B.; Ding, L.X.; Deng, J.; Zhang, H.; Zhu, M.; Yi, T.; Liu, J.; Xu, P.; Lu, F.M.; Peng, Y.H. Replication of EV71 was suppressed by MEK1/2 inhibitor U0126. *Chin. J. Biochem. Mol. Biol.* **2010**, *26*, 538–545.

29. Dimitri, C.A.; Dowdle, W.; MacKeigan, J.P.; Blenis, J.; Murphy, L.O. Spatially separate docking sites on ERK2 regulate distinct signaling events *in vivo*. *Curr. Biol.* **2005**, *15*, 1319–1324.

30. Lu, J.; He, Y.Q.; Yi, L.N.; Zan, H.; Kung, H.F.; He, M.L. Viral kinetics of enterovirus 71 in human abdomyosarcoma cells. *World J. Gastroenterol.* **2011**, *17*, 4135–4142.

31. Rodriguez, M.E.; Brunetti, J.E.; Wachsman, M.B.; Scolaro, L.A.; Castilla, V. Raf/MEK/ERK pathway activation is required for Junin virus replication. *J. Gen. Virol.* **2014**, *95*, 799–805.

32. Pleschka, S. RNA viruses and the mitogenic Raf/MEK/ERK signal transduction cascade. *Biol. Chem.* **2008**, *389*, 1273–1282.

33. Yao, Y.; Li, W.; Wu, J.; Germann, U.A.; Su, M.S.; Kuida, K.; Boucher, D.M. Extracellular signal-regulated kinase 2 is necessary for mesoderm differentiation. *Proc. Natl. Acad. Sci. USA* **2003**, *100*, 12759–12764.

34. Pages, G.; Guerin, S.; Grall, D.; Bonino, F.; Smith, A.; Anjuere, F.; Auberger, P.; Pouyssegur, J. Defective thymocyte maturation in p44 MAP kinase (Erk 1) knockout mice. *Science* **1999**, *286*, 1374–1377.

35. Nekrasova, T.; Shive, C.; Gao, Y.; Kawamura, K.; Guardia, R.; Landreth, G.; Forsthuber, T.G.

Both ERK1 and ERK2 are Required for Enterovirus 71 (EV71) Efficient Replication   239

ERK1-deficient mice show normal T cell effector function and are highly susceptible to experimental autoimmune encephalomyelitis. *J. Immunol.* **2005**, *175*, 2374–2380.

36. Yu, C. Distinct roles for ERK1 and ERK2 in pathophysiology of CNS. *Front. Biol.* **2012**, *7*, 267–276.

37. Lefloch, R.; Pouyssegur, J.; Lenormand, P. Single and combined silencing of ERK1 and ERK2 reveals their positive contribution to growth signaling depending on their expression levels. *Mol. Cell Biol.* **2008**, *28*, 511–527.

Permissions

All chapters in this book were first published by MDPI; hereby published with permission under the Creative Commons Attribution License or equivalent. Every chapter published in this book has been scrutinized by our experts. Their significance has been extensively debated. The topics covered herein carry significant findings which will fuel the growth of the discipline. They may even be implemented as practical applications or may be referred to as a beginning point for another development.

The contributors of this book come from diverse backgrounds, making this book a truly international effort. This book will bring forth new frontiers with its revolutionizing research information and detailed analysis of the nascent developments around the world.

We would like to thank all the contributing authors for lending their expertise to make the book truly unique. They have played a crucial role in the development of this book. Without their invaluable contributions this book wouldn't have been possible. They have made vital efforts to compile up to date information on the varied aspects of this subject to make this book a valuable addition to the collection of many professionals and students.

This book was conceptualized with the vision of imparting up-to-date information and advanced data in this field. To ensure the same, a matchless editorial board was set up. Every individual on the board went through rigorous rounds of assessment to prove their worth. After which they invested a large part of their time researching and compiling the most relevant data for our readers.

The editorial board has been involved in producing this book since its inception. They have spent rigorous hours researching and exploring the diverse topics which have resulted in the successful publishing of this book. They have passed on their knowledge of decades through this book. To expedite this challenging task, the publisher supported the team at every step. A small team of assistant editors was also appointed to further simplify the editing procedure and attain best results for the readers.

Apart from the editorial board, the designing team has also invested a significant amount of their time in understanding the subject and creating the most relevant covers. They scrutinized every image to scout for the most suitable representation of the subject and create an appropriate cover for the book.

The publishing team has been an ardent support to the editorial, designing and production team. Their endless efforts to recruit the best for this project, has resulted in the accomplishment of this book. They are a veteran in the field of academics and their pool of knowledge is as vast as their experience in printing. Their expertise and guidance has proved useful at every step. Their uncompromising quality standards have made this book an exceptional effort. Their encouragement from time to time has been an inspiration for everyone.

The publisher and the editorial board hope that this book will prove to be a valuable piece of knowledge for researchers, students, practitioners and scholars across the globe.

List of Contributors

Niels Piot, Simon Snoeck, Maarten Vanlede, Guy Smagghe and Ivan Meeus
Laboratory of Agrozoology, Department of Crop Protection, Faculty of Bioscience Engineering, Ghent University, 9000 Ghent, Belgium

Jose L. Nieto-Torres, Carlos Castaño-Rodriguez and Luis Enjuanes
Department of Molecular and Cell Biology, National Center of Biotechnology (CNB-CSIC), Campus Universidad Autónoma de Madrid, 28049 Madrid, Spain

Carmina Verdiá-Báguena and Vicente M. Aguilella
Laboratory of Molecular Biophysics, Department of Physics, Universitat Jaume I, 12071 Castellón, Spain

Shuang Lv, Qingyuan Xu, Encheng Sun, Tao Yang, Junping Li, Yufei Feng, Qin Zhang, Haixiu Wang, Jikai Zhang and Donglai Wu
State Key Laboratory of Veterinary Biotechnology, Harbin Veterinary Research Institute, Chinese Academy of Agricultural Sciences, Harbin 150001, China

Alexandre Carpentier, Pierre-Yves Barez, Malik Hamaidia, Hélène Gazon, Alix de Brogniez, Srikanth Perike and Nicolas Gillet
Molecular and Cellular Epigenetics (GIGA) and Molecular Biology (Gembloux Agro-Bio Tech), University of Liège (ULg), 4000 Liège, Belgium

Luc Willems
Molecular and Cellular Epigenetics, Interdisciplinary Cluster for Applied Genoproteomics (GIGA) of University of Liège (ULg), B34, 1 avenue de L'Hôpital, Sart-Tilman Liège 4000, Belgium
Molecular and Cellular Biology, Gembloux Agro-Bio Tech, University of Liège (ULg), 13 avenue Maréchal Juin, Gembloux 5030, Belgium

Colleen R. Reid and Adriana M. Airo
Department of Medical Microbiology and Immunology, University of Alberta, Edmonton, AB T6G 2E1, Canada

Tom C. Hobman
Department of Medical Microbiology and Immunology, University of Alberta, Edmonton, AB T6G 2E1, Canada
Department of Cell Biology, University of Alberta, Edmonton, AB T6G 2H7, Canada

Jason W. Rausch and Stuart F. J. Le Grice
Reverse Transcriptase Biochemistry Section, Basic Research Program, Frederick National Laboratory for Cancer Research, Frederick, MD 21702, USA

Cristina Ferrer-Orta, Diego Ferrero and Núria Verdaguer
Molecular Biology Institute of Barcelona (CSIC), Barcelona Science Park (PCB), Baldiri i Reixac 10, Barcelona E-08028, Spain

Melanie Schwarten and Silke Hoffmann
Institute of Complex Systems, Structural Biochemistry (ICS-6), Forschungszentrum Jülich, 52425 Jülich, Germany

Yu-Fu Hung, Dieter Willbold and Bernd W. Koenig
Institute of Complex Systems, Structural Biochemistry (ICS-6), Forschungszentrum Jülich, 52425 Jülich, Germany
Institut für Physikalische Biologie, Heinrich-Heine-Universität Düsseldorf, Universitätsstraße 1, 40255 Düsseldorf, Germany

Ella H. Sklan
Department Clinical Microbiology and Immunology, Sackler School of Medicine, Tel Aviv University, Tel Aviv 69978, Israel

Valerie J. Klema and Kyung H. Choi
Department of Biochemistry and Molecular Biology, Sealy Center for Structural Biology and Molecular Biophysics, University of Texas Medical Branch at Galveston, Galveston, TX 77555-0647, USA

Radhakrishnan Padmanabhan
Department of Microbiology and Immunology, Georgetown University School of Medicine, Washington, DC 20057, USA

Nicolas Albert Gillet
Molecular and Cellular Epigenetics, Interdisciplinary Cluster for Applied Genoproteomics (GIGA) of University of Liège (ULg), B34, 1 avenue de L'Hôpital, Sart-Tilman Liège 4000, Belgium
Molecular and Cellular Biology, Gembloux Agro-Bio Tech, University of Liège (ULg), 13 avenue Maréchal Juin, Gembloux 5030, Belgium

Bárbara Rojas-Araya and Ricardo Soto-Rifo
Molecular and Cellular Virology Laboratory, Program
of Virology, Institute of Biomedical Sciences, Faculty of
Medicine, University of Chile, Independencia 834100,
Santiago, Chile

Théophile Ohlmann
CIRI, International Center for Infectiology Research,
Université de Lyon, Lyon 69007, France
Inserm, U1111, Lyon 69007, France
Ecole Normale Supérieure de Lyon, Lyon 69007,
France

Université Lyon 1, Centre International de Recherche
en Infectiologie, Lyon 69007, France
CNRS, UMR5308, Lyon 69007, France

**Meng Zhu, Hao Duan, Meng Gao, Hao Zhang and
Yihong Peng**
Department of Microbiology, School of Basic Medical
Sciences, Peking University Health Science Center, 38
Xueyuan Road, Beijing 100191, China

Index

Printed in the USA
CPSIA information can be obtained
at www.ICGtesting.com
JSHW051624061123
51533JS00005B/87